DESIGN AND MEASUREMENT OF ANTENNAS AND PROPAGATION IN MOBILE CELLULAR SYSTEMS

For a complete listing of titles in the
Artech House Antennas and Propagation Library,
turn to the back of this book.

DESIGN AND MEASUREMENT OF ANTENNAS AND PROPAGATION IN MOBILE CELLULAR SYSTEMS

Hiroyuki Arai

ARTECH HOUSE
BOSTON | LONDON
artechhouse.com

Library of Congress Cataloging-in-Publication Data
A catalog record for this book is available from the U.S. Library of Congress.

British Library Cataloguing in Publication Data
A catalogue record for this book is available from the British Library.

ISBN 13: 978-1-63081-890-6

Cover design by Publishers' Design and Production Services, Inc.

Artech House
685 Canton Street
Norwood, MA 02062

10 9 8 7 6 5 4 3 2 1

CONTENTS

PREFACE *xi*

ACKNOWLEDGMENTS *xiii*

1
BASIC THEORY OF ANTENNAS 1

1.1 Maxwell's Equations and Their Solutions 1

1.2 Radiation from an Infinitesimal Current Element 3

1.3 Radiation from Infinitesimal Magnetic Current Element 7

1.4 Radiation from Dipole Antenna 8

1.5 Basic Antenna Elements 12

1.6 Antenna Gain 15

1.7 Antenna Parameters 16

1.7.1 Impedance Matching 17

1.7.2 Radiation Resistance and Input Impedance 18

1.7.3 Antenna Effective Length and Area 19

1.7.4 Cross Sectional Area and Antenna Gain 22

1.8 Array Antenna 23

1.8.1 Linear Array Antenna 24

1.9 Mutual Coupling 28

References 30

2

BASIC THEORY OF ELECTROMAGNETIC WAVE PROPAGATION 31

2.1 Classification of Propagation Paths 31

2.2 Propagation Loss in Free Space 32

2.3 Reflection and Transmission at the Boundary 34

2.4 Oblique Incidence 36

2.5 Effect of Ground Reflection 39

2.6 Knife-Edge Diffraction 42

2.7 Fresnel Zone 45

2.8 Fading Structure 46

2.9 Diversity Reception 51

2.10 Delay Profile 53

2.11 Angular Profile 56

References 58

3

FUNDAMENTALS OF ANTENNAS AND PROPAGATION MEASUREMENT 59

3.1 Impedance Measurement 59

- 3.1.1 Reflection Coefficient 60
- 3.1.2 Smith Chart 63
- 3.1.3 S-Parameter Measurement 64
- 3.1.4 Balanced and Unbalanced Feeding Lines 67

3.2 Measurement of the Antenna-Radiation Pattern 70

- 3.2.1 Far, Fresnel, and Near Fields in Antenna Measurements 71
- 3.2.2 Far-Field Measurements in Anechoic Chamber 72
- 3.2.3 Near-Field Measurements 77

3.3 Propagation Measurement 80

- 3.3.1 Propagation Loss Measurement 80
- 3.3.2 Cross Polarization and Correlation Coefficient 83
- 3.3.3 Delay Profile Measurement 85

References 86

4
ANTENNA ELEMENT AND ARRAY 87

4.1 Antenna Elements of the Base Station 87
- 4.1.1 Dipole Antenna 88
- 4.1.2 Microstrip Antenna 93
- 4.1.3 Multiband Dipole and Microstrip Antenna Elements 95

4.2 Array Antennas 97
- 4.2.1 Colinear Array 97
- 4.2.2 Linear Array 98
- 4.2.3 Antenna Pattern Synthesis 102

4.3 Antenna Elements of the Mobile Terminal 104
- 4.3.1 Monopole on a Finite Ground Plane 104
- 4.3.2 Inverted-F Shaped Antenna 107
- 4.3.3 Electrically Small Antenna 110
- 4.3.4 Theory of Electrically Small Antennas 113

References 115

5
DESIGN OF BASE STATION ANTENNAS 117

5.1 Categories of Base Stations 117
- 5.1.1 Circular Coverage Antennas 118
- 5.1.2 Sector Antennas 119
- 5.1.3 Other Pattern Shapes for Coverage Areas 121

5.2 Design of BS Antennas 123
- 5.2.1 Antenna Configuration of BS Antennas 123
- 5.2.2 Components of BS Antennas 125
- 5.2.3 Design of Array Pattern 136

5.3 Diversity and MIMO Antenna 139
- 5.3.1 Antenna Pattern Correlation 139
- 5.3.2 Butler Matrix and Multibeam 141
- 5.3.3 Active Integrated Antenna 143

References 146

6

DESIGN OF MOBILE TERMINAL ANTENNAS 149

6.1 Built-In Antenna 150

6.1.1 Antenna Location and the Effect of the Chassis 150

6.1.2 Antenna Elements 152

6.1.3 Matching Circuit 154

6.2 Multiantenna System 157

6.2.1 Chassis Mode 158

6.2.2 Antenna Position and Mutual Coupling 160

6.2.3 Decoupling Method 162

6.3 Body Effect 165

6.3.1 Numerical Phantom 165

6.3.2 Phantom in Measurement 167

6.3.3 Body Effect 171

References 177

7

MEASUREMENT OF BASE STATION AND MOBILE TERMINAL ANTENNAS 179

7.1 Base Station Antenna Measurements 179

7.1.1 Antenna Pattern Measurements in Open Sites 180

7.1.2 Near-Field and Very-Near-Field Measurements 182

7.1.3 Passive Intermodulation 187

7.2 Mobile Terminal Measurement 190

7.2.1 Measurement Parameters 190

7.2.2 OTA Measurement 192

7.2.3 MIMO Performance Measurement 196

7.3 EMC and SAR/PD Measurement 199

7.3.1 Open Site EMI Measurement 200

7.3.2 Reverberating Enclosure Measurement 203

7.3.3 SAR/PD Measurement 205

References 208

8
PROPAGATION ESTIMATION AND EMULATION OF FADING 211

8.1 Prediction of Propagation Losses in Cellular System 211
8.1.1 The Okumura-Hata Model 212
8.1.2 Physical Model 216
8.1.3 Standardization of Loss Predictions 221
8.2 Ray-Tracing Methods 223
8.2.1 The FDTD Method and the Ray-Tracing Method 223
8.2.2 Basis of the Ray-Tracing Method 224
8.2.3 Image and Ray-Launching Methods 227
8.2.4 Examples of Ray-Tracing Simulations 229
8.3 Multipath Fading Emulation 232
8.3.1 Fading Simulator 232
8.3.2 Delay Spread Simulation 234
8.3.3 Fading-Field Simulator 236
References 238

9
ANTENNA AND CHANNEL CAPACITY 241

9.1 Channel Capacity of SISO and MIMO 241
9.1.1 System Model of MIMO 242
9.1.2 Channel Capacity Under Rayleigh Environment 244
9.1.3 Channel Capacity Under Nakagami-Rice Environment 246
9.2 BS Antenna Design and Channel Capacity 247
9.2.1 BS Near Walls 248
9.2.2 BS at the Ceiling 251
9.2.3 BS in Open Site 255
9.3 Terminal Antenna Design and Channel Capacity 259
9.3.1 Mutual Coupling and Radiation Efficiency 259
9.3.2 Channel Capacity Estimation of MT Antennas 263
9.3.3 Channel Capacity of MT Antennas 265
References 269

10

PROPAGATION MEASUREMENTS 271

10.1 Propagation-Loss and Delay-Profile Measurements 271

10.1.1 Measurement System 272

10.1.2 Delay-Profile Measurement 274

10.1.3 Examples of Propagation Measurement 276

10.2 Link Budget 280

10.3 Indoor Simulator Measurement 283

10.3.1 Fading Emulation 283

10.3.2 MIMO Emulation 286

References 288

11

ANTENNA AND PROPAGATION CHALLENGES IN FUTURE MOBILE COMMUNICATION SYSTEMS 291

11.1 Overview of Beyond 5G System 291

11.2 Channel Capacity and Link Budget Estimations 293

11.3 Optical Wireless Communication 295

References 298

ABOUT THE AUTHOR 301

INDEX 303

PREFACE

This book presents a practical design method for cellular systems, especially focusing on antenna design to make coverage areas for indoor and outdoor cells. In the first part (Chapters 1–3), this book describes the basics of antennas, propagation, and measurement to simulate, design, and verify the mobile antenna systems.

The second part (Chapters 4–7) describes practical antenna elements and array structure for mobile base station and terminal antennas. The details of base station and mobile terminal antennas are presented to lay out the cell coverage area of typical site conditions. Antenna performance should be confirmed by a measurement procedure for the development and production stages, respectively. The measurement method for each stage is explained separately.

The third part (Chapters 8–10) shows measurements of mobile systems used to check their characteristics under real propagation environments. These final tests are described at the end of this part. Estimation of propagation loss under indoor and outdoor environments are explained based on empirical statistics using measured data and ray tracing simulations. High-speed data transmission in mobile systems are evaluated by site channel capacity such as multiple input multiple output (MIMO) antennas. Channel capacity estimation methods are shown theoretically and experimentally.

The last chapter (Chapter 11) presents strategies and challenges of antennas and propagation systems for the future system.

The goal of this book is to understand the practical design procedure of mobile communication antenna systems based on propagation estimation by measurements and simulation. This book provides all the required knowledge for the research and development of current and future systems.

ACKNOWLEDGMENTS

My sincere thanks are due to the reviewers for their patient and skillful advice. I am deeply grateful to the following persons for their technical information, photos, numerical data, and illustrations: Koichi Ito, Chiba University; Nobuhiro Kuga and Megumi Takahashi, Yokohama National University; Yuko Kumagai and Toshiyuki Maruyama, Nazca Industrial Technology; and Toshiyasu Tanaka and Youki Shimizu, Microwave Factory Co. I am also grateful to Merlin Fox, senior commissioning editor, and Natalie McGregor, acquisitions editorial assistant, for their encouragement and schedule planning throughout this book.

The last acknowledgment goes to my wife, Keiko, who provided the encouragement and willingness to complete this book in a timely manner.

1

BASIC THEORY OF ANTENNAS

This chapter describes the basic theory of antennas to facilitate a discussion on their performance with respect to mobile base stations and terminals. For readers who are unfamiliar with the field of antenna engineering, explanations of the technical terms associated with mobile communication systems are provided. Additionally, a basic knowledge of electromagnetics is useful.

Antennas are basic elements for converting radio frequency (RF) energy into electromagnetic waves that propagate through space. Their fundamental characteristics are described by an infinitesimal current element as an initial point source, and the radiation performances of differently shaped antenna elements evaluated with respect to this infinitesimal current element are explained using several supporting examples. The second part of this chapter describes the basics of mutual coupling and array antennas, which are used by mobile base stations to produce the desired radiation pattern for controlling field distributions inside the coverage area.

1.1 MAXWELL'S EQUATIONS AND THEIR SOLUTIONS

Electromagnetic waves are governed by Maxwell's equations. The differential forms are expressed under the assumption that only the current source $\boldsymbol{J}$ exists in free space [1]. The time-dependent factor is expressed by the angular frequency ω and time t as $e^{j\omega t}$:

$$\nabla \times \boldsymbol{H} = \boldsymbol{J} + j\omega\varepsilon_o \boldsymbol{E} \tag{1.1}$$

$$\nabla \times \boldsymbol{E} = -j\omega\mu_o \boldsymbol{H} \tag{1.2}$$

$$\nabla \cdot \boldsymbol{E} = 0 \tag{1.3}$$

$$\nabla \cdot \boldsymbol{H} = 0 \tag{1.4}$$

where $\boldsymbol{E}$ and $\boldsymbol{H}$ are the electric and magnetic field vectors, and ε_o and μ_o represent the permittivity and permeability in free space, respectively. The term $j\omega\varepsilon_o\boldsymbol{E}$ in (1.1) expresses the displacement current and is an extension of Ampere's law. Equation (1.2) is Faraday's law, whereas (1.3) and (1.4) represent Gauss's law for electric and magnetic fields.

To determine the solution of electromagnetic fields generated by a current source, vector $\boldsymbol{A}$ is introduced in (1.5) [2]:

$$\boldsymbol{H} = \nabla \times \boldsymbol{A} \tag{1.5}$$

The rotation of arbitrary vectors becomes zero by taking its divergence as $\nabla \cdot \nabla \times \boldsymbol{A} = 0$, which allows (1.5) to identically satisfy (1.4). Equation (1.6) is obtained by substituting (1.5) into (1.2), as follows:

$$\nabla \times \left(\boldsymbol{E} + j\omega\mu_o \boldsymbol{A}\right) = 0 \tag{1.6}$$

Using the vector formula of $\nabla \times \nabla\phi = 0$ for an arbitrary potential ϕ, the above equation can be rewritten as:

$$\boldsymbol{E} = \nabla\phi - j\omega\mu_o \boldsymbol{A} \tag{1.7}$$

When (1.5) and (1.7) are substituted into (1.1), (1.8) is obtained using the vector formula of $\nabla \times \nabla \times \boldsymbol{A} = \nabla\nabla \cdot \boldsymbol{A} - \nabla^2\boldsymbol{A}$ as follows:

$$\nabla\left(\nabla \cdot \boldsymbol{A} - j\omega\varepsilon_o\phi\right) - \nabla^2\boldsymbol{A} - k_o^2\boldsymbol{A} = \boldsymbol{J} \tag{1.8}$$

where $k_o = \omega\sqrt{\varepsilon_o\mu_o}$ is the wave number in free space. If the arbitrary vector $\boldsymbol{A}$ and the potential ϕ satisfy the equation of $\nabla \cdot \boldsymbol{A} - j\omega\varepsilon_o\phi = 0$, (1.8) can be rewritten as:

$$\nabla^2\boldsymbol{A} + k_o^2\boldsymbol{A} = -\boldsymbol{J} \tag{1.9}$$

It should be noted that the solution of (1.9) identically satisfies $\nabla \cdot \boldsymbol{A} - j\omega\varepsilon_o\phi = 0$ by the substitution (1.9) into (1.8). Vector $\boldsymbol{A}$ for the solution of (1.9) is given as [3]:

$$\boldsymbol{A} = \frac{1}{4\pi}\int_v \frac{\boldsymbol{J}(\boldsymbol{r}_0)}{|\boldsymbol{r} - \boldsymbol{r}_0|}\boldsymbol{e}^{-jk_o|\boldsymbol{r}-\boldsymbol{r}_0|}\,d\boldsymbol{r}_0 \tag{1.10}$$

The observation and source point vectors are denoted by $\boldsymbol{r}$ and $\boldsymbol{r}_o$, respectively, as shown in Figure 1.1, in which v is the volume region of current $\boldsymbol{J}$. Finally, the electric field generated by current $\boldsymbol{J}$ can be obtained, as follows:

$$\boldsymbol{E} = -j\omega\mu_o\left(\boldsymbol{A} + \frac{\nabla\nabla \cdot \boldsymbol{A}}{k_o^2}\right) \tag{1.11}$$

1.2 RADIATION FROM AN INFINITESIMAL CURRENT ELEMENT

Electromagnetic waves are radiated from current sources. The arbitrary current source shown in Figure 1.2(a) is decomposed into short current elements, as presented in Figure 1.2(b). To calculate

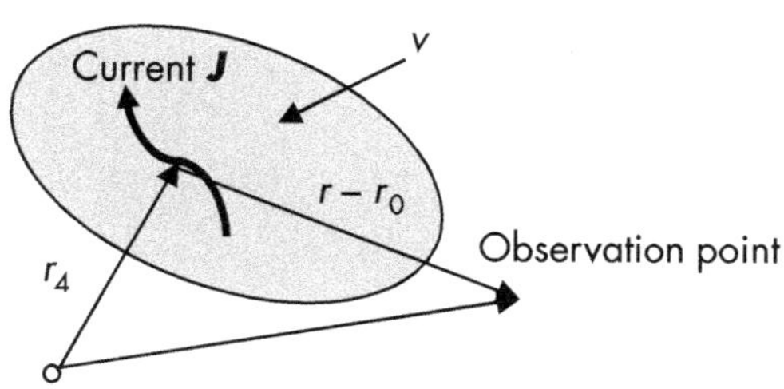

Figure 1.1 Current source and observation point.

the characteristics of the array of current elements, a short current element of length l located at the origin of the spherical coordinate system is considered (see Figure 1.2), in which the current element is the point source directed along the z axis. After the derivation of the radiation characteristics of the short currents, the fields from the arbitrary current can be obtained by a summation of the short currents.

Equations (1.9) and (1.10) provide the electric and magnetic field components (E_i, H_i, $i = r$, θ, and ϕ) from the short current at an observation point P. Then, the current vector $\boldsymbol{J}$ can be rewritten by the current amplitude I and the unit vector along the z direction of $\boldsymbol{a_z}$ as $J = I\boldsymbol{a_z}$. The current is considered a point source at $\boldsymbol{r_o} = 0$, and the integral volume in (1.9) is replaced by Il. The unit vectors of the spherical coordinate system are used to express the unit vector $\boldsymbol{a_z}$; then, the components of (1.11) can be obtained, as follows:

$$E_r = Z_o \frac{Il}{2\pi} k_o^2 \left\{ \frac{1}{(k_o r)^2} - j\frac{1}{(k_o r)^3} \right\} \cos\theta e^{-jk_o r} \tag{1.12}$$

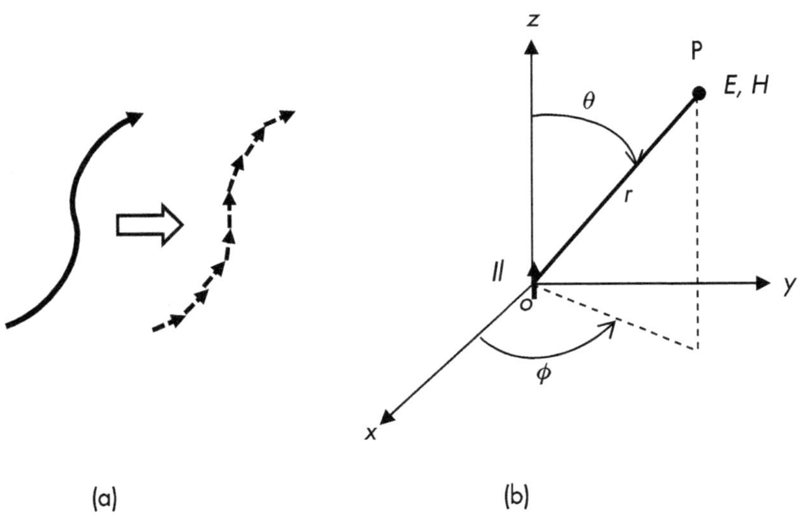

Figure 1.2 Current source and the coordinate system: (a) arbitral current, and (b) coordinate system.

$$E_\theta = Z_o \frac{Il}{4\pi} k_o^2 \left\{ j\frac{1}{(k_o r)} + \frac{1}{(k_o r)^2} - j\frac{1}{(k_o r)^3} \right\} \sin\theta e^{-jk_o r} \tag{1.13}$$

$$H_\phi = \frac{Il}{4\pi} k_o^2 \left\{ j\frac{1}{(k_o r)} + \frac{1}{(k_o r)^2} \right\} \sin\theta e^{-jk_o r} \tag{1.14}$$

$$E_\phi = H_r = H_\theta \tag{1.15}$$

where $Z_o = \sqrt{\mu_o/\varepsilon_o}$ represents the characteristic impedance and wavenumber. The wavenumber k_o is also obtained as $k_o = 2\pi/\lambda$ using the wavelength λ. The components in (1.12)–(1.14) are decomposed by $(k_o r)^{-n}$, $n = 1$, 2, and 3, and there is a decrease in each term, as can be seen in Figure 1.3. When the observation point is a function of distance from the source, the dominant term of $(k_o r)^{-1}$ in each component is designated as a radiation far field. The other two terms

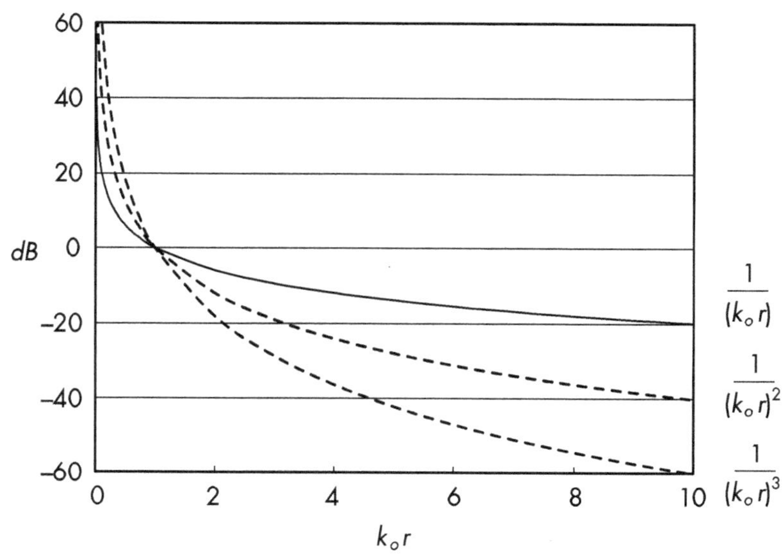

Figure 1.3 Field components.

for $k_o r < 1$ are greater than the far field, with $(k_o r)^{-2}$ and $(k_o r)^{-3}$ being characterized as an induced and a static field, respectively.

In the far-field region, radiation field components can be approximated by the following [4]:

$$E_\theta \cong jZ_o k_o \frac{Il}{4\pi} \frac{e^{-jk_o r}}{r} \sin\theta \tag{1.16}$$

$$H_\phi \cong k_o \frac{Il}{4\pi} \frac{e^{-jk_o r}}{r} \sin\theta \tag{1.17}$$

These two components are orthogonal to each other and have the relation of $E_\theta/H_\phi = Z_o$, which is the same as a plane wave. The term $e^{-jk_o r}/r$ represents both the decay in amplitude according to distance r and the phase change as a function of $k_o r$. As shown in Figure 1.4, the angular function of $\sin\theta$ is known as radiation pattern of the point source current. The radiation pattern's field strength is provided by the distance from the origin, which is the location of the point source. The plane including the z axis, which is parallel to component E_θ, is defined as E plane, and its orthogonal plane, which is parallel to component H_ϕ, is defined as the H plane.

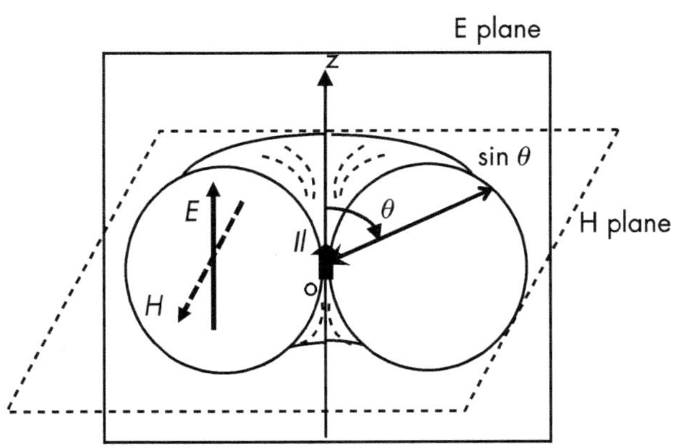

Figure 1.4 Radiation pattern infinitesimal current element.

1.3 RADIATION FROM INFINITESIMAL MAGNETIC CURRENT ELEMENT

Most antennas are characterized by the current distribution. Conversely, slot and planar antennas are characterized easily using the fictitious magnetic current $\boldsymbol{M}$. Maxwell's equations in (1.1) and (1.2) can be rewritten by introducing $\boldsymbol{M}$ into (1.2) as:

$$\nabla \boldsymbol{H}_j = \boldsymbol{J} + j\omega\varepsilon_o \boldsymbol{E}_j, \qquad \nabla \times \boldsymbol{E}_m = -\boldsymbol{M} - j\omega\mu_o \boldsymbol{H}_m \tag{1.18}$$

The electric and magnetic fields generated by the current source are denoted as $\boldsymbol{E}_j$ and $\boldsymbol{H}_j$, whereas those generated by magnetic current source are denoted as $\boldsymbol{E}_m$ and $\boldsymbol{H}_m$. A high degree of symmetry for ($\boldsymbol{E}_j$, $\boldsymbol{H}_j$) and ($\boldsymbol{E}_m$, $\boldsymbol{H}_m$) is found in (1.18), then the following relations can be interchanged in both fields:

$$\varepsilon \leftrightarrow \mu, \quad \boldsymbol{J} \leftrightarrow \boldsymbol{M}, \quad \boldsymbol{E}_j \leftrightarrow \boldsymbol{H}_m, \quad \boldsymbol{H}_j \leftrightarrow -\boldsymbol{E}_m \tag{1.19}$$

This complementarity of electric and magnetic fields is known as Babinet's principle [5]. To describe the electric and magnetic fields generated by the magnetic current, the vector equation and its solution (given in the preceding section) can be rewritten by the previous relations as follows:

$$\nabla^2 \boldsymbol{A}_m + k_o^2 \boldsymbol{A}_m = -\boldsymbol{M} \tag{1.20}$$

$$\boldsymbol{E}_m = -\nabla \times \boldsymbol{A}_m \tag{1.21}$$

$$\boldsymbol{H}_m = -j\omega\varepsilon\left(\boldsymbol{A}_m + \frac{\nabla\nabla \cdot \boldsymbol{A}_m}{k_o^2}\right) \tag{1.22}$$

Additionally, the solution of vector $\boldsymbol{A}_m$ can be obtained, as follows:

$$\boldsymbol{A}_m = \frac{1}{4\pi}\int_v \frac{\boldsymbol{M}(\boldsymbol{r}_0)}{|\boldsymbol{r} - \boldsymbol{r}_0|}\mathbf{e}^{-jk_o|r-r_0|}\, d\boldsymbol{r}_0 \tag{1.23}$$

Assuming the magnetic current at the origin directed in the z direction $\boldsymbol{M} = M\boldsymbol{a_z}$, the field components can be provided as follows:

$$E_\phi = \frac{Ml}{4\pi}k_o^2\left\{j\frac{1}{(k_or)} + \frac{1}{(k_or)^2}\right\}\sin\theta e^{-jk_or} \tag{1.24}$$

$$H_r = \frac{1}{Z_o}\frac{Ml}{2\pi}k_o^2\left\{\frac{1}{(k_or)^2} - j\frac{1}{(k_or)^3}\right\}\cos\theta e^{-jk_or} \tag{1.25}$$

$$H_\theta = \frac{1}{Z_o}\frac{Ml}{4\pi}k_o^2\left\{j\frac{1}{(k_or)} + \frac{1}{(k_or)^2} - j\frac{1}{(k_or)^3}\right\}\sin\theta e^{-jk_or} \tag{1.26}$$

$$E_r = E_\theta = H_\phi = 0 \tag{1.27}$$

E_ϕ and H_θ are the radiation field components according to the magnetic current and have complementarity with those components obtained using the current elements shown in (1.16) and (1.17).

1.4 RADIATION FROM DIPOLE ANTENNA

The infinitesimal current element in the previous section is a basic theoretical antenna. Next, a short dipole antenna is introduced as a basic practical element. To understand the radiation mechanism of a short dipole, the study considers an open-ended Lecher line excited by an RF feed (see Figure 1.5).

The forward currents flowing toward the end section are reflected to the feeding side as backward currents. The backward current cancels the forward current at the end section and creates a node of a current standing wave. The voltage standing wave that takes an antinode at the end section shifts by $\lambda/4$ compared with the current one. The currents on the top and bottom Lecher lines flow in reverse directions to each other, which cancels out the radiating field components, as shown in Figure 1.5. To radiate an electric field at the end section of a Lecher line, the top- and bottom-line currents should be aligned by bending, as shown in Figure 1.6. Consequently, the end

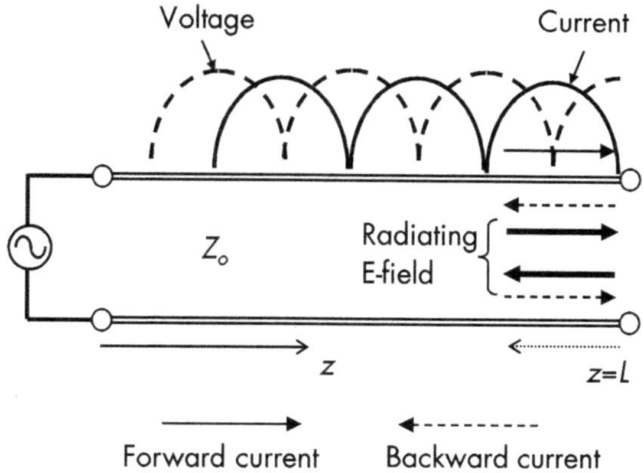

Figure 1.5 Open-ended Lecher line.

section becomes a dipole antenna. The radiation is increased by accumulating electromagnetic energy from the antenna element; this creates a standing wave of $\lambda/2$ length, which is known as a half-wavelength dipole antenna.

As Figure 1.7 shows, the radiation pattern of a $\lambda/2$ dipole antenna is calculated by the summation of the short dipole. Assuming the location of observation point P to be in the far-field region, the distance r' between point P and the short dipole on the z axis can be approximated as $r' \cong r - z \cos\theta$. This approximation of distance

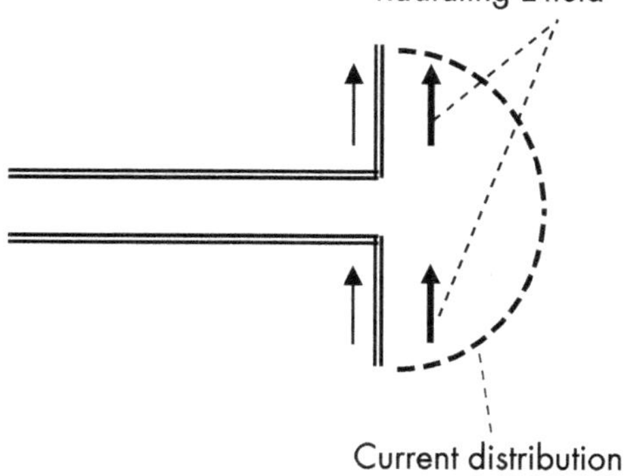

Figure 1.6 Lecher line and dipole antenna.

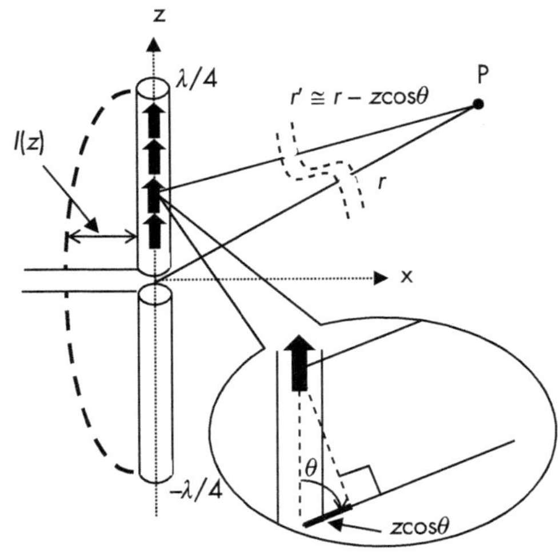

Figure 1.7 Dipole antenna and observation point.

is used only in the phase term in (1.16), and the radiation field of the $\lambda/2$ dipole antenna can be obtained by integrating (1.16) as follows:

$$E_\theta \cong \frac{jZ_o k_o}{4\pi}\sin\theta \int_{-\lambda/4}^{\lambda/4} I(z)\frac{e^{-jk_o r'}}{r}dz \tag{1.28}$$

where the current distribution of the dipole antenna of $I(z)$ is approximated by the sinusoidal function with a maximum amplitude I_o [6].

$$I(z) = I_o \cos(k_o z) \tag{1.29}$$

The radiation field can be obtained by substituting (1.29) into (1.28):

$$E_\theta = \frac{jZ_o I_o}{2\pi}\frac{e^{-j_o r}}{r}\frac{cos\left(\frac{\pi}{2}\cos\theta\right)}{\sin\theta} \tag{1.30}$$

The radiation pattern of a $\lambda/2$ dipole antenna is the function of θ and ϕ, which is expressed as:

$$E_\theta(\theta,\phi) = \frac{\cos\left(\frac{\pi}{2}\cos\theta\right)}{\sin\theta} \tag{1.31}$$

Figure 1.8 presents the E (zx) plane patterns of the $\lambda/2$ dipole antenna and the short dipole obtained by (1.16). The pattern of (1.31) is normalized by its maximum value at $\theta = \pi/2$. The half-power beamwidth of the antenna pattern is defined by the angle between the half-power (the amplitude of $1/\sqrt{2}$) of the main lobe. For the $\lambda/2$ dipole, the angle is 78°, whereas for the short dipole, the angle is 90°. This shape is known as a figure-of-eighth pattern, and the patterns of the H (xy) plane of both antennas are omnidirectional. Figure 1.9 contains a three-dimensional view of the pattern of $\lambda/2$ dipole.

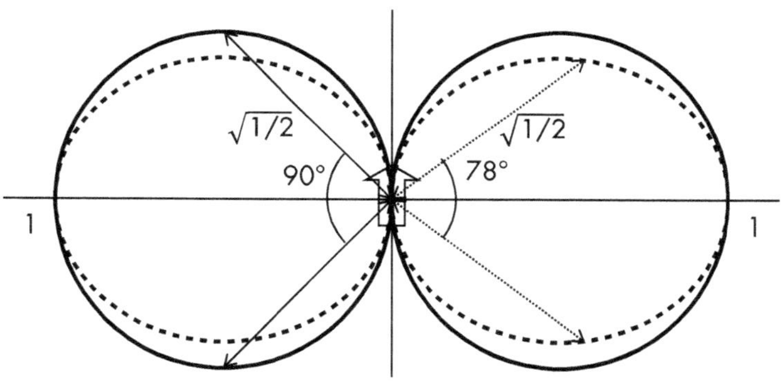

Figure 1.8 Radiation pattern of half-wavelength dipole antenna (dotted line) and infinitesimal current element (solid line).

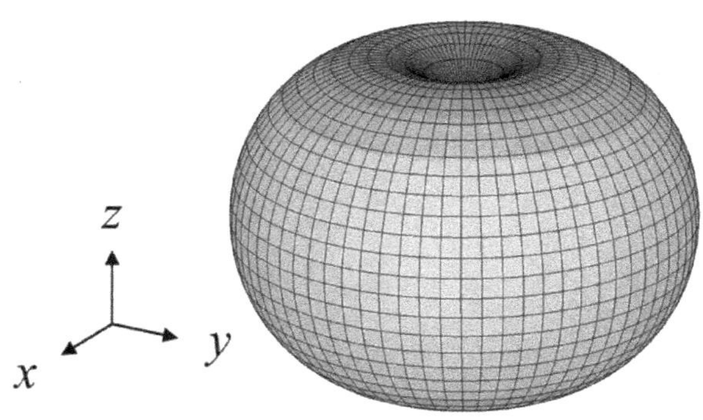

Figure 1.9 Three-dimensional view of $\lambda/2$ dipole antenna.

Typically, the radiation electric field of an arbitrary antenna is expressed as:

$$\boldsymbol{E}(r,\theta,\phi) = C\frac{e^{-j_o r}}{r}\left\{E_\theta(\theta,\phi)\boldsymbol{e}_\theta + E_\phi(\theta,\phi)\boldsymbol{e}_\phi\right\} = C\frac{e^{-j_o r}}{r}\boldsymbol{D}(\theta,\phi) \tag{1.32}$$

where C is a constant, and $\mathbf{e}_\theta$ and $\mathbf{e}_\phi$ are unit vectors in θ and ϕ directions, respectively. Figure 1.10 presents the radiation patterns produced by $E_\theta(\theta, \phi)$ and $E_\phi(\theta, \phi)$ in which the maximum radiation direction of 0° is designated by the front radiation. It can be seen in Figure 1.10 that the radiation pattern has a main lobe and several side lobes. In most cases, the front-to-back ratio can be obtained by dividing the front level by the maximum level within $180° - 60° \leq \theta \leq 180° + 60°$.

1.5 BASIC ANTENNA ELEMENTS

The principle of a dipole antenna can be explained by changing part of the Lecher line, as detailed in the preceding section. The current section uses a similar approach to describe several basic antenna elements.

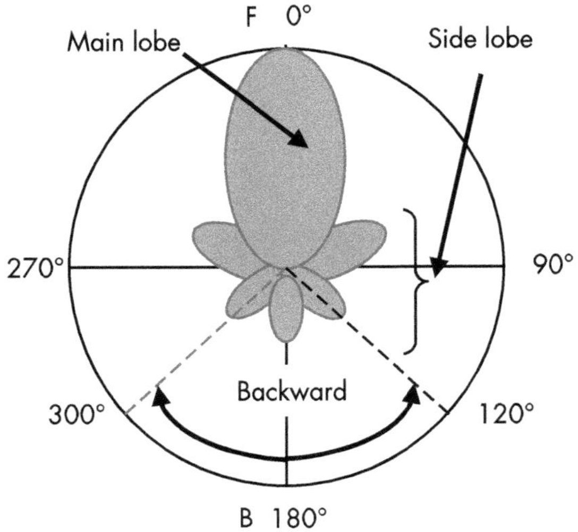

Figure 1.10 Radiation pattern.

A dipole antenna is created by bending one open end of the Lecher line in Figure 1.11(a) in the opposite direction (Figure 1.11(b)). For resonance, the length L of the bent section is almost half a wavelength. This form of antenna is known as a standard dipole antenna and is used for the antenna measurement.

Printed antennas are fabricated easily with current printing technology, and their thickness is very thin compared with their operating wavelength. This antenna is also described as a flat, planar, or patch antenna. Figure 1.12(a) shows a microstrip line etched on a dielectric substrate, which is widely used in RF circuits. After cutting out part of a strip conductor with length L, the width of the strip is extended to increase the radiation from its open edges to produce a microstrip antenna, as can be seen in Figure 1.12(b). The strip length L should be around half of the guided wavelength for the dominant mode resonance. Additionally, this type of antenna is characterized by assuming the magnetic currents at the edge of the strip.

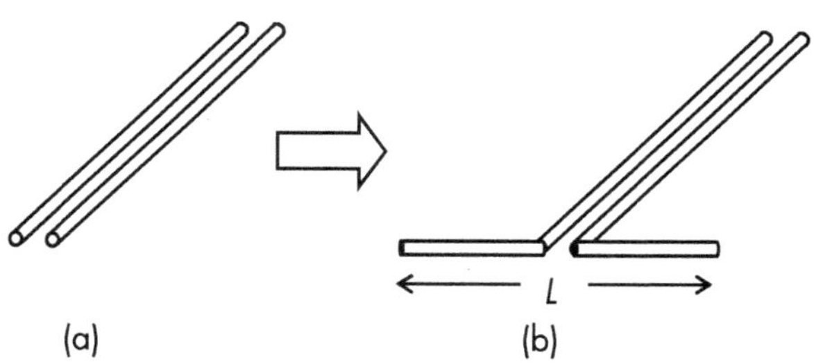

Figure 1.11 Lecher line and dipole antenna: (a) Lecher line, and (b) dipole antenna.

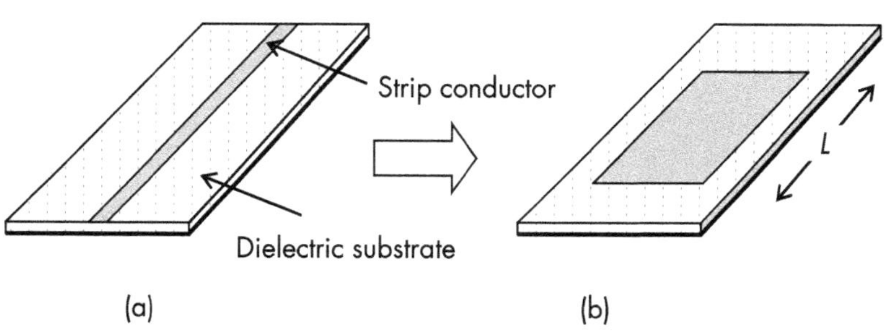

Figure 1.12 Microstrip line and microstrip antenna: (a) microstrip line, (b) microstrip antenna.

Sleeve antennas are fabricated from a coaxial cable; see Figure 1.13(a). As can been seen in Figure 1.13(b), the materials covering the inner conductor are removed to form a radiating element. To suppress current leakage on the surface of the cable, the bottom part of the antenna is overlapped by another outer conductor. This conductor is a type of balun that connects antennas with balanced feed and the transmission line, such as the coaxial cable, and its mechanism is explained in Section 3.1.4.

The final example is the horn antenna. A rectangular waveguide and a pyramidal horn antenna are presented in Figure 1.14.

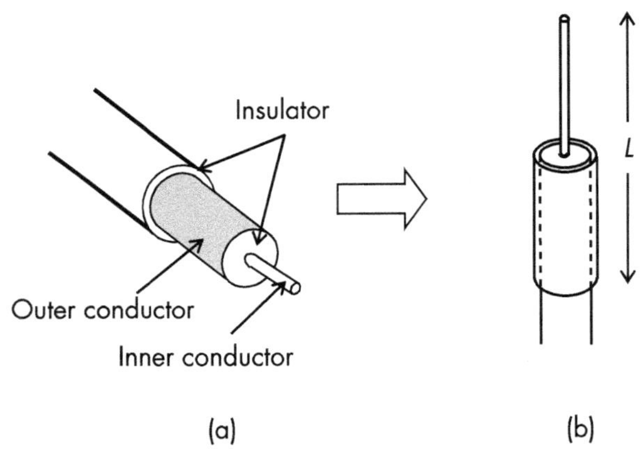

Figure 1.13 Coaxial cable and sleeve antenna: (a) coaxial cable, and (b) sleeve antenna.

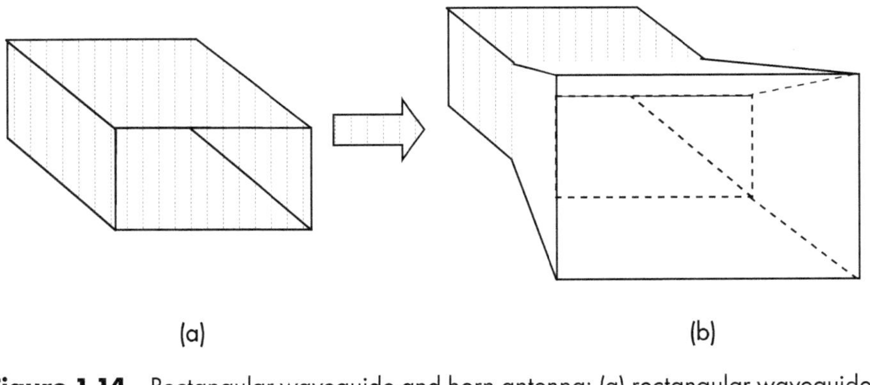

Figure 1.14 Rectangular waveguide and horn antenna: (a) rectangular waveguide, and (b) horn antenna.

The open aperture of the waveguide is widened gradually to match the aperture impedance with the characteristic impedance in free space. This geometry does not create resonance, and it has a relatively wide operating frequency band.

1.6 ANTENNA GAIN

In a transmitting antenna, the antenna gain describes the amount of power transmitted in the direction of peak radiation compared with an isotropic antenna. The opposite is true for a receiving antenna. As shown in Figure 1.15, an ideal isotropic antenna radiates uniformly in all directions.

When the same power is input to an antenna (e.g., a short dipole), the transmitting power density p_d of the peak radiation along the y axis is greater than that of an isotropic antenna p_i. As established in (1.33), the ratio between p_d and p_i is defined as the antenna gain G_i.

$$G_i = \frac{p_d}{p_i} \tag{1.33}$$

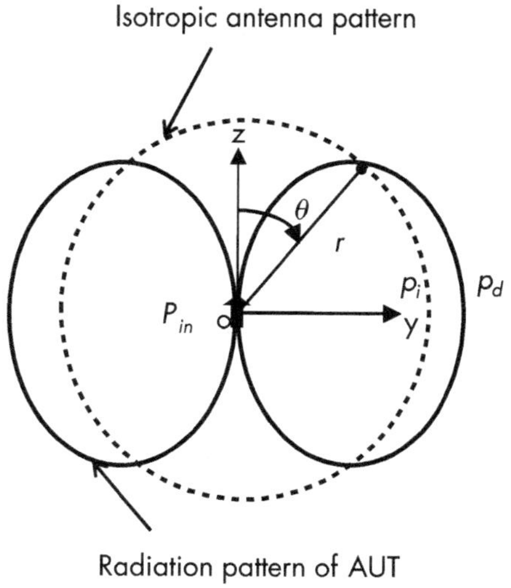

Figure 1.15 Antenna directivity gain.

Additionally, this gain can be determined by the radiation pattern of the antenna under test (AUT) using the following procedure. The power density at the observation point of the AUT can be obtained by using the radiation field in (1.16) as follows:

$$p_d = \frac{1}{2Z_o}\left|\boldsymbol{E}(r,\theta,\phi)\right|^2 \tag{1.34}$$

When deriving the antenna gain, the same power is input to the AUT, and the power density of isotropic antenna can be obtained using (1.34):

$$p_i = \frac{1}{4\pi r^2}\int_0^{2\pi}\int_0^{\pi}\frac{1}{2Z_o}\left|\boldsymbol{E}(r,\theta,\phi)\right|^2 r^2 \sin\theta \, d\theta \, d\phi \tag{1.35}$$

The peak direction of the beam is denoted as (θ_o, ϕ_o), and the antenna gain can be obtained using (1.35):

$$G_i = \frac{p_d}{p_i} = 4\pi\frac{\left|\boldsymbol{D}(\theta_o,\phi_o)\right|^2}{\int_0^{2\pi}\int_0^{\pi}\left|\boldsymbol{D}(\theta,\phi)\right|^2 \sin\theta d\theta d\phi} \tag{1.36}$$

Compared with the isotropic antenna, the antenna gain is defined as the isotropic gain G_i in units of dBi. However, isotropic antennas are theoretical and are particularly difficult to use in practical antenna measurement. Therefore, the standard for antenna measurement is the $\lambda/2$ dipole antenna, and the gain compared with a $\lambda/2$ dipole antenna is defined as the relative gain G_d in units of dBd. The G_i of a $\lambda/2$ dipole is calculated as 2.15 dBi using the previous formula and (1.31), and the conversion of G_d to G_i is realized by $G_i = G_d - 2.15$ dBi.

1.7 ANTENNA PARAMETERS

This section presents the basic parameters of antennas, including input impedance, radiation resistance, effective length, and area, and it contains specific examples of parameters for $\lambda/2$ dipole antenna.

Additionally, the relation between the effective area of the antenna and the antenna gain is presented.

1.7.1 Impedance Matching

With respect to a circuital element functioning as a radio transmitter/receiver, antennas are evaluated according to the frequency characteristics of the input impedance as a complex value of $Z_i = R + jX$. The real part of R involves the internal loss of the antenna (the antenna loss resistance R_l) and the loss factor based on electromagnetic wave radiation (antenna radiation resistance R_r). Antenna loss should be minimized for efficiency, and the ratio between R_l and R_r is defined in (1.37) by the radiation efficiency [7] as follows:

$$\eta = \frac{R_r}{R_l + R_r} \tag{1.37}$$

The reactance part of X, which represents the reactive energy stored in the vicinity of an antenna, is changed by the function of frequency. The resonance frequency is defined at $X = 0$ and $X = \pm\infty$ as series resonance and parallel resonance, respectively. The antenna input impedance must correspond with the impedance of the input and output ports of the radio equipment, which in most radio systems is $Z_o = 50\Omega$. A disparity in impedance at the antenna–radio equipment connection can be calculated using the reflection coefficient from the Z_o side, as in (1.38):

$$\Gamma = \frac{Z_i - Z_o}{Z_i + Z_o} \tag{1.38}$$

This reflected power level is described in decibels (dBs), and the frequency bandwidth is defined by the reflection at less than −10 dB (the typical example used in journal papers). Additionally, the degree of impedance mismatch is defined using the voltage standing wave ratio (VSWR), which is denoted by ρ as follows:

$$\rho = \frac{1 + |\Gamma|}{1 - |\Gamma|} \tag{1.39}$$

In practical antenna systems, typical VSWR values are $\rho \le 1.1$ for high-power broadcasting transmitting antennas, $\rho \le 1.5$ for cellular base station antennas, and $\rho \le 2$ or 3 for mobile terminal antennas.

1.7.2 Radiation Resistance and Input Impedance

The reactance part of input impedance is dominated by the stored energy in the vicinity of the antenna element. However, as it cannot be calculated using a simple formula, it is recommended that other publications are referenced. The real part of input impedance is radiation resistance, which depends on far-field radiation and can be calculated using a simple procedure as follows; let I be the input current of an antenna. The radiation power P_r is obtained using $R_r I^2/2$. Equation (1.40) can be obtained by integrating the power density over the entire solid angle:

$$P_r = \frac{1}{2} R_r I^2 = \int_0^{2\pi} \int_0^{\pi} \frac{1}{2Z_o} \left| \boldsymbol{E}(r,\theta,\phi) \right|^2 r^2 \sin\theta \, d\theta \, d\phi \tag{1.40}$$

Using the radiation far field of the infinitesimal current element in (1.16), the radiation resistance of the short current element can be given as:

$$R_r = 80\pi^2 \left(\frac{l}{\lambda} \right)^2 \tag{1.41}$$

As there is no simple analytical formulation for the $\lambda/2$ dipole antenna, the radiation resistance of 73.13Ω can be obtained using the numerical integral for (1.40). An approximation formula for the input impedance of dipole antenna depicted in Figure 1.16 can be given as [8]:

$$Z_i = R(k_o l) - j\left\{ 120\left(\ln\frac{l}{a} - 1 \right)\cot(k_o l) - X(k_o l) \right\} \tag{1.42}$$

$$R(x) = -1.636x + 28.24x^2 - 12.59x^3 + 8.985x^4 \tag{1.43}$$

$$X(x) = 9.748x + 14.23x^2 - 12.31x^3 + 6.934x^4 \tag{1.44}$$

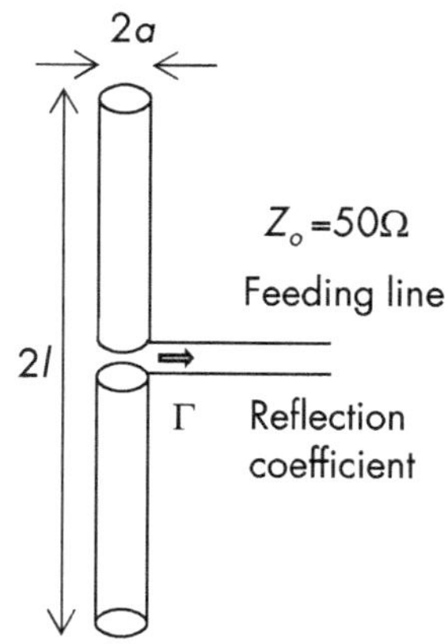

Figure 1.16 Geometry of dipole antenna.

This formula is valid for $l < \lambda/2$. In Figure 1.17, Z_i is shown as a function of frequency and the reflection coefficient for the feed line impedance of 50Ω. This dipole length of $2l$ = 30 cm is half a wavelength at 1 GHz, whereas due to the effect of the antenna thickness of a = 1 mm, the resonant frequency of X = 0 is 0.9 GHz. As shown in Figure 1.17(b), the frequency bandwidth of $|\Gamma| \leq -10$ dB is approximately 18%.

1.7.3 Antenna Effective Length and Area

In a receiving antenna, the effective length multiplies the incoming wave power density to provide the RF open circuit voltage at the antenna feed. Figure 1.18 demonstrates that by converting the real current distribution to the uniform one, the effective length can be obtained as follows:

$$l_e = \frac{1}{I_o}\int I(z)\,dz \tag{1.45}$$

Therefore, assuming the sinusoidal current distribution in (1.29), the effective length of $\lambda/2$ dipole antenna is λ/π.

Figure 1.19 depicts an equivalent circuit of the receiving antenna.

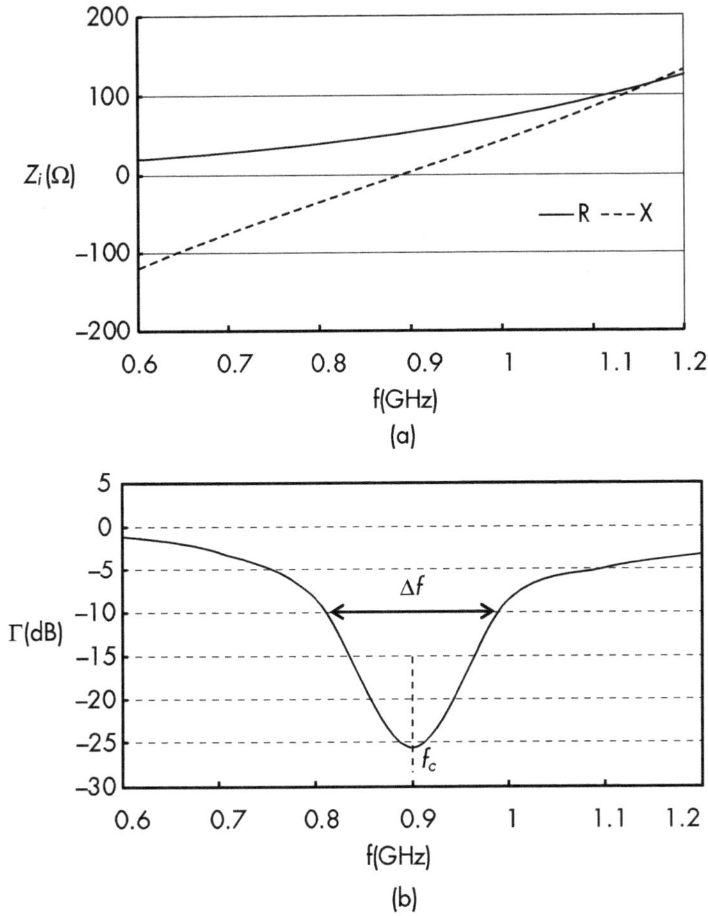

Figure 1.17 Input characteristics of dipole antenna; l = 15 cm, a = 1 mm: (a) input impedance, and (b) reflection coefficient.

The receiving voltage is given by $V_o = El_e$, where E is the electric field strength of incoming wave. The receiving power is calculated using the equivalent circuit, as follows:

$$P_r = \frac{|I_l|^2}{2} R_l = \left| \frac{V_o}{Z_i + Z_l} \right|^2 \frac{R_l}{2} \tag{1.46}$$

The condition for maximizing the receiving power is $Z_l = Z^*_i$ in which the asterisk denotes the complex conjugate of Z_i. Under the

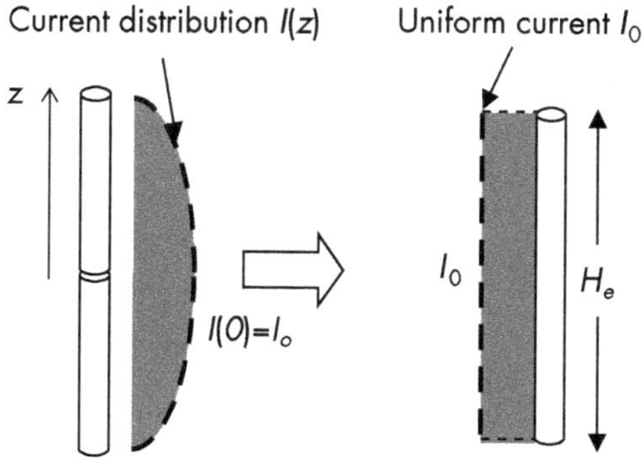

Figure 1.18 Antenna effective length.

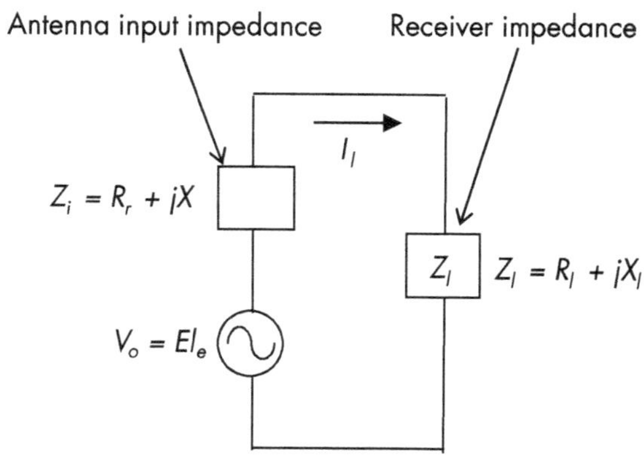

Figure 1.19 Open circuit voltage and equivalent circuit.

impedance matching condition, the available power from the antenna can be obtained, as follows:

$$P_a = \left|\frac{V_o}{2R_r}\right|^2 \frac{R_l}{2} = \frac{V_o^2}{8R_r} \tag{1.47}$$

Additionally, the receiving power at the antenna is represented by the product of the incoming plane wave with the power density of p_i and the antenna's effective area σ_e:

$$P_a = p_i \sigma_e \tag{1.48}$$

Then the effective area can be rewritten by (1.47) using $V_o = El_e$:

$$\sigma_e = \frac{P_a}{p_i} = \frac{\dfrac{V_o^2}{8R_r}}{\dfrac{1}{2Z_o}|E|^2} = \frac{Z_o}{4R_r} l_e^2 \tag{1.49}$$

From (1.49), the effective area of $\lambda/2$ dipole antenna is obtained as $\sigma_e = (0.36\lambda)^2$, which represents the electrical area of the wire antenna that is available for capturing the incoming wave. As shown in Figure 1.20, the physical area (σ_p) of aperture antennas, such as the horn or parabolic antennas, is larger than the effective area.

The ratio between σ_e and σ_p is defined as the aperture efficiency:

$$\eta = \frac{\sigma_e}{\sigma_p} \tag{1.50}$$

1.7.4 Cross Sectional Area and Antenna Gain

The effective antenna area is proportional to antenna gain, and this section presents the relation between these two values. Assuming a receiving antenna is used at the impedance-matched condition of is $Z_l = Z^*_i$, the effective area in (1.49) can be rewritten as follows:

$$\sigma_e = \frac{Z_o}{4R_r} l_e^2 \tag{1.51}$$

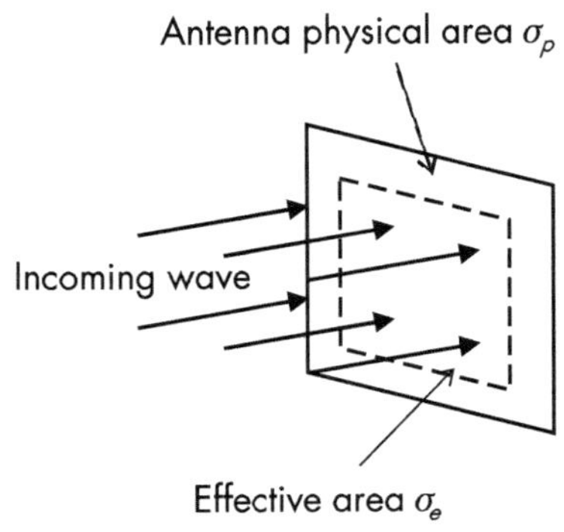

Figure 1.20 Aperture efficiency of antenna.

where the antenna input impedance is $Z_i = R_r + jX$. The peak radiation power density of the $\lambda/2$ dipole antenna is given by (1.30) at $\theta = \pi/2$

$$p_d = \frac{1}{2Z_o}\left|E_\theta\right|^2 = \frac{1}{2Z_o}\left|\frac{jZ_o I_o}{2\pi}\frac{\pi}{2}\frac{e^{-j_o r}}{r}\right|^2 = \frac{Z_o I_o^2}{8\pi^2 r^2} \tag{1.52}$$

Then the power density of the isotropic antenna in (1.35) can be rewritten using the radiation resistance:

$$p_i = \frac{1}{4\pi r^2}\left(\frac{1}{2}R_r I_o^2\right) = \frac{1}{8\pi r^2}R_r I_o^2 \tag{1.53}$$

The definition of antenna gain provides (1.54):

$$G_i = \frac{p_d}{p_i} = \frac{Z_o}{\pi R_r} \tag{1.54}$$

The replacement of the term of Z_o/R_r in (1.51) with (1.54) allows the relation between σ_e and G_i to be expressed as:

$$\sigma_e = \frac{\lambda^2}{4\pi}G_i \tag{1.55}$$

where the effective length λ/π of the $\lambda/2$ dipole antenna is used. Equation (1.55) provides the effective area of antenna from the antenna gain and is valid for other antennas.

1.8 ARRAY ANTENNA

As described earlier, antenna patterns are determined by the shape and size of antenna elements. However, controlling the shape of antenna to obtain a desired pattern is difficult. The array antenna provides another pattern synthesis by controlling the amplitude and phase of each antenna element. This section presents the principles of an array antenna.

1.8.1 Linear Array Antenna

This section considers an equally spaced linear array antenna, as shown in Figure 1.21, which is a one-dimensional array on the x axis. Assuming the in-phase array, each element is excited in phase, and the far field in the zx plane of the array can be obtained by summation, as follows [9]:

$$g(\theta) = \frac{e^{-jk_o r}}{r} \sum_{n=1}^{N} a_n e^{-j(n-1)\delta} \tag{1.56}$$

where a_n is the amplitude coefficient of nth element, and δ, which is defined as $\delta = k_o d \sin(\theta)$, shows the difference in path length of nth element by $(n - 1)\delta$. Equation (1.56) is the summation of a geometric progression in case of an equal-amplitude array, such as $a_n = 1$.

$$g(\theta) = C\frac{e^{-jk_o r}}{r} \frac{\sin\left(\frac{n\delta}{2}\right)}{n\sin\left(\frac{\delta}{2}\right)}, \quad C = ne^{-j\frac{n-1}{2}\delta} \tag{1.57}$$

The radiation pattern is determined by its function with angular variables, which is defined in (1.58) as the normalized array factor:

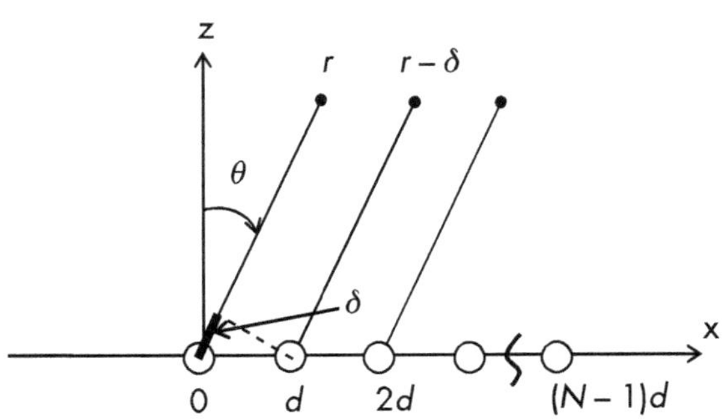

Figure 1.21 Equally spaced linear array.

$$\hat{g}(\theta) = \frac{\sin\left(\frac{n\delta}{2}\right)}{n\sin\left(\frac{\delta}{2}\right)} \tag{1.58}$$

Figure 1.22 presents the array factor dominated by the element spacing d in which the maximum value of 1 at $\delta = 0, \pm m\pi$, where m represents natural numbers. The main lobe is the center peak at $\delta = 0$, and the other lobes are known as grating lobes. To increase the gain of the array antenna, the variable range of δ is required to satisfy the condition of $|\delta| < 2\pi$. Additionally, this range is referred to as the visible region, which is observed region of the array antenna pattern. Outside of the region is the invisible region, as illustrated in Figure 1.22. The condition of the visible region, excluding the grating lobes, can be rewritten as $|\delta \sin \theta| < 1$; here, as $|\sin \theta| \le 1$; the array spacing should be $d < \lambda$.

An advantage of the array antenna is to electrically change the main beam direction. The condition of the main beam tilted to $\theta = \theta_o$ is given by:

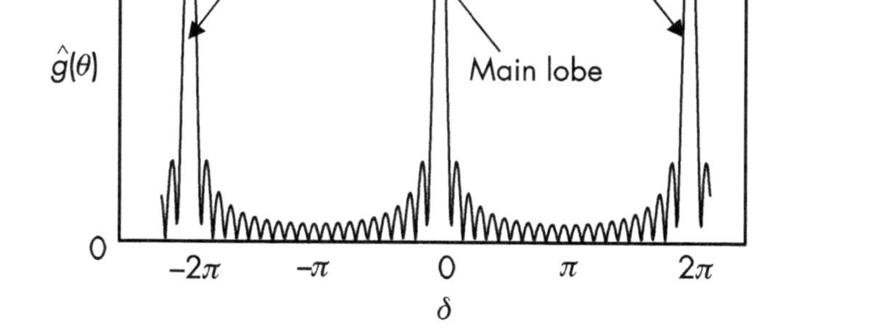

Figure 1.22 Array factor of equally spaced linear array.

$$\delta = k_o d\left(\sin\theta - \sin\theta_o\right) \tag{1.59}$$

This exciting condition is provided by adding the progressive phase difference to each element. The additional phase of each element is $(n-1)k_o d \sin\theta_o$. The variable range of δ should be $|k_o d(\sin\theta - \sin\theta_o)| < 2\pi$, excluding the grating lobes in the visible region. This formula can be rewritten as (1.60) using the condition of $|\sin\theta| \leq 1$:

$$d < \frac{\lambda}{1 + \sin\theta_o} \tag{1.60}$$

For the arbitrary tilt angle θ_o, the array spacing d should be $d < \lambda/2$. Figure 1.23 shows the array antenna patterns in dBs in which the number of elements is $n = 2$, 5, and 10. Due to the symmetry, the pattern for $0 \leq \theta \leq 90°$ is shown. Increasing the number of elements causes the main beam width to narrow; however, the first side level is a constant value of −13.2 dB. The increase in the number of elements

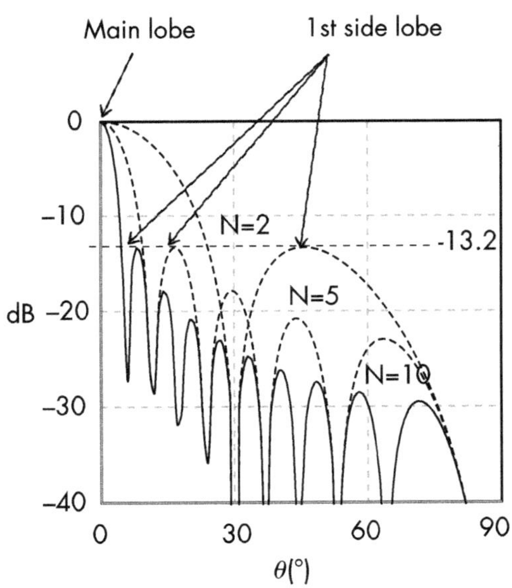

Figure 1.23 Array antenna pattern.

extends the width of visible region. It does not alter the level of the first side lobe, but it does include many side lobes (see Figure 1.23).

This formulation is discussed by ignoring the radiation pattern of the antenna element. The pattern of the array is obtained using (1.57) by introducing the element pattern $f(\theta)$:

$$g(\theta) = C\frac{e^{-jk_o r}}{r} f(\theta)\hat{g}(\theta) \tag{1.61}$$

To understand the array and element factor, a short dipole array is considered (see Figure 1.24). The radiation pattern of the short

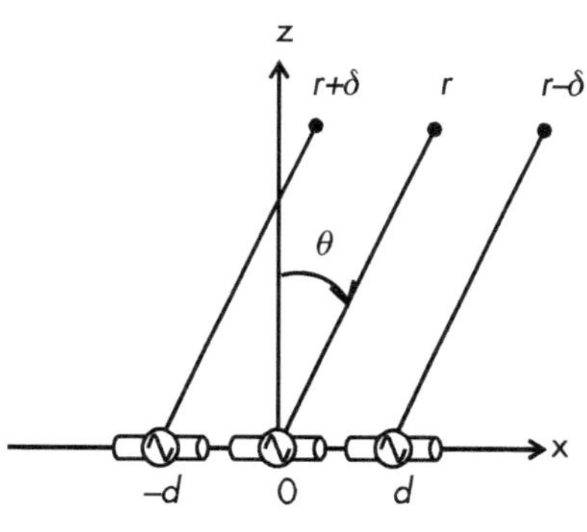

Figure 1.24 Three-element array.

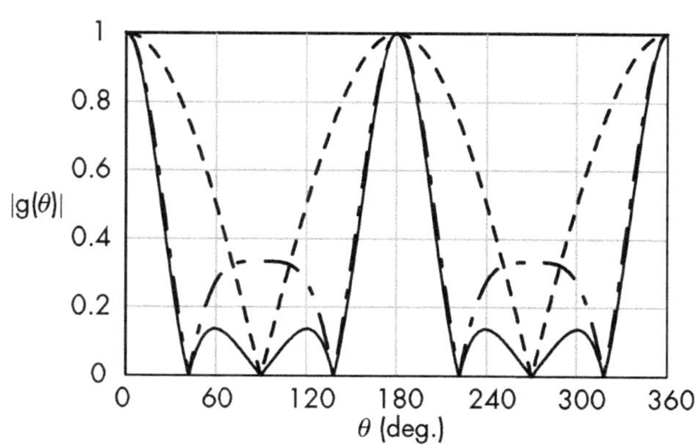

Figure 1.25 Radiation pattern of three-element array: the solid line is array pattern, the dotted line is element factor, the dashed and dotted line is array factor.

dipole is approximated as that of the infinitesimal current element. The $\sin\theta$ in (1.16) is rotated by 90° in Figure 1.24, and the element factor is $f(\theta) = \cos\theta$. It can be seen in Figure 1.25 that the main lobes of the array pattern and array factor is almost identical. The effect of the element factor appears around $\theta = 90°, 270°$, where null points of element factor reduce the level of the side lobe of the array factor. This result indicates that the array factor antenna dominates the radiation of an array antenna.

1.9 MUTUAL COUPLING

The performance of an array antenna is deteriorated by the mutual coupling between the array elements, and its effect should be reflected in the antenna's design. This section describes a simple model of mutual coupling using two dipole antennas to determine its effect according to the spacing of the elements.

Figure 1.26 shows two dipole arrays with a spacing of d. The current amplitude and length of the dipole are also shown. As is discussed in Figure 1.8, the radiation pattern of the short dipole antenna is almost the same with when using a $\lambda/2$ dipole. First, the mutual coupling between two short dipoles is examined. The electric field radiated by the short dipole 1 is expressed using (1.13) at dipole 2 after substituting $\pi/2$ into θ and d into r, respectively.

$$E_{21} = Z_o \frac{I_1 l_1}{4\pi} k_o^2 \left\{ j\frac{1}{(k_o d)} + \frac{1}{(k_o d)^2} - j\frac{1}{(k_o d)^3} \right\} e^{-jk_o d} \tag{1.62}$$

To excite the current I_2 at the dipole 2 against the above electric field by dipole 1, the induced field of $-E_{21}$ is required at the feed of dipole 2. The power density induced by this electric field is expressed by $-E_{21}I^*_2$, and the total power required at dipole 2 is obtained by its

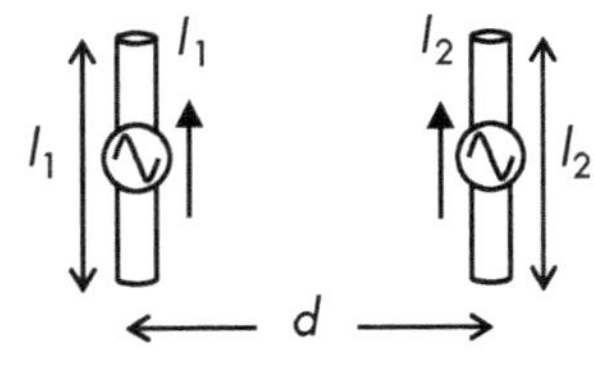

Figure 1.26 Dipole array.

integration along the length l_2. Then the mutual impedance between two dipoles is defined, as follows:

$$Z_{21} = -\frac{1}{I_1 I_2}\int E_{21} I_2^* \, dz = Z_o \frac{l_1 l_2}{4\pi} k_o^2 \left\{ j\frac{1}{(k_o d)} + \frac{1}{(k_o d)^2} - j\frac{1}{(k_o d)^3} \right\} e^{-jk_o d} \tag{1.63}$$

This procedure is known as the electromotive force method [9].

The following conversion provides the mutual coupling evaluated by S parameters:

$$[S] = \{[\hat{Z}] + [1]\}^{-1} \{[\hat{Z}] - [1]\} \tag{1.64}$$

where $[\hat{Z}]$ is the impedance matrix normalized by the value of Z_{11}, and [1] is the unit matrix. Assuming that each dipole is excited by an ideal matching circuit to cancel its reactance part, the self-impedance of Z_{11} and Z_{22} can be obtained by the radiation resistance of the short dipole.

Figure 1.27 illustrates the mutual coupling between two shorts dipoles using the previous procedure as well as the mutual coupling between two λ/2 dipoles. This calculation, which was performed by the electromagnetic simulation software FEKO [10], is in good

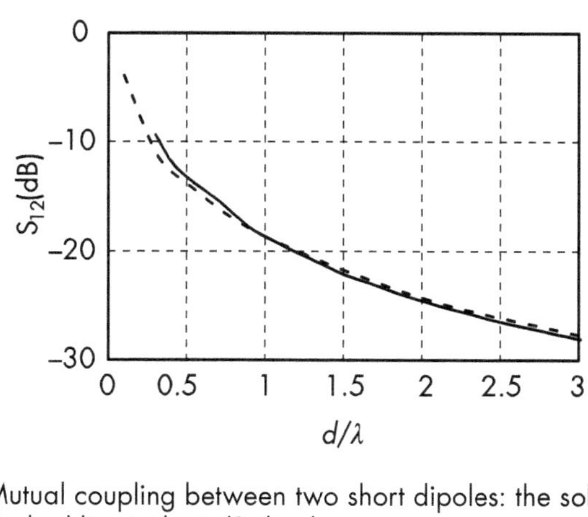

Figure 1.27 Mutual coupling between two short dipoles: the solid line is the short dipole, and the dashed line is the λ/2 dipole.

agreement with the calculation for the $\lambda/2$ dipole. In the design of the array antenna, $\lambda/2$ dipole array provides good approximation results.

The linear array discussed in Section 1.8.1 indicates that the spacing of the array element should be more than $\lambda/2$ for beam tilting. As shown in Figure 1.27, and with the exception of high-performance array antennas, the mutual coupling of −14dB for $d = \lambda/2$ is almost negligible.

References

[1] Grant, I. S., and W. R. Phillips, *Electromagnetism*, Second Edition, New York: John Wiley & Sons, 1990, pp. 356–359.

[2] Collin, R. E., *Field Theory of Guided Wave*, Second Edition, Piscataway, NJ: IEEE Press, 1991, pp. 30–34.

[3] Collin, R. E., *Field Theory of Guided Wave*, Second Edition, Piscataway, NJ: IEEE Press, 1991, pp. 34–37.

[4] Kong, J. A., *Electromagnetic Wave Theory*, Second Edition, New York: John Wiley & Sons, 1990, pp. 235–240.

[5] Collin, R. E., *Field Theory of Guided Wave*, Second Edition, Piscataway, NJ: IEEE Press, 1991, pp. 39–43.

[6] Balanis, C. A., *Antenna Theory Analysis and Design*, Third Edition, New York: John Wiley & Sons, 2005, pp. 170–171.

[7] Balanis, C. A., *Antenna Theory Analysis and Design*, Third Edition, New York: John Wiley & Sons, 2005, pp. 85–86.

[8] Johnson, R. C., *Antenna Engineering Handbook*, Third Edition, 4-2, New York: McGraw-Hill, 1993.

[9] Volakis, J. L., *Antenna Engineering Handbook*, Fourth Edition, 3-3, New York: McGraw-Hill, 2007.

[10] https://altair.com/feko-applications.

2

BASIC THEORY OF ELECTROMAGNETIC WAVE PROPAGATION

This chapter explains the fundamentals of electromagnetic (EM) wave propagation to understand the phenomena for mobile communication systems. The first part of this chapter describes the basic characteristics of the plane wave, such as propagation, reflection, transmission, and diffraction. Electromagnetic waves radiated from the base station arrive at a receiving point through several paths, which causes serious fluctuation known as fading. Its fundamental characteristics, derived by statistical models, and the diversity technique to overcome fading are described in the subsequent sections. Additionally, the paths of different lengths arrive at the receiving position with a time difference. The delay profile and angular profile of electromagnetic waves are used to determine the properties of digital mobile communication systems.

2.1 CLASSIFICATION OF PROPAGATION PATHS

Electromagnetic waves transmitted from a base station arrive at a receiving terminal, propagating through several paths. A typical propagation path model from the base station installed at the top of a building is shown in Figure 2.1. At the receiving position A, the base station antenna can be seen, and a direct path is dominant. This is

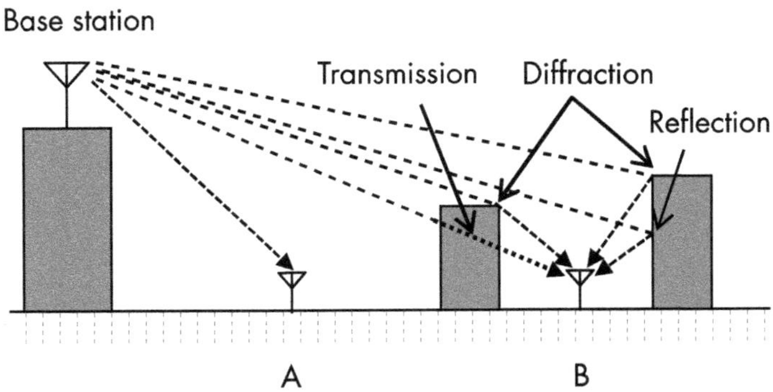

Figure 2.1 Paths from base station to receiving point.

known as the line-of-sight (LOS) propagation. Position B is placed at the shadow region from the base station and is defined as non–line of sight (NLOS). As shown in Figure 2.1, microwaves used in the current cellular system reach position B via reflection from the building wall, diffraction at the top corner of the building, and transmission through the building.

The strength of electromagnetic waves decreases as a function of the propagation distance because of the expansion of waves. This attenuation is known as the propagation loss. The received signal is decreased furthermore by the transmission, reflection, and diffraction. These factors are explained using simple models of wave propagation in a later section.

2.2 PROPAGATION LOSS IN FREE SPACE

The transmitting and receiving positions are assumed to be placed in free space to obtain the propagation loss, and no obstacles exist along the propagation path. As shown in Figure 2.2, the transmitting power P_t is radiated uniformly into space, which is equivalent to use of an isotropic antenna defined in Section 1.6. The power density p_d at the distance d from the transmitting position is given as follows:

$$p_d = \frac{P_t}{4\pi d^2} \tag{2.1}$$

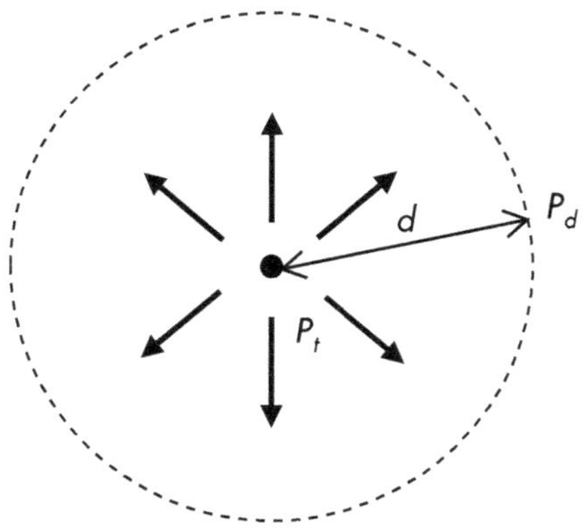

Figure 2.2 Transmitting power and power density.

The next step is to obtain the receiving power by introducing the antenna gain. The models to derive the receiving power using the propagation loss are shown in Figure 2.3. The transmitting antenna gain G_t increases the power density, and the receiving power P_r can be obtained using the cross-sectional area of the receiving antenna defined by (1.55) as follows [1]:

$$P_r = G_t \frac{P_t}{4\pi d^2} \frac{\lambda^2}{4\pi} G_r = \left(\frac{\lambda}{4\pi d} \right)^2 G_r G_t P_t \tag{2.2}$$

where G_r is the antenna gain of the receiving antenna and d is the distance between the transmitting and receiving antennas. The term $(\lambda/4\pi d)^2$ in (2.2), the loss factor due to propagation distance, is designated as a propagation loss.

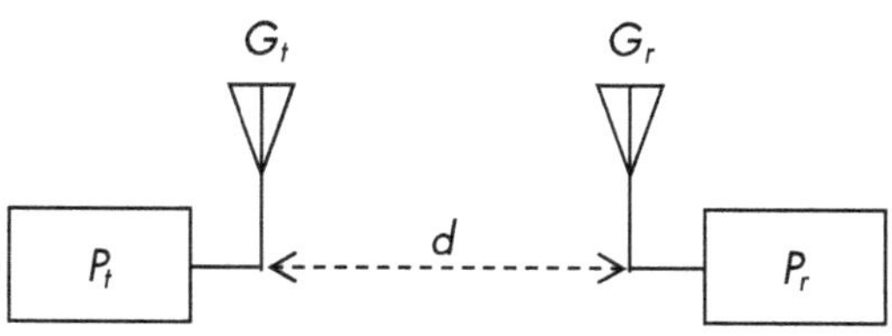

Figure 2.3 Propagation distance and propagation loss.

2.3 REFLECTION AND TRANSMISSION AT THE BOUNDARY

The boundary value condition determines how electromagnetic waves are reflected and transmitted through obstacles. The condition is that the tangential components of electric and magnetic fields are continuous across a boundary when no charges and current sources exist on the boundary. Figure 2.4 shows a simple model in which the wave is perpendicularly incident on the boundary between two consecutive obstacles.

The boundary separates regions 1 and 2, where the characteristic impedance in each region is denoted as Z_1 and Z_2, respectively. The reflected wave has the opposite magnetic field to that of the incident one using the definition of Poynting vector as follows [2]:

$$\boldsymbol{S} = \frac{1}{2}\text{Real}\left(\boldsymbol{E} \times \boldsymbol{H}^*\right) \tag{2.3}$$

where the asterisk denotes the complex conjugate of the magnetic field of the incident wave. The Poynting vector defines the direction of the propagating wave, and its amplitude is the power density of the wave. The subscript in each field indicates the incident, reflected, and transmitted waves, respectively.

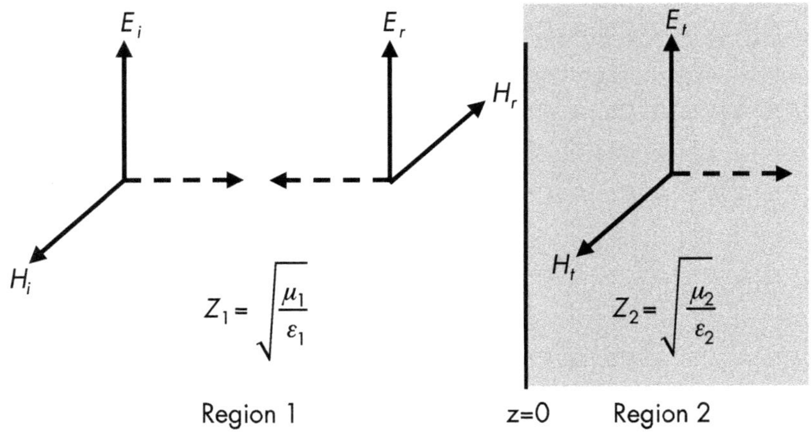

Figure 2.4 Reflection and transmission at the boundary.

Tangential field components at the boundary should satisfy the following boundary value conditions:

$$(1+\Gamma)E_i = E_t, \quad (1-\Gamma)H_i = H_t \tag{2.4}$$

where Γ is the reflection coefficient of an electric field, and the reflected wave is described by $E_r = \Gamma E_i$ and $H_r = -\Gamma H_i$. The magnetic field in each region may be expressed using the characteristics impedance and the electric field, and the reflection coefficient can be obtained by solving (2.4) as follows [3]:

$$\Gamma = \frac{Z_2 - Z_1}{Z_2 + Z_1} \tag{2.5}$$

It should be noted that the reflection coefficient is $\Gamma = -1$, assuming that region 2 is the perfect conductor, ($Z_2 = 0$).

Then, the oblique incidence is considered using the wavenumber vectors as shown in Figure 2.5. The incident and reflection angles are denoted as θ_i and θ_r in region 1 and the transmission angle is θ_t in region 2. The wavenumber vector describes the wavelength of the electromagnetic waves. Since the waves are continuous, they should

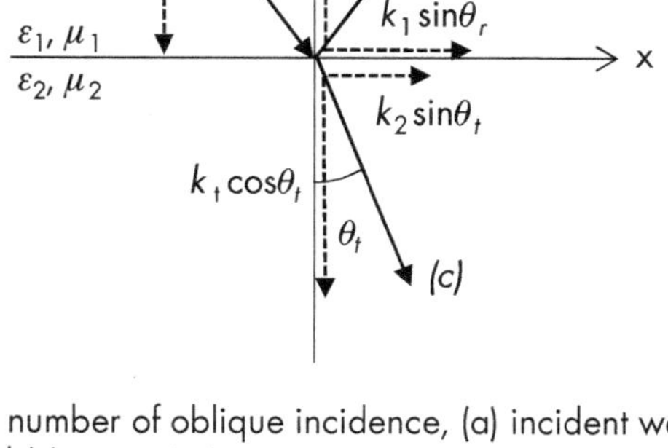

Figure 2.5 Wave number of oblique incidence, (a) incident wave vector, (b) reflection wave vector, and (c) transmission wave vector.

follow the following boundary condition to avoid the discontinuity [4].

$$k_1 \sin\theta_i = k_1 \sin\theta_r, \quad k_1 \sin\theta_i = k_2 \sin\theta_t \tag{2.6}$$

The first equation provides the law of reflection, which states that the incident and reflection angles are identical. The second is the same as Snell's law. The refractive index n_i in each region is described in (2.7):

$$n_i = \frac{c}{v_i} = \sqrt{\frac{\varepsilon_i \mu_i}{\varepsilon_o \mu_o}}, \quad i = 1,2 \tag{2.7}$$

where c is the velocity of light and ε_o, μ_o, is the permittivity and permeability in free space. Using (2.6) and (2.7), a well-known formula of Snell's law is obtained as follows:

$$\frac{\sin\theta_i}{\sin\theta_t} = \frac{n_1}{n_2} \tag{2.8}$$

Snell's law defined by (2.8) is a formula of the relationship between the incident and reflection angles, referring to waves passing through a boundary between two different regions.

2.4 OBLIQUE INCIDENCE

This section describes the oblique incidence of electromagnetic waves on the boundary to obtain the reflection and transmission coefficient in general. Figures 2.6 and 2.7 show two incident cases, parallel and orthogonal polarization, respectively. The boundary plane is placed at $y = 0$, and the xy plane is known as the incident plane. Assuming that the incident electric and magnetic fields are orthogonal to one another, as in Section 2.2, the parallel polarization of the electric field parallel to the incident plane is defined as shown in Figure 2.6. The magnetic field parallel to the incident plane is specified by the orthogonal polarization (Figure 2.7). Additionally, the parallel and orthogonal polarization is known as the transverse magnetic (TM) and transverse electric (TE) polarization. In TM and

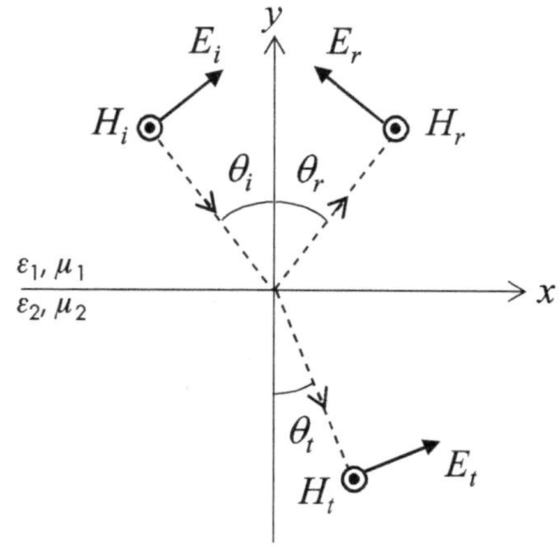

Figure 2.6 Parallel polarization (transverse magnetic).

TE cases, the normal component to the incident plane is selected to designate the polarization.

The reflection and transmission coefficients are obtained using the same procedure as in Section 2.2, solving the simultaneous equations using the tangential components at the boundary surface. The results for the parallel polarization are as follows:

$$R_p = \frac{E_r}{E_i} = \frac{Z_1 \cos\theta_i - Z_2 \cos\theta_t}{Z_1 \cos\theta_i + Z_2 \cos\theta_t} = \frac{\mu_1 n^2 \cos\theta_i - \mu_2 \sqrt{n^2 - \sin^2\theta_i}}{\mu_1 n^2 \cos\theta_i + \mu_2 \sqrt{n^2 - \sin^2\theta_i}} \quad (2.9)$$

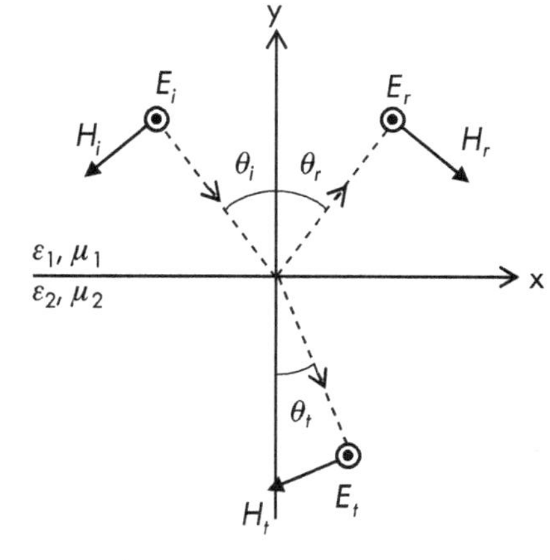

Figure 2.7 Orthogonal polarization (transverse electric).

$$T_p = \frac{E_t}{E_i} = \frac{2Z_2 \cos\theta_i}{Z_1 \cos\theta_i + Z_2 \cos\theta_t} = \frac{2\mu_2 n \cos\theta_i}{\mu_1 n^2 \cos\theta_i + \mu_2 \sqrt{n^2 - \sin^2\theta_i}} \tag{2.10}$$

And those of the orthogonal polarization are as follows:

$$R_o = \frac{E_r}{E_i} = \frac{Z_2 \cos\theta_i - Z_1 \cos\theta_t}{Z_2 \cos\theta_i + Z_1 \cos\theta_t} = \frac{\mu_2 \cos\theta_i - \mu_1 \sqrt{n^2 - \sin^2\theta_i}}{\mu_2 \cos\theta_i + \mu_1 \sqrt{n^2 - \sin^2\theta_i}} \tag{2.11}$$

$$T_o = \frac{E_t}{E_i} = \frac{2Z_2 \cos\theta_i}{Z_2 \cos\theta_i + Z_1 \cos\theta_t} = \frac{2\mu_2 \cos\theta_i}{\mu_2 \cos\theta_i + \mu_1 \sqrt{n^2 - \sin^2\theta_i}} \tag{2.12}$$

where the subscript p or o denotes the parallel or orthogonal polarization, respectively, and the ratio of the refractive index is denoted as $n = n_2/n_1$.

Figure 2.8 shows an example of the reflection and transmission coefficient as a function of the incident angle for the parallel polarization, where the permeability in regions 1 and 2 is the same and $n^2 = 2$. It should be noted that the reflection coefficient becomes zero at a specific incident angle known as Brewster's angle around $\theta_i = 55°$

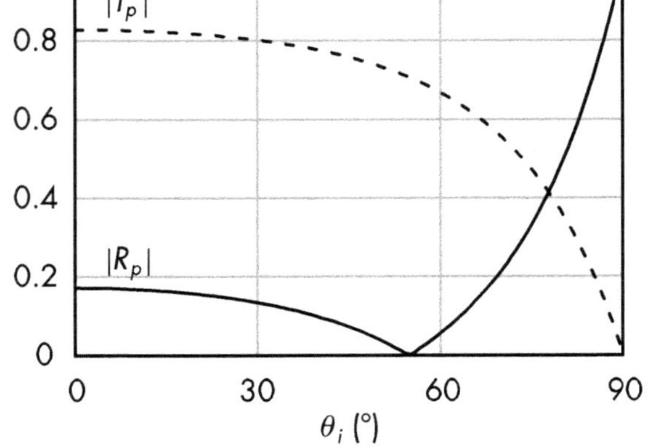

Figure 2.8 Reflection and transmission coefficient of parallel polarization for $\mu_1 = \mu_2$, $n^2 = 2$.

in Figure 2.8. This angle can be obtained from (2.9) for $\mu_1 = \mu_2$ as follows:

$$\cos\theta_i = \frac{1}{\sqrt{n^2 + 1}} \tag{2.13}$$

Brewster's angle is obtained only for the parallel polarization and not for the orthogonal polarization, assuming the same permeability in both regions [4]. This is because (2.11) shows that the condition for $R_o = 0$ can be given by $n^2 = 1$, which indicates identical electrical parameters in both regions.

2.5 EFFECT OF GROUND REFLECTION

This section describes the effect of reflection from the ground. Figure 2.9 shows that electromagnetic waves transmitted from point O arrive at the receiving point A via a direct path $\overline{\text{OA}}$ and ground reflected path $\overline{\text{OP}} + \overline{\text{PA}}$. The intersection of the path can obtain reflection point P, $\overline{\text{OA}'}$, and the ground level line, assuming an image of point A′. This is based on Fermat's principle that a ray between two given points is taken the path traveling in the least time [5]. The received electric field can be given by the addition of two paths as follows:

$$E_A = E_o\left(\frac{e^{-jk_od_1}}{d_1} + R\frac{e^{-jk_od_2}}{d_2}\right) \tag{2.14}$$

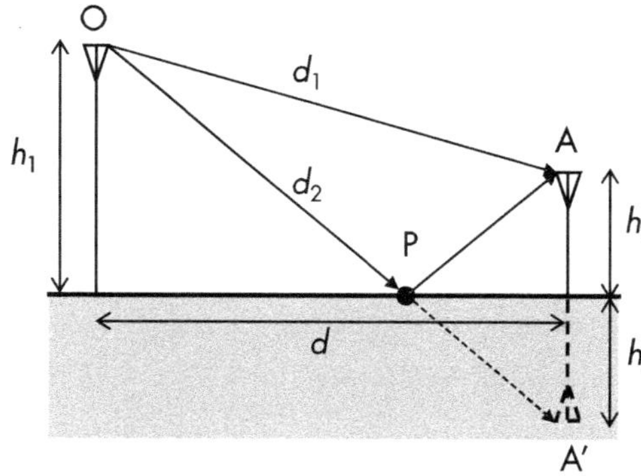

Figure 2.9 Direct patch and reflection from the ground.

where E_o and R are the amplitude of the transmitted wave and the reflection coefficient of the ground, respectively. Equation (2.15) approximates the difference between the two paths, assuming the distance d is much larger than h_1 and h_2.

$$d_2 - d_1 \cong \frac{2h_1h_2}{d} \tag{2.15}$$

The receiving electric field can be obtained by substituting (2.15) into (2.14) and by approximating the denominator in (2.14) as $d_1 = d_2 \cong d$:

$$E_A \cong E_o \frac{e^{-jk_od_1}}{d}\left\{1 + \mathrm{R}e^{-j\frac{2k_oh_1h_2}{d}}\right\} \tag{2.16}$$

Additionally, the ground can be regarded as the perfect conductor ($R = -1$) and the absolute value of (2.16) is rewritten as follows [6]:

$$\left|E_A\right| \cong \frac{E_o}{d}\left|1 - e^{-j\frac{2k_oh_1h_2}{d}}\right| = \frac{2E_o}{d}\left|\sin\left(\frac{k_oh_1h_2}{d}\right)\right| \tag{2.17}$$

The receiving electric field strength at point A is changed periodically by increasing A's height (see Figure 2.10), which is known as the height pattern. Figure 2.11 presents the variation of $|E_A|$ as a function of the distance d, where the heights h_1 and h_2 are taken as fixed values. The strength of $|E_A|$ in (2.17) is changed to the specified distance as shown in Figure 2.11, and its periodical variation is expired after that. This position, known as the breakpoint, can be obtained from the difference between two paths being less than $\lambda/2$ as follows:

$$d_b = \frac{4h_1h_2}{\lambda} \tag{2.18}$$

When the distance d is much larger than h_1 and h_2 and satisfies the condition of $d \gg h_1h_2$, (2.17) can be rewritten as:

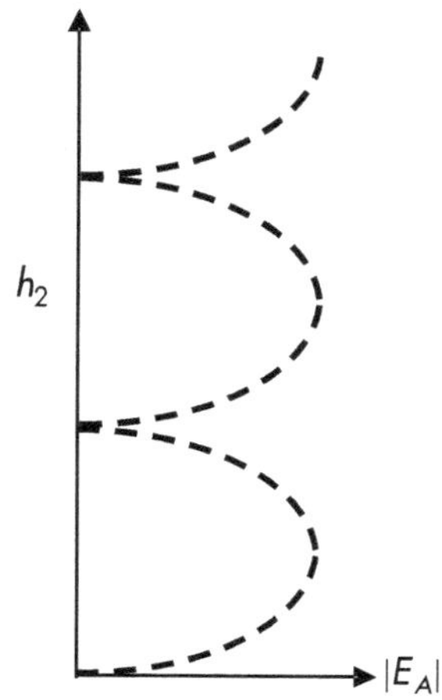

Figure 2.10 Height pattern.

$$|E_A| \cong \frac{2E_o k_o h_1 h_2}{d^2} \tag{2.19}$$

The electric field strength is proportional to $1/d^2$, and then the power level decreases by $1/d^4$ beyond the breakpoint. In Figure 2.11, the

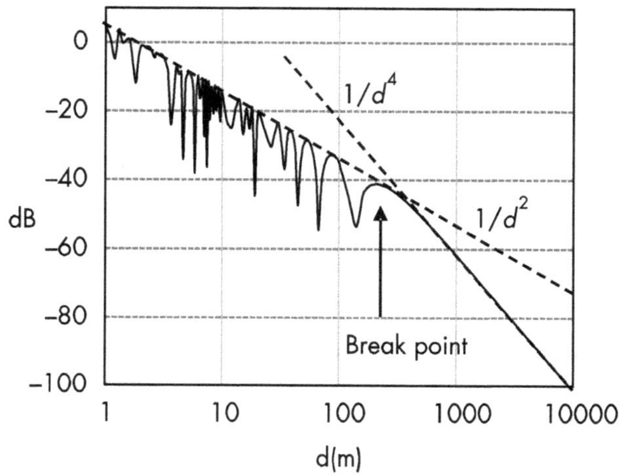

Figure 2.11 Break point; f = 2GHz, h_1 = 10, and h_2 = 1m.

distance of the breakpoint is $d_b = 266$m. This is a simple propagation model, and the power decrease beyond d_b is larger than $1/d^4$ in a real environment.

2.6 KNIFE-EDGE DIFFRACTION

Electromagnetic waves can be received in the shadow region via reflection and diffraction as described in Section 2.1. The reflection characteristics are explained in the preceding sections, then knife-edge diffraction is discussed in this section. As Figure 2.12 shows, mountains and buildings are approximated as a knife-edge. T and R denote the transmitting and receiving positions, and u is the length from the straight line of $\overline{\text{TR}}$ to the edge top. The LOS environment is obtained from $u < 0$, and the NLOS by $u > 0$.

Diffracted waves from the knife-edge can be obtained by assuming the source sheet extension in the semi-infinite region S above the edge, which is designated as $-\infty < x < \infty$, $y > H$ and $z = 0$, as shown in the coordinate system in Figure 2.13. Based on Huygens' principle, the secondary sources can be given by the electromagnetic waves at the source sheet region from the transmitting point T. At the $P(x, y)$ on the source sheet, the electromagnetic wave is expressed as:

$$E_P = E_o \frac{e^{-jk_o r_1}}{r_1} \tag{2.20}$$

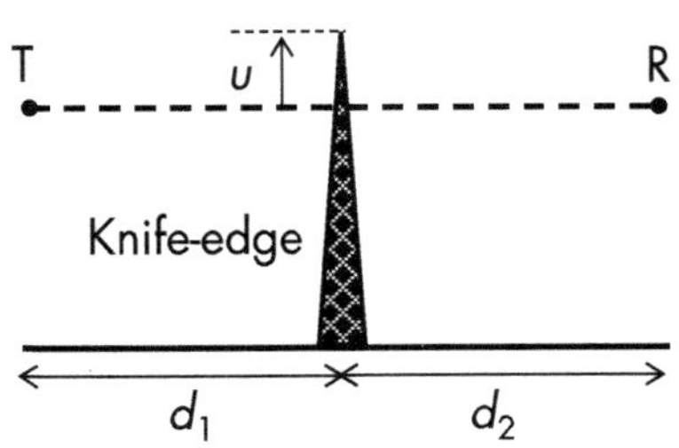

Figure 2.12 Knife-edge diffraction.

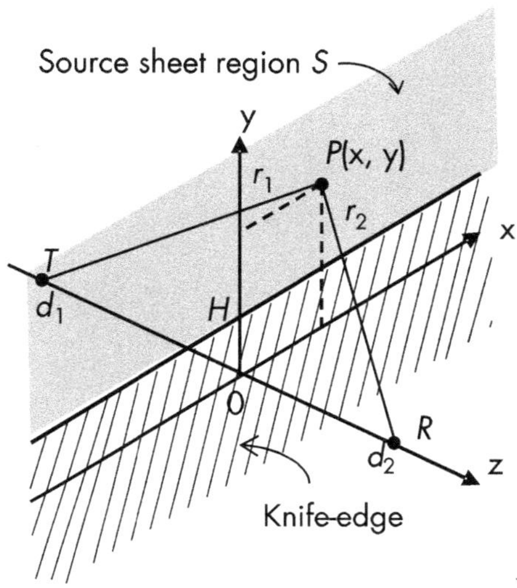

Figure 2.13 Coordinate system of knife-edge.

where E_o is the amplitude coefficient. The receiving electric field can be obtained by integrating the source sheet as follows:

$$E_R = \frac{j2k_oE_o}{4\pi}\int_S \frac{e^{-j_or_1}}{r_1}\frac{e^{-j_or_2}}{r_2}dS \tag{2.21}$$

Assuming $x,y \gg d_1,d_2$, the distances r_1 and r_2 can be approximated by (2.22) as:

$$r_i \cong d_i + \frac{x^2+y^2}{2d_i}, \quad i = 1,2 \tag{2.22}$$

This approximation is effective because most electromagnetic waves are concentrated around the edge top at $x = 0$, $y = H$.

Substituting (2.22) into (2.21), the receiving field E_R can be rewritten as follows:

$$E_R \cong \frac{j2k_oE_o}{4\pi}\frac{e^{-jk_o(d_1+d_2)}}{d_1d_2}\int_{-\infty}^{\infty}\int_{H}^{\infty} e^{-jk_o\frac{x^2+y^2}{2d_1}}e^{-jk_o\frac{x^2+y^2}{2d_2}}dx\,dy \tag{2.23}$$

where the integrals in (2.23) are given by the Fresnel integral [7]:

$$C(s) = \int_0^s \cos\left(\frac{\pi}{2}t^2\right)dt, \quad S(s) = \int_0^s \sin\left(\frac{\pi}{2}t^2\right)dt \tag{2.24}$$

Introducing a new variable u for edge height H by (2.25), the diffracted fields are expressed as follows [8]:

$$u = H\sqrt{\frac{2}{\lambda}\left(\frac{1}{d_1} + \frac{1}{d_2}\right)} \tag{2.25}$$

No knife-edge case (free space propagation) is given by $u \to -\infty$:

$$E_{Ro} = E_o \frac{e^{-jk_o(d_1+d_2)}}{d_1 + d_2} \tag{2.26}$$

The receiving point in the shadow region by knife-edge is designated by $u \geq 0$ and the diffraction coefficient $S(u)$ can be obtained from the diffracted field normalized by (2.26) as follows:

$$S(u) = \left|\frac{E_R}{E_{Ro}}\right| = \frac{1}{\sqrt{2}}\left|\frac{1}{2} - j\frac{1}{2} + \{C(u) - jS(u)\}, \quad u \geq 0\right| \tag{2.27}$$

The direct path from T to R is not shadowed for $u < 0$, and the $S(u)$ is given by replacing u with $u' = -u$:

$$S(u') = \left|\frac{E_R}{E_{Ro}}\right| = \frac{1}{\sqrt{2}}\left|\frac{1}{2} - j\frac{1}{2} - \{C(u') - jS(u')\}\right|, \quad u < 0 \tag{2.28}$$

Figure 2.14 shows the diffraction coefficient of the knife-edge. In the LOS environment for $u < 0$, the $S(u)$ varies periodically and converges to the free space level of E_{Ro}. The $S(u)$ often exceeds the free space case, which is as the diffraction gain. This phenomenon is remarkable at $u \cong 0$ as shown in Figure 2.14. In the shadow region, $u > 0$, the diffracted field decreases exponentially.

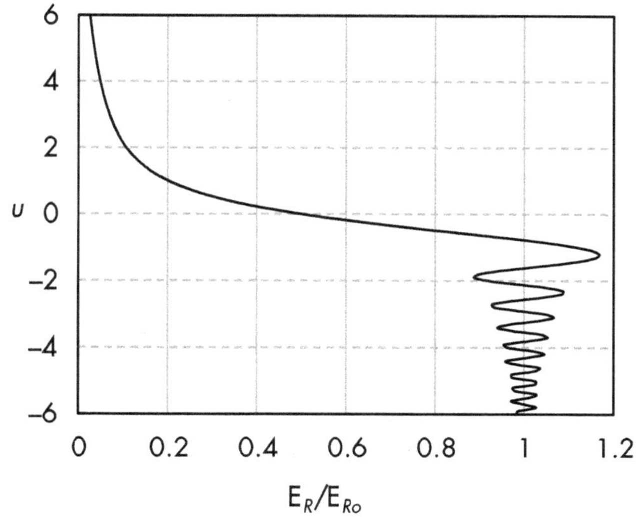

Figure 2.14 Diffraction coefficient of knife-edge.

The knife edge diffraction is also given by the Vogler solution and is extended to the case of multiple edges using a formula expressed as multiple diffraction integrals. The method is obtained by the use of the repeated integrals of the error function, which can improve the computation accuracy [9, 10].

2.7 FRESNEL ZONE

Due to the phase mismatch between the direct and diffracted paths, the diffracted field by the knife edge becomes greater or smaller than the free space level. This phase difference is described in detail in this section. Figure 2.15 shows the geometry to consider the location of the obstacle to create diffracted waves. An interference object Q is located on the z-axis, and the distance of the transmission T and receiving R positions from the origin is d_1 and d_2, respectively. The difference between the direct and diffracted paths is approximated by assuming that the Q position is much smaller than d_1 and d_2 as follows:

$$\Delta d = \sqrt{\left(d_1^2 + z^2\right) + \left(d_2^2 + z^2\right)} - d_1 - d_2 \cong \frac{z^2}{2}\left(\frac{1}{d_1} + \frac{1}{d_2}\right) \tag{2.29}$$

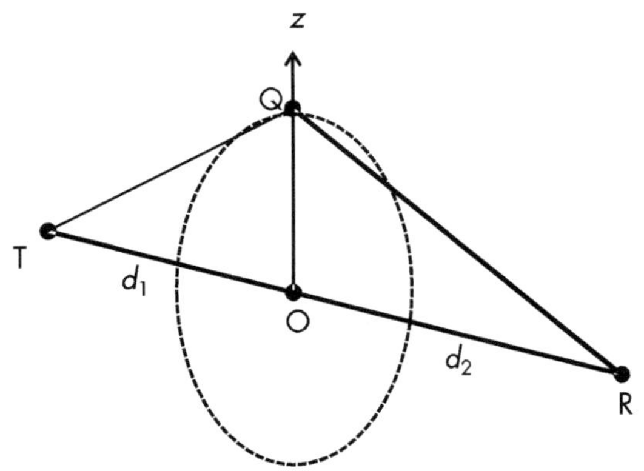

Figure 2.15 Fresnel zone.

This distance difference is examined using the detracted coefficient in Figure 2.14. When the difference is $\lambda/2$, the object position can be obtained as follows:

$$z = \sqrt{\frac{\lambda}{\left(\frac{1}{d_1} + \frac{1}{d_2}\right)}} \tag{2.30}$$

Substituting (2.30) into (2.25), the variable $u = \sqrt{2}$ is obtained. In this case the edge is below the line of TR, and the diffracted wave enhances the direct path and is defined as the first Fresnel zone. In the next region of $\lambda/2 < \Delta d < \lambda$ ($\sqrt{2} < u < 2$), the direct path is weakened by the diffracted one and is defined as the second Fresnel zone [11]. Similarly, the diffracted wave enhances the direct path in the odd number zone and vice versa in the even number zone.

The Fresnel zone exists as a ring around the line of TR and, when the origin moves along the TR, it becomes a spheroid. To avoid variations in the direct path between the transmitting and receiving positions, the interference object should not be present inside the first Fresnel region.

2.8 FADING STRUCTURE

Multipath fading occurs when electromagnetic waves emitted by a transmitting antenna reach the receiving point P through various reflections and diffractions. This section describes the fading structure from a statistical perspective. The signals arriving at point P are concentrated in a horizontal plane when the transmitting antenna is located far from the point P. Figure 2.16 shows that the receiving point P on a vehicle moves with a velocity of v (m/s). For this situation, N waves arrive from the Φ_i direction, each with an amplitude A_i and phase ψ_i. The time-varying electric field $E_r(t)$ at point P is expressed in terms of the angular frequency, ω_c as [12]:

$$E_r(t) = \sum_{i=1}^{N} A_i \cos\left(\omega_c t + \psi_i\right), \quad \psi_i = \zeta_i + 2\pi \frac{V}{\lambda} \cos \Phi_i \tag{2.31}$$

where the phase ψ_i consists of the phase delay of the receiving signal ζ_i and the frequency shift due to the Doppler effect caused by the movement of the vehicle. Using standard trigonometric expressions to expand (2.31), E_r(t) can be rewritten as follows:

$$\begin{aligned} E_r(t) &= \sum_{i=1}^{N} x_i(t) \cos \omega_c t - \sum_{i=1}^{N} y_i(t) \sin \omega_c t \\ &= x(t) \cos \omega_c t - y(t) \sin \omega_c t \end{aligned} \tag{2.32}$$

where $x_i(t) = A_i \cos \psi_i$ and $y_i(t) = A_i \sin \psi_i$.

When the amplitudes of the individual waves A_i are almost the same and the phase ζ_i changes randomly, the coefficients of $x(t)$ and $y(t)$ are given by the Gaussian distributions:

$$x(t) = \frac{1}{\sqrt{2\pi}\sigma} \exp\left(-\frac{x^2}{2\sigma^2}\right), \quad y(t) = \frac{1}{\sqrt{2\pi}\sigma} \exp\left(-\frac{y^2}{2\sigma^2}\right) \tag{2.33}$$

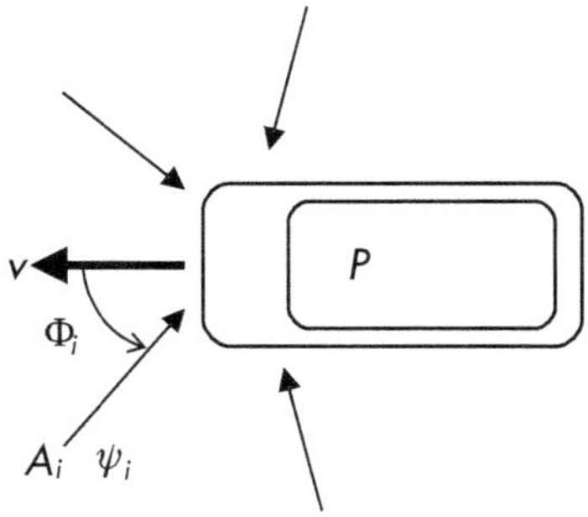

Figure 2.16 Arrival waves at moving vehicle.

where the mean values of both $x_i(t)$ and $y_i(t)$ are 0, with the same variance σ as:

$$\sigma^2 = \frac{1}{N}\sum_{i=1}^{N} x_i(t) = \frac{1}{N}\sum_{i=1}^{N} y_i(t) \tag{2.34}$$

It should be noted that $\sigma_r = 2\sigma^2$ is defined as the average of receiving power.

The Gaussian distributions $x(t)$ and $y(t)$ are replaced by the following probability density function, assuming $x(t)$ and $y(t)$ are independent of each other as:

$$f(x, y) = \frac{2}{\pi\sigma_r}\exp\left(-\frac{x^2 + y^2}{\sigma_r}\right) \tag{2.35}$$

In (2.35), the amplitude of the total receiving field strength is defined $r = \sqrt{x^2 + y^2}$ as shown in Figure 2.17, then the amplitude probability in the range r to $r + dr$ is obtained by integrating (2.35) over the region $0 \leq \theta \leq 2\pi$ after employing the transformations $x = r\cos\theta$ and $y = r\sin\theta$ as:

$$p(r) = 2\pi r f(x, y) = \frac{2r}{\sigma_r}\exp\left(-\frac{r^2}{\sigma_r}\right) \tag{2.36}$$

Equation (2.36) is known as the Rayleigh distribution [13]. From a practical viewpoint, it is useful to express the receiving field

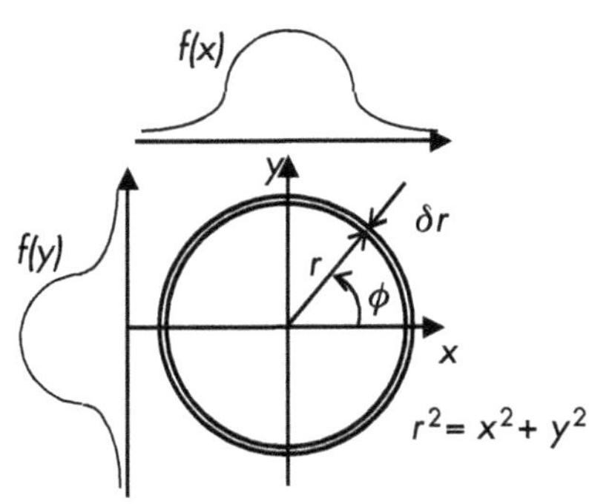

Figure 2.17 Rayleigh distribution.

strength as the probability function to determine whether the field strength is less than a critical level R. Then, the cumulative probability of a received signal level less than R is given as follows:

$$P(R) = \int_{o}^{R} p(r)\,dr = 1 - \exp\left(-\frac{R^2}{\sigma_r}\right) \tag{2.37}$$

The Rayleigh distribution is the summation of independent waves with almost the same amplitude. As shown in Figure 2.18, if the signal is composed of a dominant wave r_0 and several weak scattered waves, the receiving signal is given by a Gaussian distribution as the following probability density function:

$$f(r,\phi) = \frac{r}{2\pi\sigma^2}\exp\left(-\frac{r_o^2 - 2r_o r\cos\phi + r^2}{2\sigma^2}\right) \tag{2.38}$$

where the parameters in (2.38) are defined in Figure 2.19. Using the same transformations in (2.35), the probability function can be obtained, known as Nakagami-Rice distribution [14, 15]:

$$p(r) = \int_{0}^{2\pi} f(r,\phi)\,d\phi = \frac{r}{\sigma^2}\exp\left(-\frac{r_o^2 + r^2}{2\sigma^2}\right) I_o\left(\frac{r_o r}{\sigma^2}\right) \tag{2.39}$$

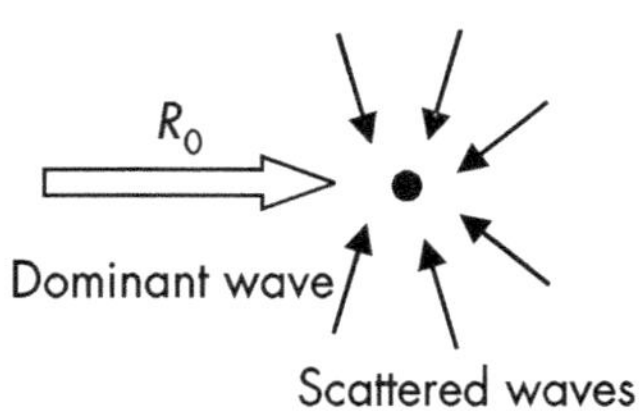

Figure 2.18 Dominant and scattered waves.

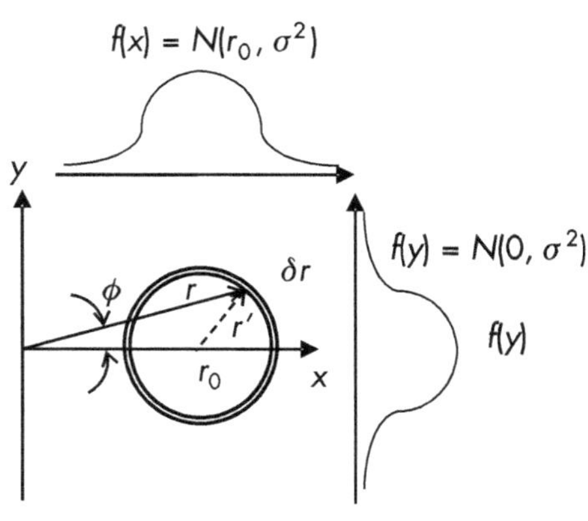

Figure 2.19 Nakagami-Rice distribution.

The $I_o(u)$ is the zero-order modified Bessel function. The ratio between the dominant and scattered waves is defined by the K-factor as follows:

$$K = \frac{r_o^2}{2\sigma^2}, \quad r_o^2 + 2\sigma^2 = 1 \tag{2.40}$$

Figure 2.20 shows the probability function of (2.39). Its peak level is increased for the strong dominant wave case, that is, the large K-factor. The probability of $K = 0$ corresponds to the Rayleigh distribution because only the scattered waves exist.

The distribution defined by (2.37) is known as the cumulative distribution function. Figure 2.21 shows Nakagami-Rice distribution substituting (2.39) into (2.37). The probability takes 0.5 for all K-factors at $r = 0$ dB, which is characterized as a medium value. These distributions show the probability of appearance for the specific fading depth (the abscissa in Figure 2.21). The $P(r) = 0.01$ for $K = 0$ and $r = -20$ dB means that the fading depth of less than -20 dB indicates the probability of 0.01 or less. In other words, the receiving electric field strength of more than -20 dB indicates the probability of 0.99 or more.

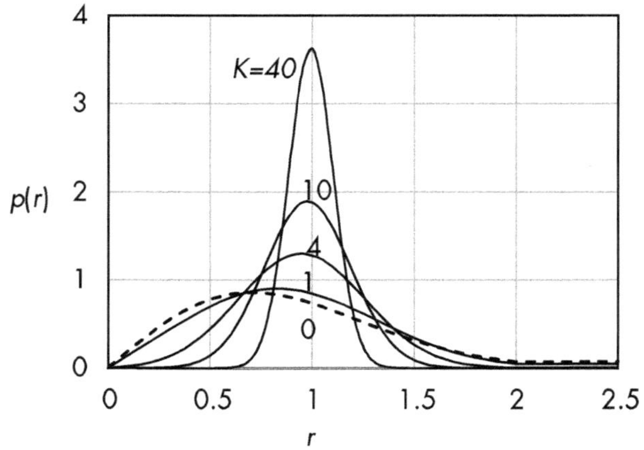

Figure 2.20 Probability of Nakagami-Rice distribution.

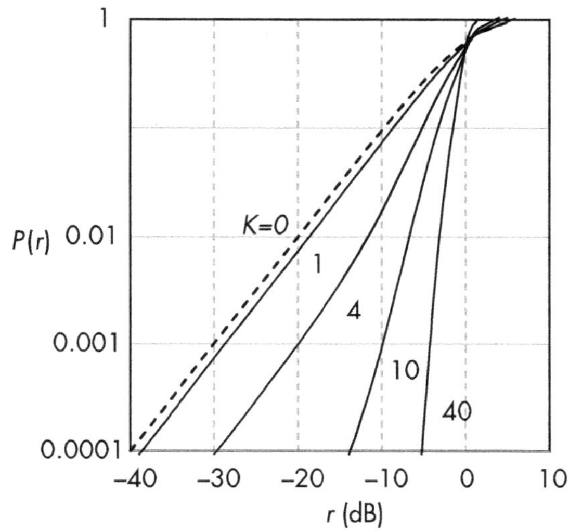

Figure 2.21 Cumulative distribution function of Nakagami-Rice distribution.

2.9 DIVERSITY RECEPTION

The receiving electric field at the mobile terminal varies with the multipath fading described in Section 2.7. Figure 2.22 shows an example of receiving field strength along the moving direction, where seven random incoming waves are assumed in Figure 2.16.

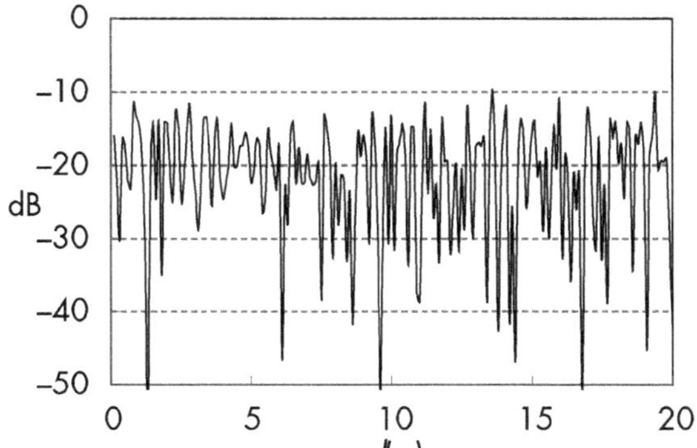

Figure 2.22 Electric field distribution along moving position; $f = 2$GHz, seven random waves.

One technique to reduce the influence of deep fading is diversity reception using several simultaneously received independent signals and synthesizing them to compensate for fading. Figure 2.23 shows a simple fading model using two forward and backward waves, producing a standing wave with a period of $\lambda/2$. When two receiving points are placed with $\lambda/4$ spacing, either e_1 or e_2 avoids the minimum electric field by the fading. This technique is known as space diversity.

Antenna, frequency, and time are the three basic types of diversity systems. The antenna diversity approach is independent of the type of communication system, whereas the other two methods depend on it. To help prevent fading, antenna diversity strategies employ various methods such as using multiple independent transmissions from various antennas spaced apart and using two orthogonal polarization components to create polarization variety. The antenna pattern diversity is another strategy that uses signals from antennas with varied antenna emission patterns. The antenna pattern diversity, also known as angular diversity, has been utilized to decrease fading caused by time changes from signals reflected from the ionosphere for overseas wireless communications in the high frequency (HF) band (3–30 MHz).

The envelope of the receiving signal for each diversity branch is combined to obtain the diversity effect by selecting the branch at a higher level. Figure 2.24 shows the cumulative probability distributions in a Rayleigh environment, where the receiving fields are shown with and without the diversity reception. An evaluation factor for diversity effect is the diversity gain. In Figure 2.24, the diversity

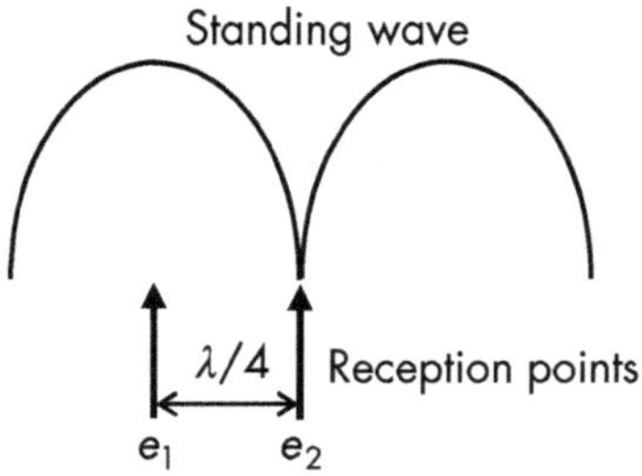

Figure 2.23 Space diversity reception.

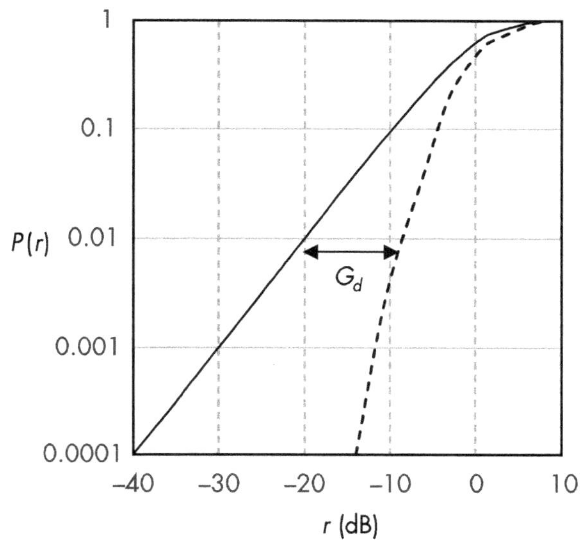

Figure 2.24 Diversity gain; solid line without diversity, dashed line with diversity.

gain G_d is defined by the difference in the receiving level at $P(r)$ = 0.01 between the two curves. G_d = 12 dB is obtained in this example.

2.10 DELAY PROFILE

In wireless high-speed digital transmission, the previously transmitted data packet can arrive at the receiving point along with the currently transmitted data packet. The techniques to eliminate these effects are diversity reception, adaptive antenna pattern forming, and signal processing system. To select an appropriate method for designing a communication system, the characteristics of the delayed signal should be estimated in advance.

The base station transmits a pulse signal for the delayed signal measurement, and the mobile terminal receives an echo. Figure 2.25 shows an example of an actual delayed signal profile received at a fixed observation point. To make it easier to understand the definition of the delay profile, the actual received signal is modeled using a simple waveform, as shown in Figure 2.26. The distribution function

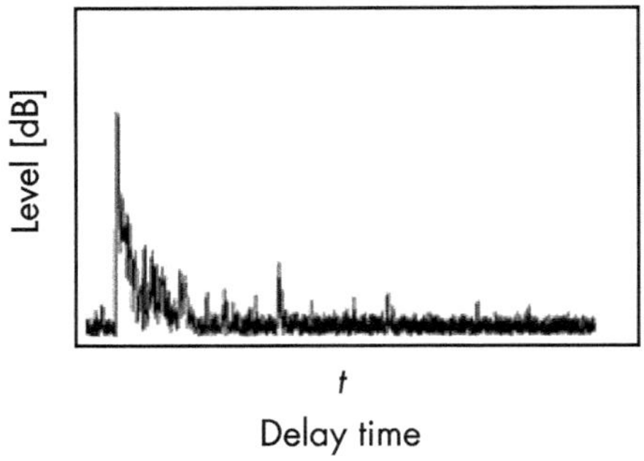

Figure 2.25 Measured delay profile.

of the delay profile is denoted via $f(t)$. Thus, the received power P_m is given by:

$$P_m = \int_{t_0}^{t_1} f(t)\,dt \tag{2.41}$$

where t_0 is the time at which $f(t)$ first exceeds the noise level, L_0, and t_1 is the time at which $f(t)$ next drops down to the noise level L_0.

The second evaluation factor T_D is defined as the mean of the delay time as [16]:

$$T_D = \frac{1}{P_m}\int_{t_0}^{t_1} (t - t_0) f(t)\,dt \tag{2.42}$$

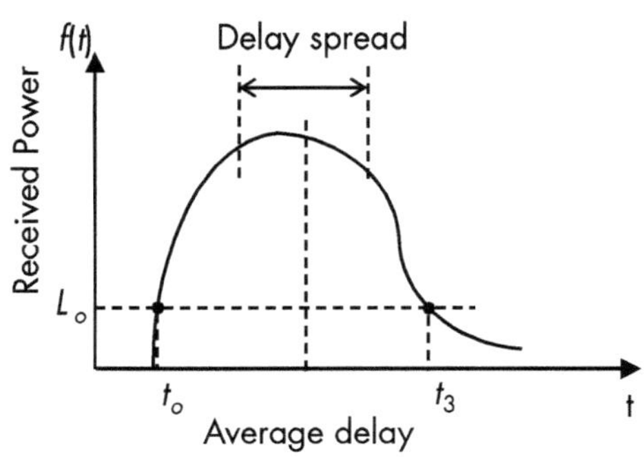

Figure 2.26 Delay profile model.

from which the delay spread or standard deviation of $f(t)$ may be defined as follows:

$$S = \sqrt{\frac{1}{P_m}\int_{t_0}^{t_1}(t - t_0)^2 f(t)\,dt - T_D^2} \tag{2.43}$$

Several simple mathematical models can express the delay profile. Figure 2.27 shows the double spike model in which direct and other paths arrive at receiving points at $t = t_1$ and t_2, respectively. The delay profile function of this model can be given using the delta function $\delta(x)$ as:

$$p(t) = \frac{1}{P_1 + P_2}\{P_1\delta(t - t_1) + P_2\delta(t - t_2)\} \tag{2.44}$$

Substituting (2.44) into (2.42) and (2.43), the mean of the delay and the delay spread can be obtained as follows:

$$T_D = \frac{t_1 P_1 + t_2 P_2}{P_1 + P_2}, \quad S = \frac{\sqrt{P_1 P_2}}{P_1 + P_2}(t_2 - t_1) \tag{2.45}$$

Figure 2.28 shows the exponential function model in which the delay profile decays exponentially as:

$$p(t) = \frac{1}{\sigma_0}\exp\left(-\frac{t - t_0}{\sigma_0}\right)U(t - t_0) \tag{2.46}$$

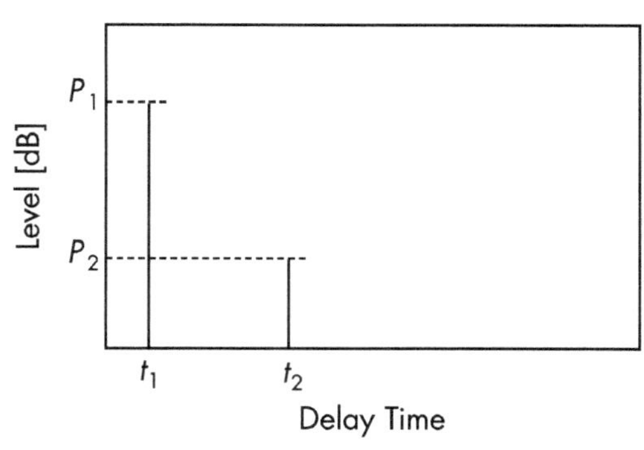

Figure 2.27 Double spike model.

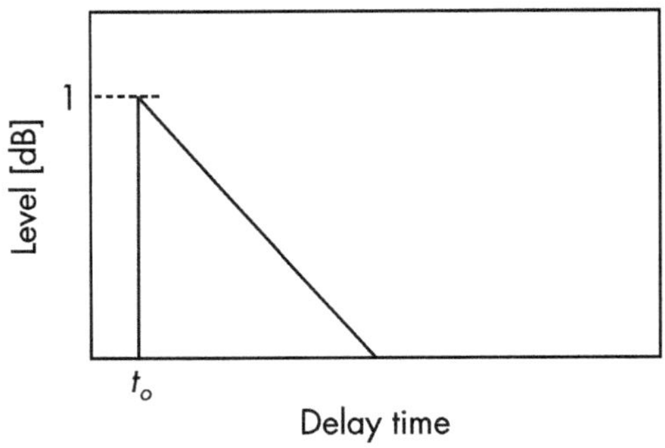

Figure 2.28 Exponential function model.

where $U(x)$ is the unit function defined: $U(x) = 1$, $x \geq 0$, and $U(x) = 0$, $x < 0$. The mean of the delay and the delay spread can be obtained as:

$$T_D = \sigma_0, \quad S = \sigma_0 \tag{2.47}$$

2.11 ANGULAR PROFILE

The electromagnetic waves radiated from the base station arrive at the mobile terminal in several different paths, which causes the time delay described in Section 2.10 and the spread of arrival angles. The angular profile is defined in the same manner as the delay profile. The reference angle fixed at the mobile terminal side by $\phi = 0°$ is usually the direction of the base station. Let the arrival power level as $P(\phi)$ in the horizontal plane, then the angular spread is defined as:

$$S_\phi = \sqrt{\frac{\int_{-\pi}^{\pi} P(\phi)(\phi - \phi_0)^2 \, d\phi}{\int_{-\pi}^{\pi} P(\phi) \, d\phi}} \tag{2.48}$$

where the average angle ϕ_o is given as:

$$\phi_0 = \frac{\int_{-\pi}^{\pi} P(\phi)\phi \, d\phi}{\int_{-\pi}^{\pi} P(\phi) \, d\phi} \tag{2.49}$$

The angular spread in the vertical plane is defined in the same manner. Several simple mathematical models can express the angular profile. Figure 2.29 shows the Gaussian and Laplacian distributions and are defined as follows:

$$P(\phi) = \exp\left(-\frac{\phi^2}{2S_\phi}\right) \tag{2.50}$$

$$P(\phi) = \exp\left(-\frac{\sqrt{2}|\phi - \phi_0|}{S_\phi}\right) \tag{2.51}$$

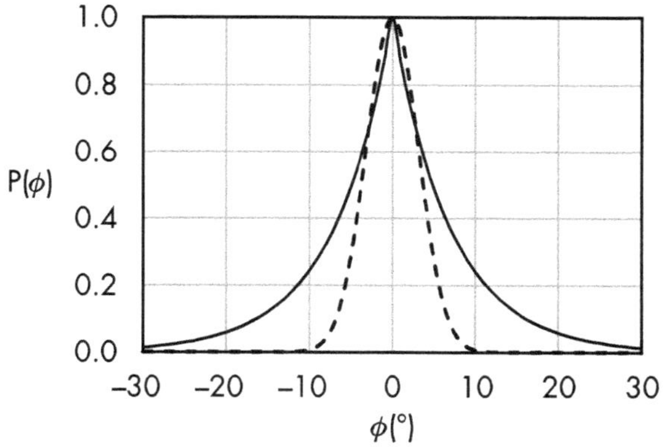

Figure 2.29 Angular profile model, solid line is Laplacian distribution, and dotted line is Gaussian distribution.

References

[1] Stuszman, W. L., and G. A. Thiele, *Antenna Theory and Design*, Third Edition, New York: John Wiley & Sons, 2012, pp. 109–110.

[2] Stratton, J. A., *Electromagnetic Theory*, New York: McGraw-Hill, 1941, pp. 132–137.

[3] Kong, J. A., *Electromagnetic Wave Theory*, Second Edition, New York: John Wiley & Sons, 1990, pp. 118–122.

[4] Kong, J. A., *Electromagnetic Wave Theory*, Second Edition, New York: John Wiley & Sons, 1990, pp. 122–126.

[5] Kong, J. A., *Electromagnetic Wave Theory*, Second Edition, New York: John Wiley & Sons, 1990, pp. 100–101.

[6] Collin, R. E., *Antennas and Radiowave Propagation*, New York: McGraw-Hill, 1985, pp. 341–349.

[7] Wolfram MathWorld, https://mathworld.wolfram.com/FresnelIntegrals.html (accessed February 12, 2022).

[8] Sizun, H., *Radio Wave Propagation for Telecommunication Applications*, Paris: Springer-Verlag, 2003, pp. 49–51.

[9] Vogler, L. E., "An Attenuation Function for Multiple Knife-Edge Diffraction," *Radio Science*, Vol. 17, No. 6, November–December 1982, pp.1541–1546.

[10] Nguyen, V. D., et. al., "On the Proof of Recursive Volger Algorithm for Multiple Knife-Edge Diffraction," *IEEE Trans. on Antennas Propag.*, Vol. 69, No. 6, June 2021, pp. 3617–3622.

[11] Sizun, H., *Radio Wave Propagation for Telecommunication Applications*, Paris: Springer-Verlag, 2003, pp. 53–54.

[12] Jakes, W. C., *Microwave Mobile Communications*, Piscataway, NJ: IEEE Press, 1974, pp. 13–16.

[13] Jakes, W. C., *Microwave Mobile Communications*, Piscataway, NJ: IEEE Press, 1974, pp. 17–19.

[14] Nakagami, M., "Some Statistical Characteristics of Short-Wave Fading," *Journal of IEICE*, Vol. 27, No. 2, 1943, pp. 145–150 (in Japanese).

[15] Rice, S. O., "Mathematic Analysis of Random Noise," *Bell System Tech. Journal*, Vol. 23, No. 3, July 1944, pp. 282–332.

[16] Sizun, H., *Radio Wave Propagation for Telecommunication Applications*, Paris: Springer-Verlag, 2003, pp. 205–209.

3

FUNDAMENTALS OF ANTENNAS AND PROPAGATION MEASUREMENT

This chapter describes the fundamentals of antennas and propagation measurement to verify the performance of antennas under test. The antenna performance is characterized using the input impedance and radiation pattern. The first part describes the measurement of the input impedance using a standing wave in the transmission line. A network analyzer can easily determine the frequency characteristics of the antenna impedance. The theory and calibration techniques are presented, which include the introduction of the Smith chart to easily understand the impedance characteristics. The second part describes the antenna pattern measurements in the far and near fields. Evaluation parameters using the antenna pattern are also defined. The last part of this chapter presents the propagation measurements, which depend on the statistical analysis of huge, corrected data using the technical terms for propagation.

3.1 IMPEDANCE MEASUREMENT

The antenna characteristic is defined as an input impedance of electrical circuits. Initially, the reflection coefficient measured in the feeding transmission line connected to the test antennas is presented. The reflection coefficient can be transformed into antenna input impedance, which can be easily observed using the Smith chart. The theory of transformation and how to use the chart are explained. The performance of several closely located antennas is characterized using the S-parameter network. The definition of an S-parameter and its relationship with the impedance matrix are described. In practical measurements, the connection of the test antenna to the feeding line must be considered according to the transmission line type, such as balanced and unbalanced lines. The use of a balun to connect different line types is presented in the last part of this section.

3.1.1 Reflection Coefficient

Most RF output and input ports are designed to have an impedance of 50Ω. Therefore, the antenna input impedance should match this value. Figure 3.1 shows the model used to determine the impedance mismatch between a transmission line with impedance Z_o and terminating load impedance Z_L. The reflection coefficient at $z = 0$ can be defined as [1]:

$$\Gamma(0) = \frac{V_2}{V_1} = \frac{Z_L - Z_c}{Z_L + Z_c} \tag{3.1}$$

where V_1 and V_2 are the amplitudes of the forward- and backward-propagated voltages in the transmission line. The propagation constant is denoted as $j\beta$, assuming a lossless transmission line.

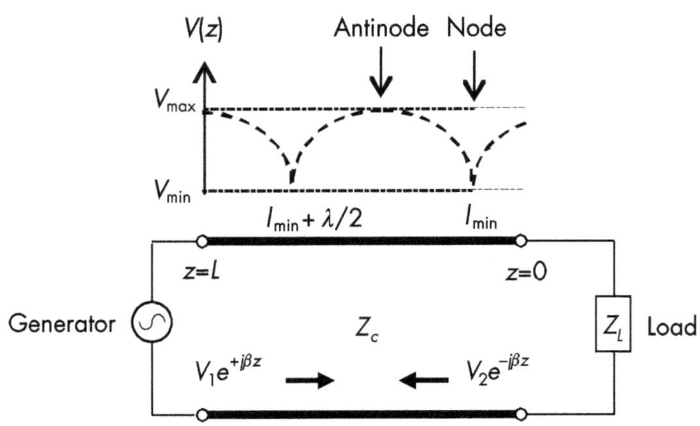

Figure 3.1 Reflection form load and impedance on transmission line.

Then, the voltage and current are varied according to the following functions:

$$V(z) = V_1\left(e^{+j\beta z} + \Gamma(0)e^{-j\beta z}\right) \tag{3.2}$$

$$I(z) = \frac{V_1}{Z_c}\left(e^{+j\beta z} - \Gamma(0)e^{-j\beta z}\right) \tag{3.3}$$

The impedance at an arbitrary position in the transmission line is given by (3.2) and (3.3) as:

$$Z(z) = \frac{V(z)}{I(z)} = \frac{Z_L + jZ_c\tan(\beta z)}{Z_c + jZ_L\tan(\beta z)} \tag{3.4}$$

where $\Gamma(0)$ is rewritten using (3.1).

Figure 3.1 shows that the voltage standing wave changes with period $\lambda/2$, and the maximum and minimum voltages are known as antinode and node, respectively. Measurement of first node distance $l_{\min}$ from the load and the ratio of $V(l_{\min})$ and $I(l_{\min})$ allows the load impedance to be calculated by solving (3.4). Simultaneous measurement of the voltage and current distributions in the transmission line is difficult; thus, voltage measurement is only performed with the aid of the Smith chart [2].

Figure 3.2 shows two impedance-mapping cases. The normalized impedance in the transmission line can be obtained using (3.1) as follows:

$$\hat{Z}(z) = \frac{Z(z)}{Z_c} = r + jx = \frac{1+\Gamma}{1-\Gamma} \tag{3.5}$$

where real part r is positive (i.e., $r \geq 0$), whereas imaginary part x varies from $-\infty$ to $+\infty$, as shown in Figure 3.2(a). The reflection coefficient takes the value of $|\Gamma| \leq 1$ and is expressed as follows:

$$\Gamma = u + jv = \frac{r - 1 + jx}{r + 1 + jx} \tag{3.6}$$

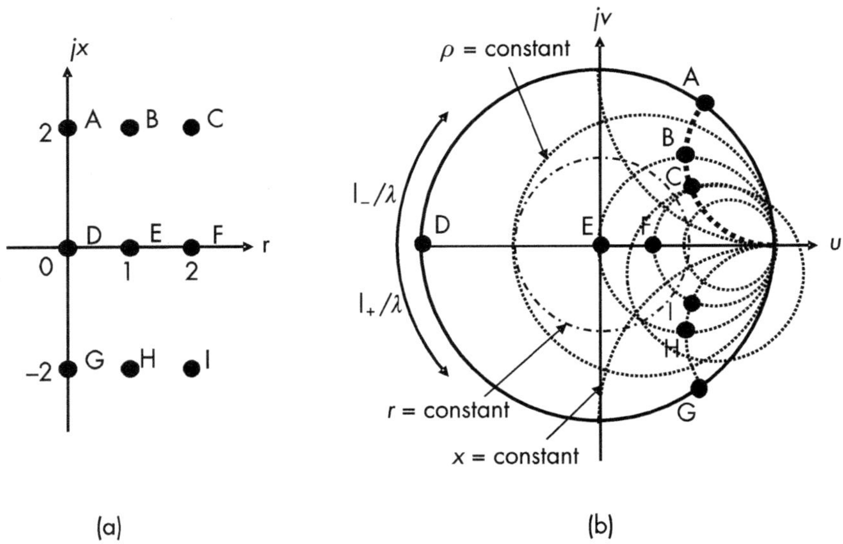

Figure 3.2 Impedance mapping: (a) impedance on *rx* plane, and (b) impedance on *uv* plane.

After this transformation, three infinite points, namely, $r = \infty$, $x = -\infty$, and $x = +\infty$, converge to a single point, as shown in Figure 3.2(b). Markers A to I, which indicate the impedance positions shown in Figure 3.2(a), are transformed on the *uv* plane, as shown in Figure 3.2(b). Any r and x constant lines are given by the following circles:

$$\left(u - \frac{r}{r+1}\right)^2 + v^2 = \left(\frac{1}{r+1}\right)^2, \quad (u-1)^2 + \left(v - \frac{1}{x}\right)^2 = \left(\frac{1}{x}\right)^2 \tag{3.7}$$

A circle with radius ρ is shown in Figure 3.2(b) to represent the voltage standing wave ratio (VSWR) as follows:

$$\rho = \frac{1+|\Gamma|}{1-|\Gamma|} = \frac{V_{max}}{V_{min}} \tag{3.8}$$

The impedance in the transmission line moves along the circle with radius ρ according to (3.4). The clockwise and counterclockwise motion directions are denoted as l_-/λ and l_+/λ, respectively. Figure 3.1 shows that l_-/λ and l_+/λ are known as the directions toward the generator and load, respectively. The standing wave period is $\lambda/2$. Then, the measurement point in the Smith chart returns to the same position after one rotation.

3.1.2 Smith Chart

Figure 3.2(b) shows the Smith chart, which is utilized to determine the value of the load impedance by measuring the VSWR in the transmission line and to determine the impedance-matching condition of the load. A classical VSWR measurement is performed using a coaxial waveguide terminated with a load. The distributions of the standing wave and node positions are measured through a longitudinal slit in the outer conductor of the coaxial waveguide. Figure 3.3 shows the procedure for determining the load impedance using the Smith chart.

Figure 3.1 shows that the first node position from the load is denoted as l_{min}. The measured VSWR, which is the ratio of the maximum and minimum voltage of the standing wave, defines a circle with radius ρ, as determined by (3.8). The two-point intersections in the u axis and VSWR circle represent a node at the left and an antinode at the right, respectively. The node corresponds to the position of the minimum voltage at distance l_{min} from the load. The load impedance can be obtained from the Smith chart by rotating the measurement point from the node by an angle equivalent to $\beta l_{\mathrm{min}}/\lambda$ toward the load side of the chart, as shown in Figure 3.3. The scales of r_l and x_l that cross the circle with radius ρ indicate the value of the normalized impedance of the load.

The aforementioned impedance is investigated at a fixed frequency. To find the matching condition of the load, frequency characteristics are often used in the measurement. Figure 3.4 shows a typical locus of the measured load impedance. The measurement point rotates clockwise from low frequency f_l to high frequency f_h. The center of the chart shows the impedance-matching point between the load and transmission line, and its frequency is denoted as f_o, as shown in Figure 3.4. The matching condition is defined by a VSWR value that is less than the specific value of ρ in which the impedance locus is inside the circle with radius ρ. The gray area in Figure 3.4 shows this region at VSWR $\leq \rho_1$.

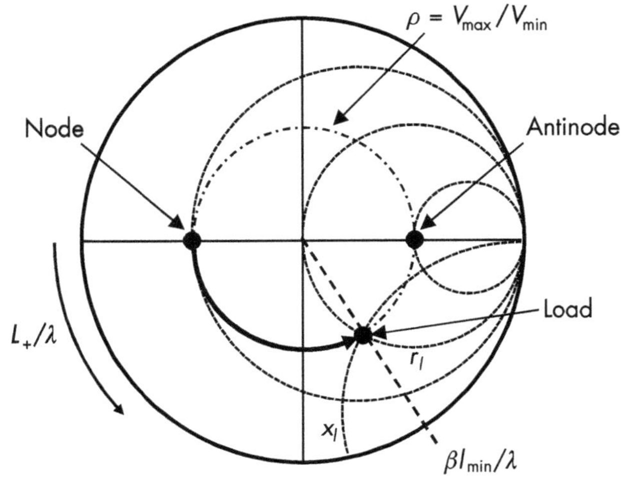

Figure 3.3 Impedance locus on Smith chart.

3.1.3 S-Parameter Measurement

The load impedance characteristics are usually measured using the scattering parameters or S-parameters by a network analyzer. The S-parameters describe the input and output relationship of an electrical circuit, as shown in Figure 3.5. Ports 1 and 2 of a two-port

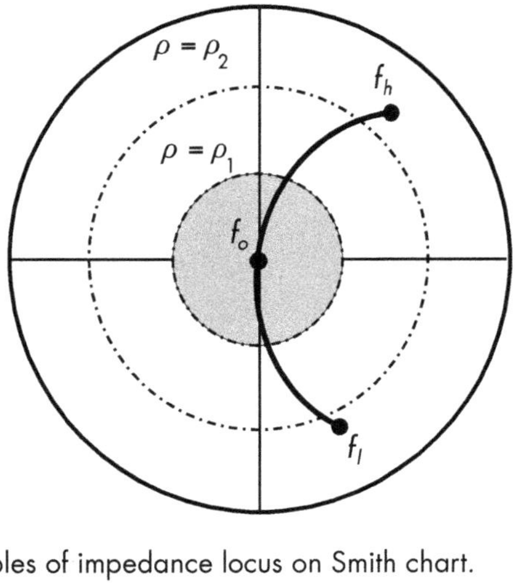

Figure 3.4 Examples of impedance locus on Smith chart.

network are connected to the transmission line with characteristic impedance Z_c. By denoting the port voltage and current as V_i and I_i ($i = 1,2$), respectively, the normalized waves inward and outward the network are expressed as follows:

$$a_i = \frac{V_i + Z_c I_i}{2\sqrt{Z_c}}, \quad b_i = \frac{V_i - Z_c I_i}{2\sqrt{Z_c}} \tag{3.9}$$

The matrix of the S-parameters is defined as a transformation from an inward to an outward wave as follows [3]:

$$\begin{pmatrix} b_1 \\ b_2 \end{pmatrix} = \begin{pmatrix} S_{11} & S_{12} \\ S_{21} & S_{22} \end{pmatrix} \begin{pmatrix} a_1 \\ a_2 \end{pmatrix}, \quad (\boldsymbol{b}) = [S]\,(\boldsymbol{a}) \tag{3.10}$$

$S_{ii} = b_i/a_i$ denotes the reflection coefficient at each port, and $S_{ij} = b_i/a_j$, $(i \neq j)$ represents the coupling coefficient from port j to i. Scales $-10\log_{10}|S_{ii}|$ and $-10\log_{10}|S_{ij}|$ are known as the return loss and isolation, respectively.

Additionally, the two-port network shown in Figure 3.5 is defined by the impedance matrix as:

$$\begin{pmatrix} V_1 \\ V_2 \end{pmatrix} = \begin{pmatrix} Z_{11} & Z_{12} \\ Z_{21} & Z_{22} \end{pmatrix} \begin{pmatrix} I_1 \\ I_2 \end{pmatrix}, \quad (\boldsymbol{V}) = [Z]\,(\boldsymbol{I}) \tag{3.11}$$

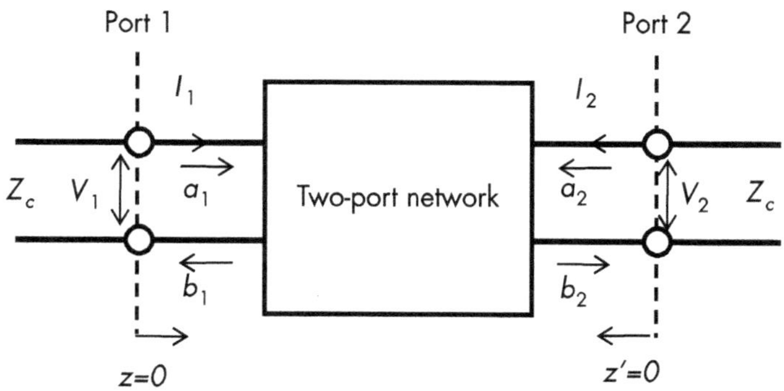

Figure 3.5 Two-port network.

The transformations from [S] to [Z] and from [Z] to [S] can be performed using unit matrix [1] as follows:

$$[S] = \left\{\left[\hat{Z}\right] + [1]\right\}^{-1}\left\{\left[\hat{Z}\right] - [1]\right\}, \quad \left[\hat{Z}\right] = \frac{1}{Z_c}[Z] \tag{3.12}$$

$$\left[\hat{Z}\right] = \{[1] - [S]\}^{-1}\{[1] + [S]\} \tag{3.13}$$

The impedance matrix is often used in computer simulations, which is verified by measuring the S-parameters and this transformation. The S-parameters are directly measured by the network analyzer. Figure 3.6 shows that the network analyzer has two measurement ports. A signal generator is connected to port 1 or 2 via a directional coupler (DC), and the input and output signals are sent to the comparator after the down-conversion of the RF signals. The scattering matrix of the device under test (DUT) is obtained as follows:

$$\begin{pmatrix} S_{11} & S_{12} \\ S_{21} & S_{22} \end{pmatrix} = \begin{pmatrix} \dfrac{C_2}{C_1} & \dfrac{C_2}{C_4} \\ \dfrac{C_3}{C_1} & \dfrac{C_3}{C_4} \end{pmatrix} \tag{3.14}$$

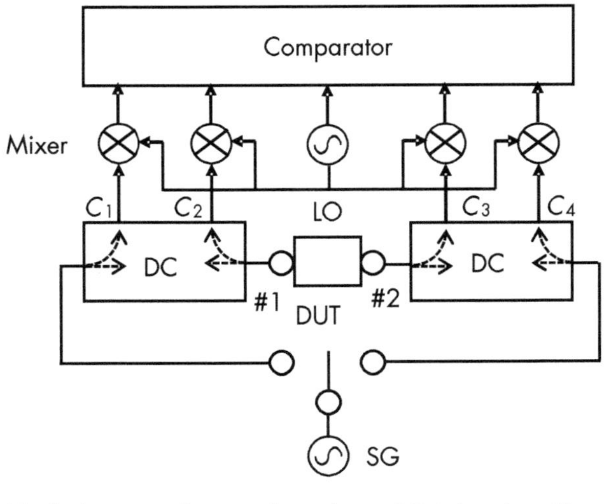

Figure 3.6 Block diagram of network analyzer, LO is local oscillator.

To correctly measure the S-parameters of the DUT, we need sufficient time to allow the instrument to warm up before the measurement and correct the calibration between the ports using a calibration kit supplied by the instrument manufacturer. We should note that the position of the reference plane of the measurement is correctly considered to obtain the impedance value. The reference plane is the contact surface of the port connector.

3.1.4 Balanced and Unbalanced Feeding Lines

The feeding lines of antennas are classified into balanced or unbalanced lines. The Lecher line shown in Figure 3.7(a) is composed of a pair of parallel conducting wires to supply out-of-phase current and voltage.

The coaxial cable shown in Figure 3.7(b) represents an unbalanced line because the voltage in the outer conductor of the cable is always zero. Another name for this coaxial cable is a shield cable because the outer conductor, which is connected to electrical earth, effectively shields the cable from the effects of external RF noise.

These two different feeding lines cannot be directly connected together but must be connected via a voltage transformer known as a balun (see Figure 3.8) [4], which consists of a pair of inductors and a pair of capacitors connected in the form of a bridge circuit.

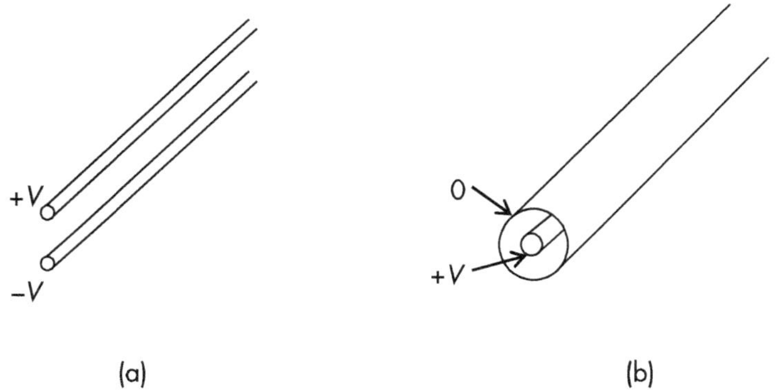

Figure 3.7 Balanced and unbalanced lines: (a) Lecher line, and (b) coaxial cable.

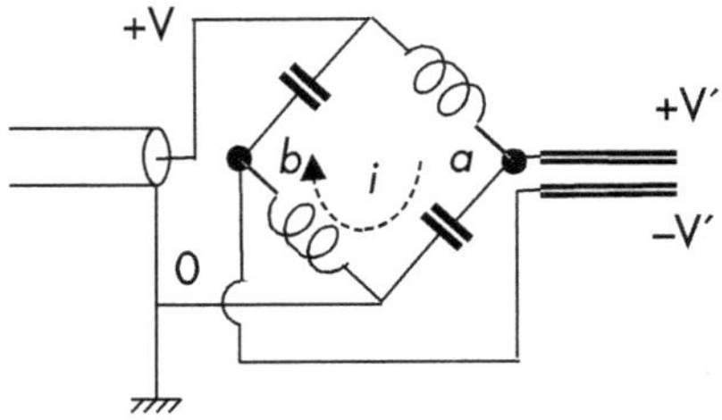

Figure 3.8 Balun circuit.

The capacitor voltage at point *a* provides a phase lead of $\pi/2$ for current *i* that flows in the bridge, and the inductor voltage at point *b* provides a phase delay of $\pi/2$. These phase lead and delay allow conversion from an unbalanced input to a balanced output. An autotransformer balun consists of primary and secondary wires wound on a ferrite rod or toroid. Figure 3.9 shows the circuit diagram. This balun is made of two or more electrically connected coils. The current in the primary coil induces a magnetic field inside the core and generates current in the secondary coil. A tap point between the two coils is grounded to connect the unbalanced voltage. The output voltage becomes double. Then, the impedance transformer ratio becomes 1:4.

Another type of balanced antenna feeding is a sleeve or bazooka balun, which is shown in Figure 3.10. If a dipole antenna is fed by a coaxial cable, undesired currents flow in the outer conductor of the cable. This leakage current can be suppressed using a double coaxial cable. The balun has an electrical length of $\lambda/4$, producing infinite

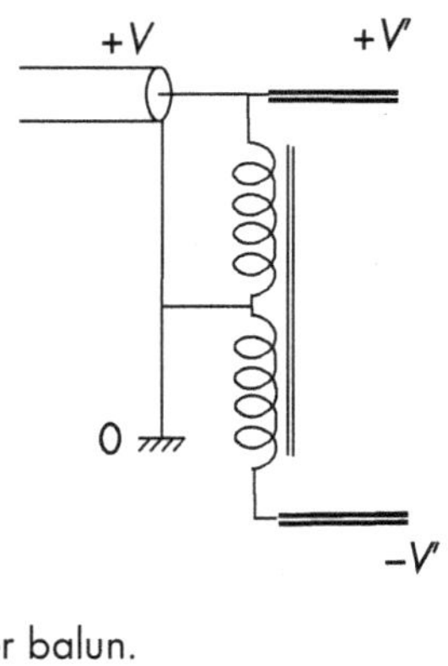

Figure 3.9 Autotransformer balun.

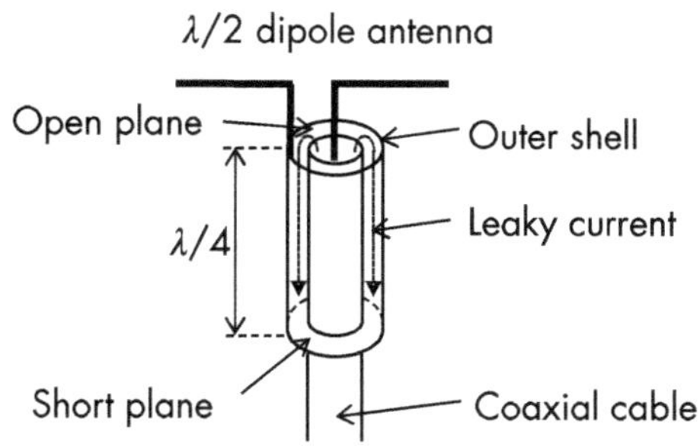

Figure 3.10 Sleeve balun.

impedance at the open end of the cable to prevent the flow of leakage current.

To measure the input impedance of an antenna with a balanced feeding line, a method based on image theory is used. Figure 3.11 shows a typical measurement setup for a $\lambda/2$ dipole antenna. The antenna structure is divided into two halves on a symmetrical plane. One-half of the antenna is mounted on an electrically large ground plane and is fed by a coaxial connector. Physically, the measured antenna is only one-half of the actual antenna structure, and, thus, the input impedance of the original $\lambda/2$ dipole antenna is twice the measured impedance value.

The S-parameter method measures the input impedance of a balanced antenna viewed as a two-port network system. This method can be extended to measure asymmetrical antenna structures such as monopole antennas on a finite-sized ground plane. The input port of these antennas is viewed as a two-port network, as shown in Figure 3.12(a), assuming a virtual short circuit at the center of the

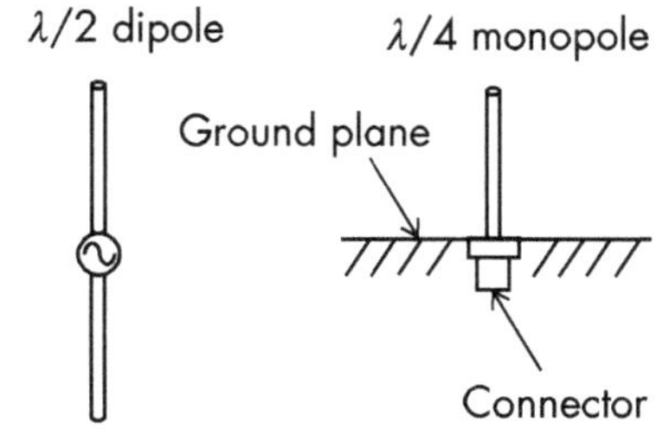

Figure 3.11 Image method.

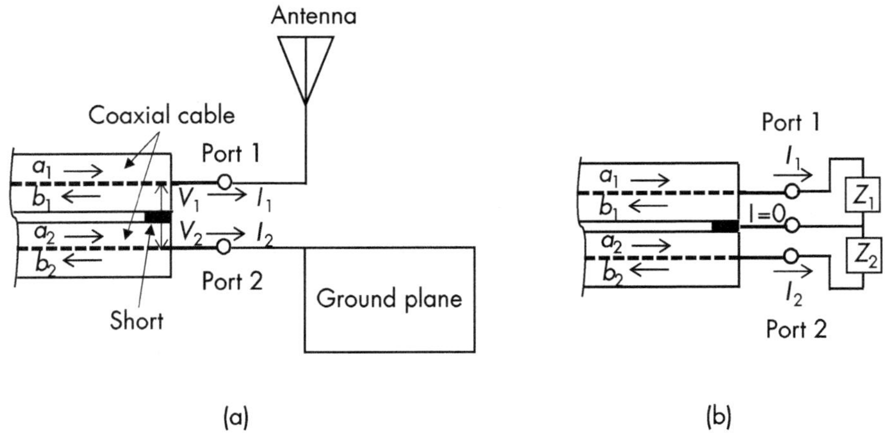

Figure 3.12 S-parameter method: (a) measurement setup, and (b) equivalent circuit.

antenna input port. A symmetrical current condition of $I_1 = I_2$ yields reflection coefficients S_1 and S_2, respectively, as:

$$S_i = \frac{b_i}{a_i}, \qquad i = 1, 2 \tag{3.15}$$

This symmetrical condition also provides the equivalent circuit shown in Figure 3.12(b) for the two-port model. Finally, antenna input impedance Z_i is given by series impedances Z_1 and Z_2.

$$Z_i = Z_1 + Z_2 = Z_c\left(\frac{1+S_1}{1-S_1} + \frac{1+S_2}{1-S_2}\right) \tag{3.16}$$

where Z_c is the characteristic impedance of the coaxial cable in this measurement.

3.2 MEASUREMENT OF THE ANTENNA-RADIATION PATTERN

The antenna-radiation performance is characterized by pattern measurement. The antenna pattern varies according to the distance from the antenna to the observation point, and three fields are defined. The far-field measurement in an anechoic chamber is defined, and the measurement parameters are explained. To measure electrically large-sized antennas under indoor environment, the near field measurement is often used. The theory and practical method are described in this section.

3.2.1 Far, Fresnel, and Near Fields in Antenna Measurements

Antenna-radiation patterns are measured in the far-field region described in Section 1.2 in which an infinitesimal current element is used. In practical antenna pattern measurement, the antenna size should be considered to determine the required distance of the far field. Figure 3.13 shows an antenna element with length D on the z axis, and point P indicates the observation point. The distance of point P from the origin is denoted as r, and that from arbitral position z' on the antenna is denoted as R. To find the difference between r and R, distance R is defined using binomial expansion as follows [5]:

$$R = \sqrt{r^2 + z'^2 - 2rz'\cos\theta} = r - z'\cos\theta + \frac{z'^2 \sin^2\theta}{2r} + \frac{z'^3 \sin^2\theta\cos\theta}{2r^2} + \cdots \tag{3.17}$$

When the fourth and higher terms in (3.17) are neglected in the far-field approximation, the third term gives error factor Δr_3. To define

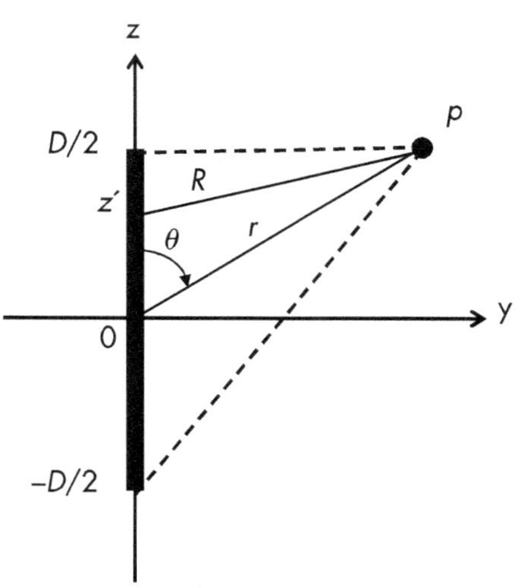

Figure 3.13 Antenna size and observation point.

the far-field region, distance r should be $\Delta r_3 < \lambda/16$ in which causes amplitude difference in Δr_3 less than 0.1 due to the phase difference. This term takes the maximum of $\Delta r_3 = D^2/8r$ at $\theta = \pi/2$ and $z' = D/2$. Then, the far-field condition for distance r is obtained as follows:

$$r > \frac{2D^2}{\lambda} \tag{3.18}$$

Additionally, following the same explanation for the fourth term in (3.17), the following condition is obtained for the distance r:

$$r = \sqrt{\frac{2D^3}{3\sqrt{3}\lambda}} = 0.62\sqrt{\frac{2D^3}{\lambda}} \tag{3.19}$$

The distance range between (3.18) and (3.19) is known as the Fresnel region, where the field distributions vary according to the distance. The region closer than the Fresnel region is known as the near-field region. The field distributions in the near field have several peaks, depending on the source distributions, and gradually converge to the field distributions in the Fresnel region. Figure 3.14 shows three regions defined in the previous discussions as a function of the antenna size.

3.2.2 Far-Field Measurements in Anechoic Chamber

The most common type of RF anechoic chamber has a rectangular cross section with an electromagnetic wave absorber installed on the chamber inner walls, as shown in Figure 3.15. Transmitting and test antennas are placed near the side walls of the chamber to satisfy a reflection-wave level of less than −20 or −30 dB. This region is the quiet zone of the chamber and is defined as a sphere whose given radius is centered on the antenna, as shown in Figure 3.15. In the microwave frequency band, the wave absorber consists of urethane foam filled with carbon particles to convert the electromagnetic wave into a thermal loss. For normally incident waves, the reflection level is reduced to less than −20 dB. To suppress the reflection of obliquely incident waves, a pyramidal-shaped absorber is used to scatter the incident waves.

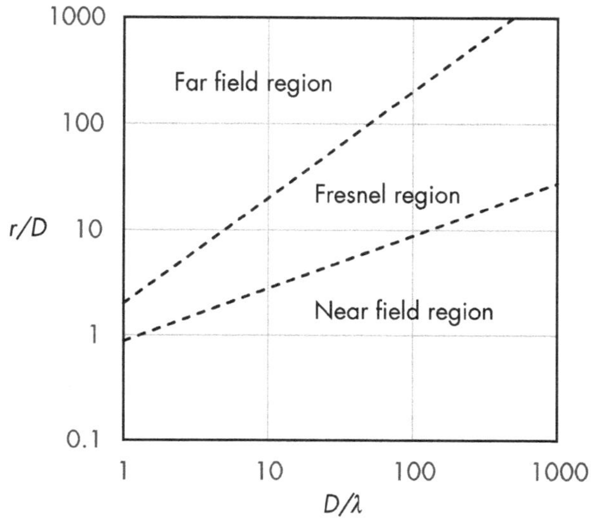

Figure 3.14 Near-field, Fresnel, and far-field regions.

An example of the measurement setup is shown in Figure 3.16. The transmitting antenna is installed at the center of the quiet zone, and the test antenna is mounted on a rotating table at the opposite side of the chamber. The test antenna can be incrementally rotated using a stepper motor. A signal generator is used to excite the transmitting antenna when the test antenna is connected to a receiver. To construct a phase-locked-loop receiver system, installing DC and an

Figure 3.15 Anechoic chamber. (Courtesy of Microwave Factory.)

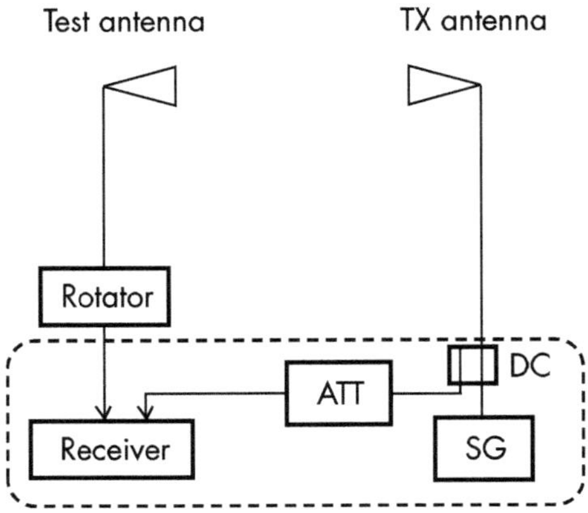

Figure 3.16 Diagram of chamber measurement.

attenuator (ATT) in the transmitting signal line to supply the phase reference is necessary.

The transmitting antennas used in the measurement system are a dipole antenna, a Yagi–Uda array, a pyramidal horn, a double-ridged horn, and a log periodic antenna. The frequency bandwidth characteristics and cross polarization level of these antennas must be considered to determine the specifications of the measurement system. For example, the frequency ranges of standard antennas can be categorized as follows: dipole (<1 GHz), double-ridged horn or log periodic (1–20 GHz), and pyramidal horn (>10 GHz). These are all examples of antennas with linearly polarized radiation patterns. If a circularly polarized radiation pattern is required, a helical or loop antenna can be employed.

For measurements of low-gain antennas where the distance between the antennas is relatively short, a network analyzer can be used. The two antennas are connected to the two ports of the network analyzer, and S_{21} is measured. This option is comparatively low-cost; however, the dynamic range is limited.

For the pattern measurements of linearly polarized antennas, the transmitting and test antennas are first aligned so that the polarization vectors are in the same plane. On the other hand, for circularly polarized antennas, two measurement processes are required. Before proceeding to the description of the circular-polarization measurement, we first need to define the cross-polarization ratio (XPR) and axial ratio (AR).

An arbitrary polarized radiation wave is expressed as follows:

$$\boldsymbol{E} = E_\theta \mathbf{e}_\theta + jE_\phi \mathbf{e}_\phi \tag{3.20}$$

where E_θ and E_ϕ are the complex amplitudes of the electromagnetic wave in the θ and ϕ directions, respectively, and $\mathbf{e}_\theta$ and $\mathbf{e}_\phi$ are the unit vectors in the θ and ϕ directions, respectively. If $E_\theta = E_\phi$, then the wave is called a circularly polarized wave. If $E_\theta \neq E_\phi$, the wave is called an elliptically polarized wave. Equation (3.20) can be decomposed into right-handed ($\boldsymbol{E}_R$) and left-handed ($\boldsymbol{E}_L$) circularly polarized components as follows:

$$\boldsymbol{E}_R = E_R\left(\mathbf{e}_\theta - j\mathbf{e}_\phi\right), \quad \boldsymbol{E}_L = E_L\left(\mathbf{e}_\theta + j\mathbf{e}_\phi\right) \tag{3.21}$$

The XPR of a circularly polarized wave is defined with respect to $\boldsymbol{E}_R$ as:

$$XPR = \frac{|\boldsymbol{E}_R|}{|\boldsymbol{E}_L|} \tag{3.22}$$

For the elliptical polarization case shown in Figure 3.17, AR is defined as the ratio of the maximum and minimum amplitudes [6].

$$AR = \frac{|\boldsymbol{E}_R + \boldsymbol{E}_L|}{|\boldsymbol{E}_R - \boldsymbol{E}_L|} \tag{3.23}$$

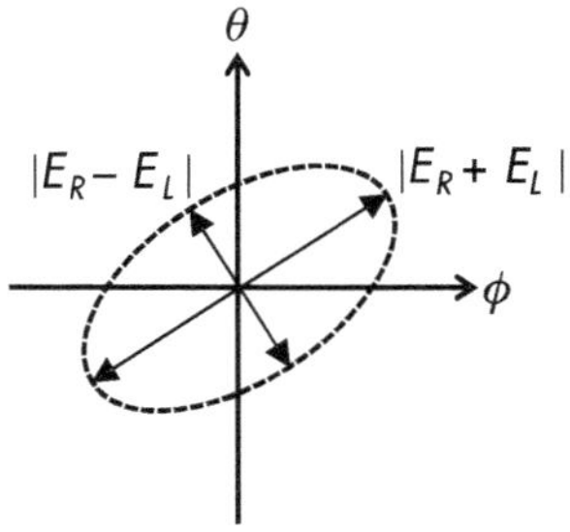

Figure 3.17 Definition of polarization.

AR is measured using the spin linear technique in which a linearly polarized standard antenna is rotated much faster than the azimuth rotation of the test antenna. A radiation pattern measured using this method is shown in Figure 3.18. The peak-to-peak level of the ripple in the pattern yields the AR of the test antenna.

Antenna gain is classified into directive and absolute gains. The directive gain does not include the input impedance mismatch and antenna internal losses. It only indicates the directivity of the test antennas (i.e., the spatial distribution of the radiated power from the antenna). The absolute gain includes all losses of the antenna, and it is the term most often used when specifying the transmitting power and link budget margin in an actual operating environment. Absolute antenna gain G_s is measured by comparing the receiving electric field strength between the standard-gain and test antennas. An example of the absolute gain of a standard antenna is $G_s = 2.15$ dBi for a $\lambda/2$ dipole antenna, which is used to measure a test antenna with expected gain G_d of less than 15 dBi. However, a standard high-gain pyramidal horn antenna can be used to measure antennas with an empirically expected gain in the range from 15 to 35 dBi.

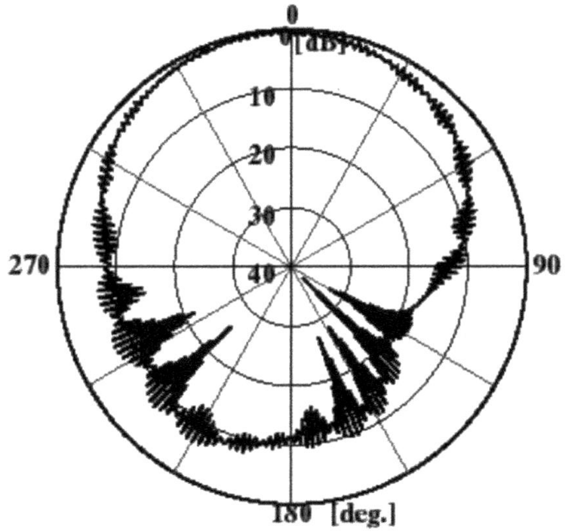

Figure 3.18 Measured spin linear radiation pattern.

Figure 3.19 shows a typical measurement setup. ATT is installed between the standard antenna and the receiver and is adjusted until the received signal levels from both the standard and test antennas become equal. When the attenuation level is equal to L_α, the antenna gain may be obtained as follows:

$$G_d(\mathrm{dB}) = G_s(\mathrm{dB}) - L_a(\mathrm{dB}) \tag{3.24}$$

Gain measurement using ATT is a recommended precise procedure. However, the following describes a simple relative-gain measurement.

Received electric field strength E_s is measured using a standard antenna with antenna gain G_s measured without inserting ATT in advance. Using the same experimental setup, the test antenna replaces the standard antenna, and, once again, electric field strength E_d is measured. The gain of the measured antenna G_d is thus given as:

$$G_d(\mathrm{dB}) = E_d(\mathrm{dB}) - E_s(\mathrm{dB}) + G_s(\mathrm{dB}) \tag{3.25}$$

3.2.3 Near-Field Measurements

The distance of the far-field measurement defined by (3.18) often becomes very large in the measurement of high-gain antennas. This section describes the near-field measurement for these antennas in

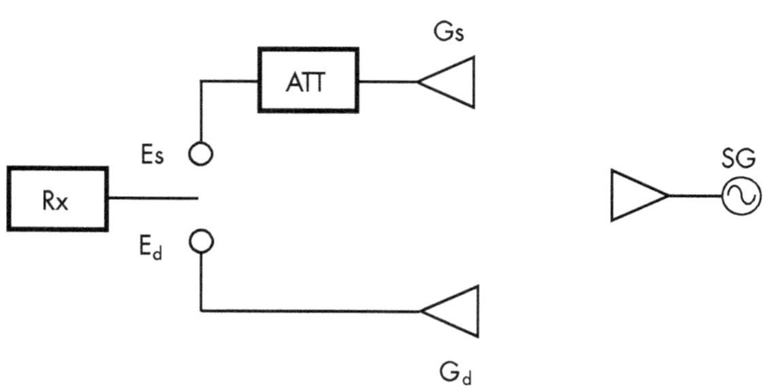

Figure 3.19 Antenna gain measurement.

indoor facilities. First, the theory of near- to far-field transformation is briefly presented.

If virtual boundary S encloses a radiating test antenna, equivalent electric and magnetic currents $\boldsymbol{J}$ and $\boldsymbol{M}$ are provided by tangential magnetic and electric fields $\boldsymbol{H}_t$ and $\boldsymbol{E}_t$, respectively, on surface S as:

$$\boldsymbol{J} = \boldsymbol{n} \times \boldsymbol{H}_t, \qquad \boldsymbol{M} = \boldsymbol{E}_t \times \boldsymbol{n} \tag{3.26}$$

where $\boldsymbol{n}$ is the outward normal vector on surface S. Equation (3.26) is rewritten using the equivalent theorem. The vector Kirchhoff integral yields the antenna far-field using (3.26) as follows:

$$\boldsymbol{E}(\boldsymbol{r}) = \frac{jke^{-jkr}}{4\pi r}\boldsymbol{r} \times \int_S \left(\boldsymbol{M}(\boldsymbol{r}_o) + Z_o \boldsymbol{r} \times \boldsymbol{J}(\boldsymbol{r}_o)\right) e^{-jkr_o}\, dS \tag{3.27}$$

where Z_o is the characteristic impedance in free space and $\boldsymbol{r}$ and $\boldsymbol{r}_o$ are the observation and source vectors, respectively.

As an example of a large aperture size such as that in a parabolic reflector antenna, the electromagnetic fields are differently defined according to the distance in the near-field, Fresnel, and far-field regions. For field measurements near the antenna, the mutual coupling between the measurement probe and test antenna becomes high and alters the electromagnetic field radiated from the antenna. The near-field to the region near the antenna is used for antenna measurement. A suitable probe for near-field measurement consists of an open-ended rectangular waveguide, which does not have an ideal omnidirectional pattern. The nonideal performance can be easily rectified using probe correction [1], and the far field can then be evaluated using the electromagnetic field in the near-field region.

Although the near-field measurements require a space that encloses the antenna, a closed-edge plane is not always necessary when the measured field strength is sufficiently small to be neglected. The scanning surface is categorized to be planar, cylindrical, or spherical, as shown in Figure 3.20. A planar scan covers only half the space and is appropriate for high-gain antenna measurements, such as parabolic and planar-array antennas. Cellular base station

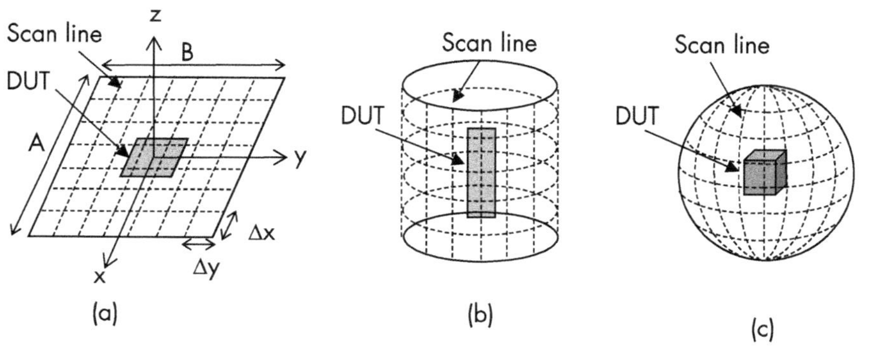

Figure 3.20 Scans are of near-field measurement: (a) planar scan, (b) cylindrical scan, and (c) spherical scan.

and marine radar antennas have a fan-shaped pattern with a narrow beam in one plane and a wide beam in the other plane, and a cylindrical scan is appropriate for near-field measurements. The spherical scan is advantageous in measurements of low-gain antennas.

The scanning area ($A \times B$) and sampling interval (Δx, Δy) are described in the planar scanning method and shown in Figure 3.20(a). The size of the scanning area is determined according to the requirement for obtaining more than 35-dB dynamic range in the scanning area without the need to extrapolate the measured data. The 35-dB value is an empirical one. The dynamic range is defined by the difference between the maximum electric field and edge-field strengths of the scanning area. Maximum data-point spacing should be made such that $\Delta x = \lambda_o/2$ and $\Delta y = \lambda_o/2$ for planar scanning using the sampling theorem [7].

If the x and y components of the measured electric field are denoted as $E_x(x,y)$ and $E_y(x,y)$, respectively, then the far-field radiation patterns are given by the coordinate transformation from (x,y,z) to (r,θ,ϕ) as:

$$E_\theta = jk\cos\theta\frac{e^{-jkr}}{r}\left(S_x\cos\theta\cos\phi + S_y\cos\theta\sin\phi\right) \tag{3.28}$$

$$E_\phi = -jk\cos\theta\frac{e^{-jkr}}{r}\left(S_x\sin\phi - S_y\cos\phi\right) \tag{3.29}$$

$$S_i = \frac{1}{2\pi}\int_{-B/2}^{B/2}\int_{-A/2}^{A/2} E_i(x,y)\exp\{jk(x\sin\theta\cos\phi + y\sin\theta\sin\phi)\}\,dx\,dy,\ i = x,y \tag{3.30}$$

The total numbers of observed discretized electric fields are $2M$ and $2N$ in the x and y directions, respectively, and S_i is expressed as follows:

$$S_i = \frac{1}{2\pi}\sum_{-M}^{M}\sum_{-N}^{N} E_i(x_m, y_n)\exp\{jk(x_m\sin\theta\cos\phi + y_n\sin\theta\sin\phi)\}\Delta x\Delta y \tag{3.31}$$

where $x_m = m\Delta x$ and $y_n = n\Delta y$. In the aforementioned calculations, the fast Fourier transform can be used to reduce the computation time.

3.3 PROPAGATION MEASUREMENT

This section describes the basics of propagation measurement. In mobile communication systems, the propagation loss is increased by reflection, scattering, and diffraction from obstacles that exist at the site, which causes fading at the mobile terminals. The propagation loss factors measured under a real propagation environment are characterized using a statistical approach. This section presents the basic evaluation method of the parameters to analyze the fading structures. Additionally, the measurement procedure to determine the delay spread is described.

3.3.1 Propagation Loss Measurement

Propagation loss is measured within a range of a few hundred meters to tens of kilometers. However, the received signal level frequently drops due to multipath fading. To overcome these signal fluctuations, an appropriate fading margin is required to guarantee system quality. Knowledge of the fading characteristics of a site allows selection of a suitable fading margin for the system design.

The basic terms of the parameters for narrow-band propagation measurements are propagation loss, cross polarization, and correlation factor between the diversity or MIMO branches.

Attenuation coefficient is a basic evaluation factor for propagation-characteristic measurement. The received signal levels are plotted as a function of distance from the transmitting station. The measured data vary over a wide range; thus, the propagation-decay profile is determined using a statistical process.

Figure 3.21 shows an example of an electric field within a range from a few meters to a few kilometers, which is known as long-range distribution. When the electric field is measured in terms of the propagation distance, the amplitude decreases according to $1/r$ under the condition where no reflected or diffracted wave is present. For practical measurement of the electric field where the relative power of the field is measured in decibel units, the measured data decrease according to $1/r^n$ ($n > 2$). This approximation is quite rough, and the measured data are quite scattered. This macroscopic-data treatment is commonly used in the design of a base station coverage area.

Another propagation measurement is carried out near the transmitting base station for indoor-propagation measurements or micro/pico cellular-system designs. Figure 3.22 shows an example of a short-range distribution of the received electric field near a transmitting station. To determine the average signal level at the measurement site, a short-term median is used. The definition of *short term* varies depending on the measurement range. When the total measurement length is a few hundred meters, the median value of the collected data of a few meters is used as the short-term median.

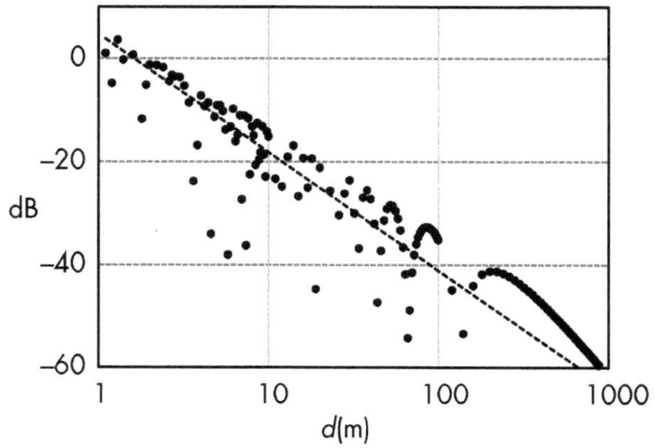

Figure 3.21 Long-range field distribution.

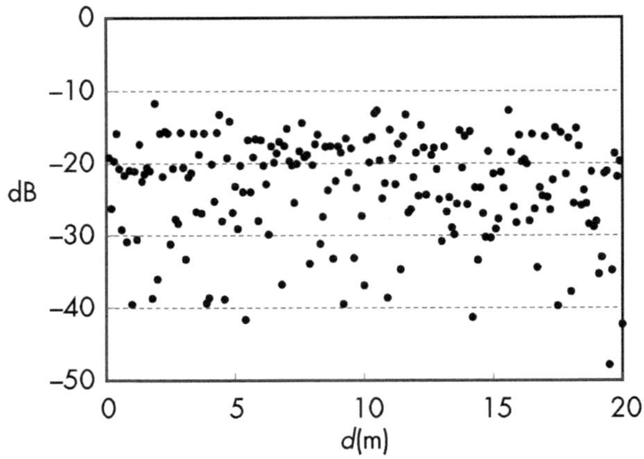

Figure 3.22 Short-range field distribution.

However, for measurements of more than tens of kilometers, the short term is defined as a few kilometers. Short term is defined as being approximately 1/100 of the total measurement length.

The standard measurement system shown in Figure 3.23 consists of a signal generator connected to a low-gain antenna and a receiver with a standard antenna with known radiation pattern and antenna gain. The envelope-signal level at the receiver output after detection is sampled using an analog-to-digital converter (ADC). For

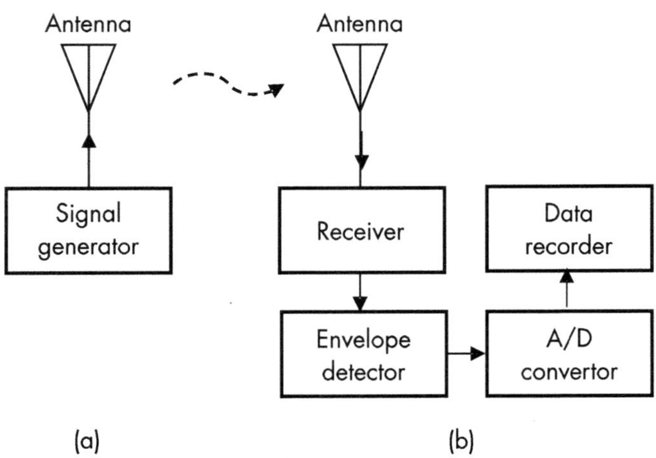

Figure 3.23 Measurement system: (a) transmitter, and (b) receiver.

long-range measurements, the ADC sampling rate is selected to sample the data at less than a quarter-wavelength spacing because the period of the standing wave created at the measurement site is $\lambda/2$, as shown in Figure 3.24.

Thus, sampling frequency f_s is determined by the relationship $f_s = 4v/\lambda$ Hz, where the receiving-system speed is v (m/s). To record the fading structure in detail without using any interpolation techniques, the sampling space should be less than $\lambda/20$ for short-range measurements [8]. To correctly measure the Rayleigh distribution data, finely spaced sampling is required to obtain more than a −30 dB fading-depth accuracy.

To determine the average received signal level from the measured data, which is normally recorded in decibel units, the measured data must first be transformed into linear units before the mean value of the data is determined. To eliminate this process, a median is used as an average value. The received data, which consist of a total number of N data points, are sorted according to the amplitude strength, and the $(N/2)$th amplitude is chosen as the median. The median value eliminates the deep fading in the measurement and provides an average value for the measurement site.

3.3.2 Cross Polarization and Correlation Coefficient

In terrestrial mobile communication systems, a set of orthogonal polarizations is employed in the diversity and MIMO schemes. The polarization combination is ±45° polarization or vertical (V) and horizontal (H) polarization. Measuring the XPR at the mobile terminal is important in the propagation measurement.

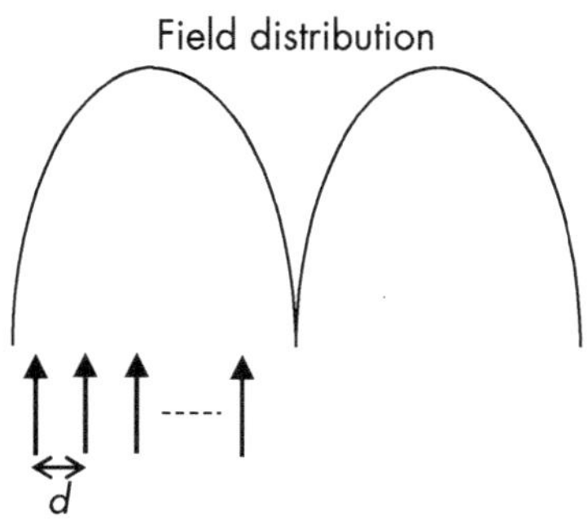

Figure 3.24 Sampling interval.

XPR is defined as the ratio of the principal and orthogonal polarizations as follows:

$$XPR = \frac{P_p}{P_x} \tag{3.32}$$

where P_p and P_x are the short-range medians for the principal and orthogonal polarizations, respectively, at the measured site. The transmitting polarization is defined as the principal polarization, which is chosen as the vertical polarization for the V/H combination, and +45° or −45° is selected for the ±45° case.

The performance of diversity antennas is evaluated by testing them in a practical environment. Generally, the term diversity branch denotes the multiple outputs from the antenna after the modulated signal is detected; however, it refers to different antenna output ports. When two output branches are present in a particular diversity antenna, the correlation coefficient of the detected signal envelope of e_1 and e_2 is defined as:

$$\rho = \frac{\frac{1}{2}\left\langle \left(e_1 - \langle e_1 \rangle\right)^* \left(e_2 - \langle e_2 \rangle\right) \right\rangle}{\frac{1}{2}\sqrt{\left\langle \left(e_1 - \langle e_1 \rangle\right)^2 \left(e_2 - \langle e_2 \rangle\right)^2 \right\rangle}} \tag{3.33}$$

where $\langle x \rangle$ denotes the mean value of measured parameter x and $(y)^*$ denotes the complex conjugate of measured parameter y.

In propagation measurements, the correlation coefficient is calculated every 100 or 200 samples, and a correlation histogram is obtained, as shown in Figure 3.25. The correlation coefficient is obtained using the average of this histogram in which the median of the received electric field level is almost constant. In practical diversity measurements, more than tens of thousands of data samples are collected in one measurement. Thus, for a single measurement, hundreds of correlation coefficients must be calculated. Because of the large number of data samples required for collection, the total number of calculations is reduced. The correlation coefficient is used for macroscopic evaluation of diversity antennas.

To measure the diversity performance, a multichannel receiver and ADC are required. For a two-branch diversity system, two

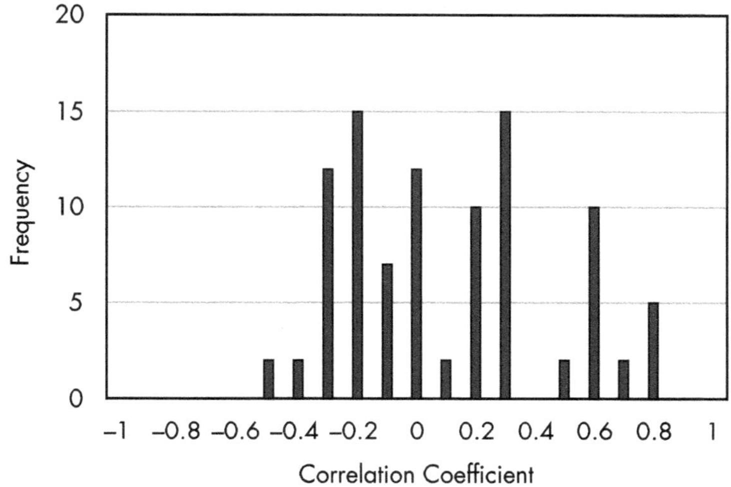

Figure 3.25 Histogram of correlation coefficient.

identical receiving subsystems are required for simultaneous measurement of the received signals from the two antennas. A portable spectrum analyzer can be used as a receiver for the measurement system, or a commercially available wideband receiver can be used for the same purpose. The two independent receiver subsystems should be calibrated using a known power-level standard as an input to each channel. A discrepancy of 1 to 2 dB can often occur between the two receivers, which need to be calibrated.

To simplify the measurement system, one receiver can be used with multiple antennas. For this measurement method, the technique of switching between different antennas at a specified switching rate is used. An interval of $\lambda/40$ is sufficiently short; thus, the received signal envelope can be considered constant between measurements [2].

3.3.3 Delay Profile Measurement

The delay profile measurement is performed by adding a precise timing signal generator to both the transmitter and receiver, as shown in Figure 3.23. When the receiving points are located close to the transmitter, a common signal generator source can provide the timing signal to both the transmitter and receiver using a coaxial cable. For these measurements, however, the use of cables is not feasible. A rubidium oscillator in a standard signal generator synchronizes the oscillators used in the transmitter and receiver. The stability of the rubidium oscillator is one part per billion (10^{-9}), and it is a cost-effective oscillator. An example of the delay profile measurement is shown in Figure 3.26. The detail of delay profile measurement is explained in Section 10.1.2.

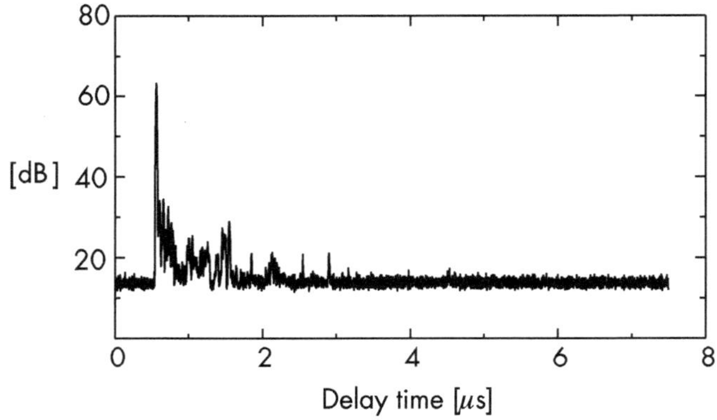

Figure 3.26 Measured delay profile.

References

[1] Collier, R. J., and A. D. Skinner, *Microwave Measurements*, Third Edition, Stevenage, U.K.: The Institution of Engineering and Technology, 2007, p. 5.

[2] Tiwari, A. K., "Smith Chart and Its Applications," *International Journal of Electronic and Electrical Engineering*, Vol. 4, No. 1, 2011, pp. 75–94.

[3] Collier, R. J., and A. D. Skinner, *Microwave Measurements*, Third Edition, Stevenage, U.K.: The Institution of Engineering and Technology, 2007, pp. 22–24.

[4] Kock, L., "Analysis and Performance of Antenna Balun," University of Stellenbosch, April 2005.

[5] Balanis, C. A., *Antenna Theory Analysis and Design*, Third Edition, New York: John Willey & Sons, 2005, pp. 168–170.

[6] Balanis, C. A., *Antenna Theory Analysis and Design*, Third Edition, New York: John Wiley & Sons, 2005, pp. 74.

[7] Yaghjian, A. D., "An Overview of Near-Field Antenna Measurement," *IEEE Trans. AP*, Vol. AP-34, No. 1, 1986, pp. 30–45.

[8] Murase, M., Y. Tanaka, and H. Arai, "Propagation and Antenna Measurements Using Antenna Switching and Random Field Measurements," *IEEE Trans. Veh. Technol.*, Vol. 43, No. 3, 1994, pp. 537–541.

4

ANTENNA ELEMENT AND ARRAY

In this section, we present the antenna elements of the base station (BS) and the mobile terminal by using practical examples in the latest mobile communication systems. Two basic antenna elements, namely, the dipole and microstrip antennas, are described. To support current and future systems, it is essential for the antenna elements to have multiband or wideband characteristics. The electrical features of those elements are shown. The antenna pattern of a BS is synthesized by adjusting the amplitude and phase of each antenna element of the array. This chapter presents the array antenna for the base station and the typical pattern synthetization techniques. The antenna elements for mobile terminals are miniaturized built-in structures and are different from those of base stations. This chapter also describes basic small antenna elements and the theory of antenna miniaturization.

4.1 ANTENNA ELEMENTS OF THE BASE STATION

Antenna elements derived from $\lambda/2$ dipole antennas are widely used for cellular base stations. Practical systems use numerous elements derived from dipole antennas. In this section, the basic dipole antenna elements are described. Another BS antenna element is the microstrip antenna. The theory of this antenna and the applications of the BS antenna are also presented. The cellular BS antennas are required to have multiband or wideband frequency characteristics for use in the frequency band assigned for the cellular system. This section also describes the techniques used for multiband and wideband operations in the antenna systems.

4.1.1 Dipole Antenna

The dipole antenna is the most basic antenna element and is widely used in practical radio systems. The principle of the $\lambda/2$ dipole antenna was described in Sections 1.4 and 1.5. The feeding mechanism and the use of the balun were presented in Section 3.1.4, and this section describes the dipole antennas used in cellular BS antennas and propagation measurements.

A standard $\lambda/2$ dipole antenna consists of a pair of radiating elements and a balun to connect a coaxial cable (see Figure 4.1). When the absolute antenna gain of the $\lambda/2$ dipole antenna is theoretically obtained (see Section 1.6), it is known as the standard dipole. This dipole is used in propagation measurements to find the electric field distributions produced by the testing base stations.

The radiation pattern of the dipole antenna is a figure-of-eight pattern in the E plane and is omnidirectional in the H plane. The feeding structure often disturbs the omnidirectional pattern, and then a sleeve dipole antenna is used to keep a uniform pattern in the H plane. Figure 4.2 shows the sleeve dipole antenna consisting of a radiation element and a balun. The leaky current on the surface of the coaxial cable is suppressed by the outer shell using the mechanism with the sleeve balun, as shown in Figure 3.10. The difference between the sleeve antenna and the balun is the plate position for connecting the outer shell with the cable surface. The top shorting

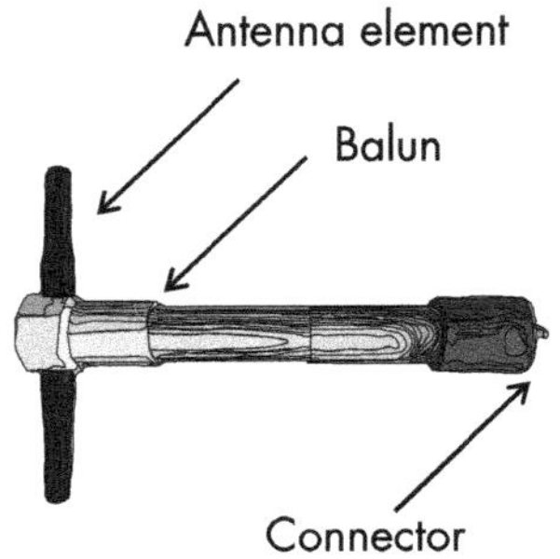

Figure 4.1 Standard dipole antenna.

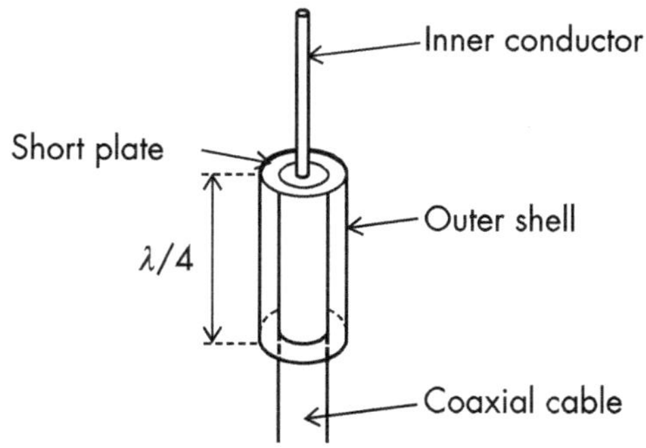

Figure 4.2 Sleeve antenna.

plate is used for the sleeve dipole, and the bottom plate is used for the balun.

A printed dipole antenna consists of an antenna element etched on a dielectric substrate, as shown in Figure 4.3, where the bottom ground plane is removed. The dominant resonance is obtained by $L = \lambda_e/2$, where the effective wavelength λ_e is approximated as $\lambda_e \cong \lambda/\sqrt{\varepsilon_r + 1}$. The effective relative dielectric constant is the average of ε_r and 1 (air value).

This antenna is usually installed on the ground plane and is fed by a coaxial cable from the back. Figure 4.4 shows a modified printed dipole antenna with a taper balun. The antenna element is etched on the front and back of a substrate to connect with a microstrip

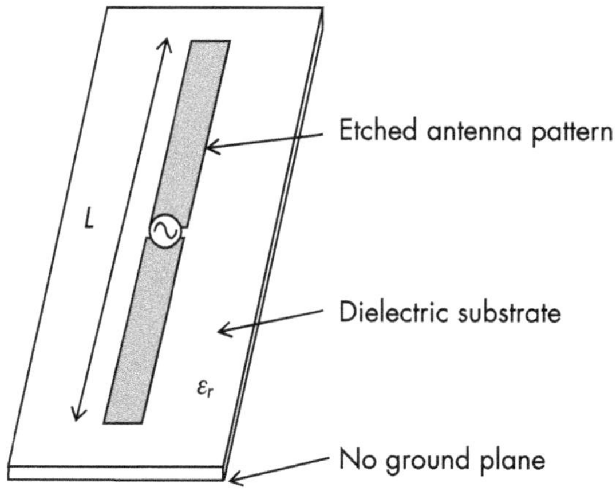

Figure 4.3 Printed dipole antenna.

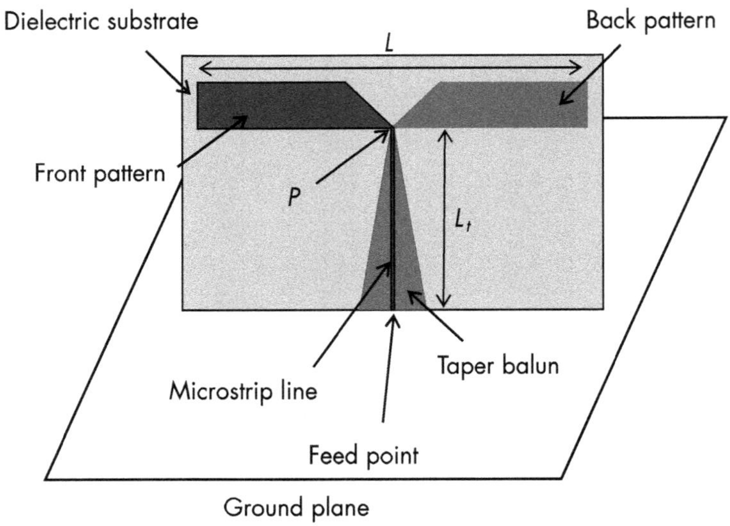

Figure 4.4 Modified printed dipole with taper balun.

line. The front antenna pattern is connected with a microstrip feeding line, and the back pattern is attached to the triangular pattern connected with the ground plane at the bottom. The bottom feeding point can be excited by an unbalanced feeding line, such as the coaxial cable, and the top point *P* is connected with the balanced antenna element. The microstrip line backed by the triangular ground plane converts the unbalanced mode into the balanced mode, and this is known as the taper balun. The resonant frequency of the antenna in Figure 4.4 is determined by the length L_t of the taper balun rather than the antenna length *L*. The dominant resonance is given by $L_t = \lambda_g/4$, where λ_g is the guided wavelength of the microstrip line.

An open stub-loaded feeding line is used for the printed dipole antenna as shown in Figure 4.5. A T-shaped back pattern of the substrate is composed into a radiating dipole antenna element and the ground plane of the feeding line. Both parts are denoted by dotted lines. The bottom rectangular pattern is the reflector for the unidirectional pattern in the *H* plane. This reflector can be replaced by the ground plane, as shown in Figure 4.4. The U-shaped feeding line on the top surface has the open stub on the right side, and the length of the feeding line is approximately $\lambda_g/4$. The point *P*, which is the top position of the feeding line, is the antinode of the current

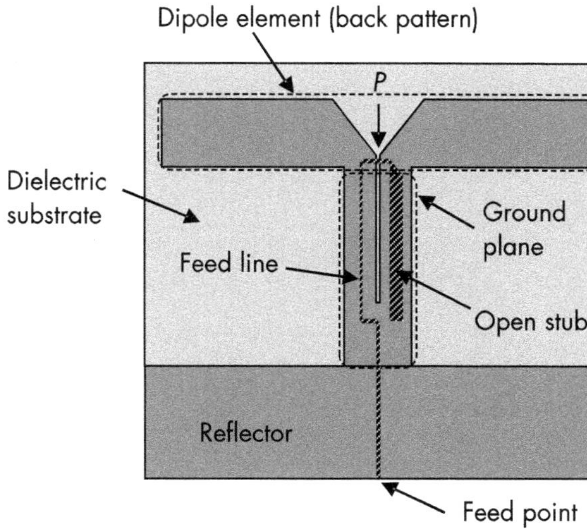

Figure 4.5 Printed dipole with merchant balun.

distribution on the feeding line because of the open stub. The dipole element is then excited. Impedance matching at the feeding point at the bottom is obtained by adjusting the characteristic impedance of the stub and the length of the feeding line. This kind of printed dipole is widely used in practical applications.

The radiation pattern of the $\lambda/2$ dipole antenna is omnidirectional in the H plane and has a figure-eight pattern along the E plane. To use the dipole antenna for the polarization diversity of the base stations, the two plane patterns should have the same beamwidth; vertical and horizontal polarization diversity is assumed. The H-plane radiation pattern of a twin-dipole antenna (see Figure 4.6) is reduced by the array factor as follows:

$$E_\phi = \cos\left(\frac{\pi}{2}\sin\phi\right) \tag{4.1}$$

The half power beamwidth given by (4.1) is 90°, which is larger than the beamwidth of the E plane by 12°. The base-station antenna elements are backed by the reflector to produce a sector pattern in the horizontal plane. Then, the beamwidth is adjusted by the size

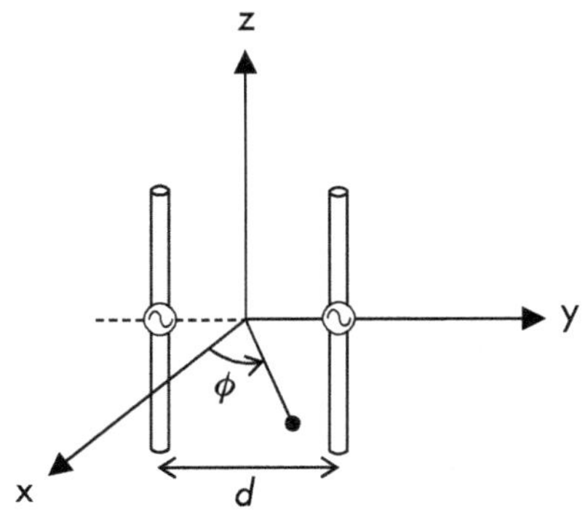

Figure 4.6 Twin dipole antenna.

and shape of the ground plane. This difference in the beamwidth is acceptable for the V/H-polarization diversity antennas.

Figure 4.7 describes a crossed dipole antenna for another diversity scheme used in cellular base stations. The antenna consists of a set of two identical dipole antennas mounted at right angles to each other. The symmetry about the *x*, *y*, and *z* axes produces identical antenna patterns in the *E* and *H* planes; this pattern is one of the advantages of this diversity scheme. Note that this antenna radiates ±45° polarization in the front and back of the antenna (along the *x* axis). The radiated electric field from −45° polarized antenna is shown in Figure 4.7. The tilt angle of the radiated field is changed by the ϕ value in the *xy* plane, and it is aligned with the *z* axis at $\phi = \pm 90°$. The tilt angle α is given as follows:

$$\alpha = \tan^{-1}(\cos\phi) \tag{4.2}$$

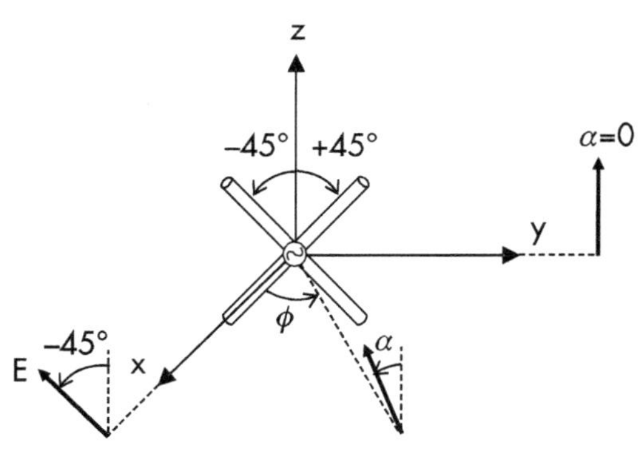

Figure 4.7 Crossed dipole antenna.

The radiated field rotated by (4.2) causes polarization mismatch when the orientation of the receiving measurement antenna is fixed at the front polarization angle. The observed radiation pattern in the *xy* plane is given by (4.3) for the measured polarization fixed at −45° in the front direction.

$$f(\phi) = \frac{1}{\sqrt{2}}(1 + \cos\phi) \tag{4.3}$$

To find the radiation pattern correctly, two orthogonal polarizations should be measured to observe the rotation of the polarization.

4.1.2 Microstrip Antenna

Microstrip antennas are basic antenna elements used for mobile communication systems. Figure 4.8(a) shows the geometry of this antenna. A rectangular or square antenna element pattern is etched on the top surface of the dielectric substrate with the thickness h. The element resonant length L is approximately $\lambda_g/2$ (see Section 1.5) for the current flowing in the pattern shown in Figure 4.8(a). The precise guide wavelength λ_g is given by introducing the effective dielectric constant ε_e as follows:

$$\lambda_g = \frac{\lambda}{\sqrt{\varepsilon_e}} \tag{4.4}$$

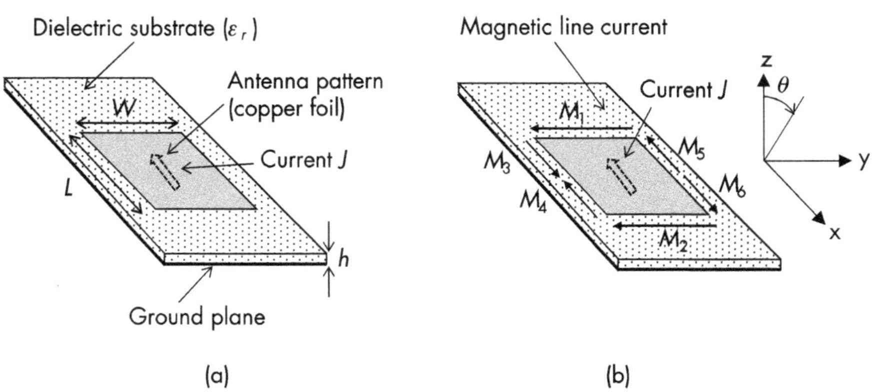

Figure 4.8 Microstrip antenna: (a) antenna geometry, and (b) magnetic line current.

$$\varepsilon_e = \frac{\varepsilon_r + 1}{2} + \frac{\varepsilon_r - 1}{2\sqrt{1 - 10\left(\dfrac{h}{W}\right)}} \tag{4.5}$$

where ε_r is the relative dielectric constant of the substrate. The antenna resonates by using the stored energy accumulated under the antenna pattern in Figure 4.8(a); however, some part of the energy leaks out from antenna pattern edges, and the resonant frequency is shifted to the lower side. This phenomenon is equivalent to seeing the increase in the relative dielectric constant, as given by (4.5).

The radiation pattern of the microstrip antenna is calculated by the current on the antenna element and the displacement current in the substrate. It is not easy to include the displacement current. The pattern is approximately calculated by fictitious magnetic line currents at the edge of the element pattern, as shown in Figure 4.8(b). The magnetic line current is given by the edge electric field $\boldsymbol{E}_g$ and the outward normal vector $\boldsymbol{n}$ at the edge as $\boldsymbol{M} = \boldsymbol{E}_n \times \boldsymbol{n}$, which are presented by the arrows $M_1,\ldots,M_6$ in Figure 4.8(b). The dominant magnetic currents producing the radiation are M_1 and M_2, and the other currents are aligned in the reverse direction to cancel the radiation along the z axis. The $E(zx)$ plane pattern is approximated by the M_1 and M_2 arrays as follows:

$$D(\theta) = \cos\left(\frac{k_o L}{2}\cos\theta\right) \tag{4.6}$$

The $H(yz)$ plane pattern is close to the H-plane pattern of the dipole antenna.

Figure 4.9 shows the feeding structures of the microstrip antenna. A coaxial connector attached to the bottom of the substrate excites the antenna by connecting the inner conductor of the connector with the antenna pattern. The cross section of the feeding point is shown in Figure 4.9(a). The position of the feeding point is adjusted to obtain impedance matching at the connector that is excited by the feeding cable.

The microstrip line is another feeding structure (see Figure 4.9(b)). The feeding point, which is the length s from the pattern edge, is adjusted to obtain impedance matching. The impedance becomes

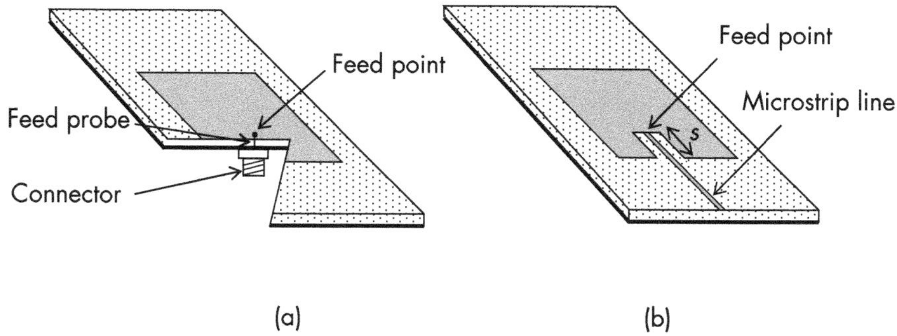

Figure 4.9 Feeding of microstrip antenna: (a) coaxial feed, and (b) microstrip line feed.

the maximum value at the antenna pattern edge and is decreased by moving the feeding position to the antenna center.

Dual polarization is obtained by two feeding points, as shown in Figure 4.10. Two feeds excite the orthogonal polarizations; the feeding point #1 radiates E_1, and the feeding point #2 radiates E_2.

4.1.3 Multiband Dipole and Microstrip Antenna Elements

A multiband or wideband antenna element is used if the frequency bandwidth of a single antenna element is not wide enough to cover the required bandwidth. A parasitic element in the vicinity of the exciting antenna provides the additional resonance for the multiband

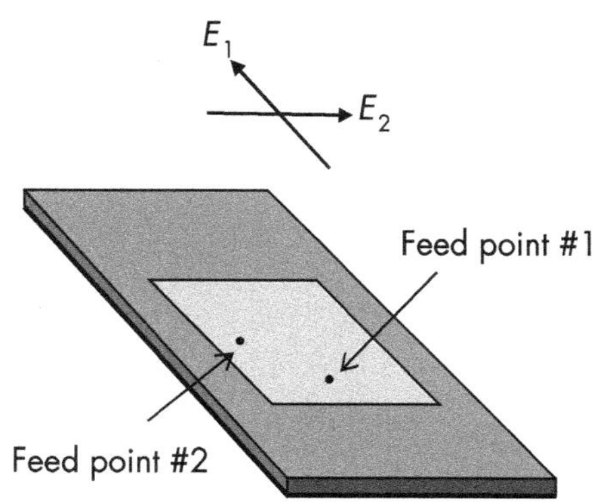

Figure 4.10 Dual polarized microstrip antenna.

antenna. Figure 4.11 shows the geometry of a dual-band dipole antenna consisting of a driven and a parasitic element. If the length of the parasitic element is less than that of the exciting element, the resonant frequency f_1 of the exciting element should be less than that of the parasitic element f_2. If the parasitic element is longer than the exciting element, it does not produce another resonance but is operated upon as a reflector.

A dual resonant microstrip antenna has a parasitic element of the same size as the exciting element (see Figure 4.12). The air gap between the parasitic and exciting elements increases the resonant frequency of the parasitic element f_2. The resonant frequency of the exciting element is determined by the guide wavelength in the dielectric substrate, and the upper frequency is given by the average wavelength of the substrate and the air gap. Then, the resonant frequency has a similar relationship of $f_1 < f_2$ with the dipole antenna in Figure 4.11.

The parasitic element is effective for adding higher resonance to the exciting element, and multiresonance can be obtained by using multiple parasitic elements. This method is easy to obtain for a multiband antenna; however, the radiation pattern and the antenna gain should be considered in practical applications.

The microstrip antenna made by the dielectric substrate has narrow frequency band characteristics because the substrate is thin. To increase the bandwidth, we use a thick antenna consisting of the

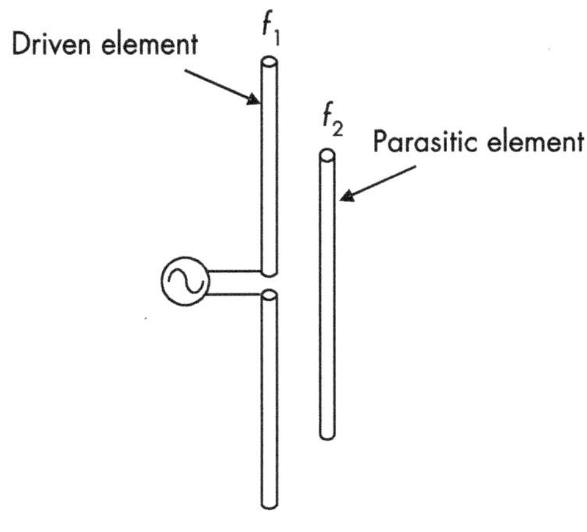

Figure 4.11 Dual resonant dipole antenna.

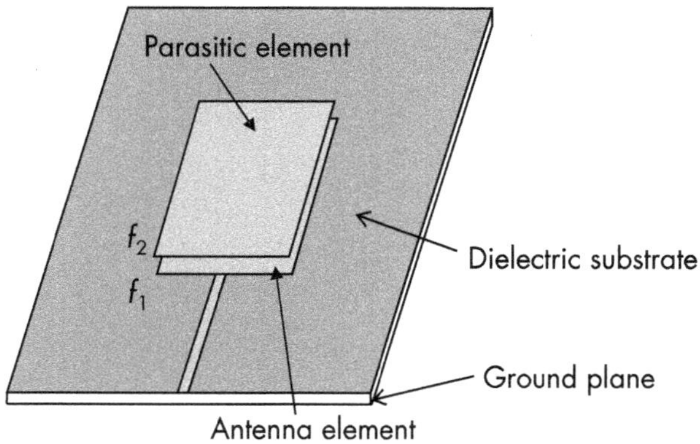

Figure 4.12 Dual resonant microstrip antenna.

conducting antenna plate without the substrate. Dual resonance is also obtained using a parasitic element as shown in Figure 4.12. The parasitic element placed above a driven element gives another resonance at f_2 and extends the frequency bandwidth.

4.2 ARRAY ANTENNAS

The array antenna is used to increase the antenna gain and to synthesize the desired radiation pattern. This section describes a colinear array antenna that is omnidirectional in the horizontal plane and has a narrow beamwidth in the vertical plane. We also present the array feeding structure and the principle of the phase shifter for the cellular base-station antennas. In addition, the beam tilt and pattern synthetization are explained.

4.2.1 Colinear Array

Figure 4.13 shows the colinear array antenna that consists of several coaxial cables with the length of $\lambda/2$. A bottom cable is excited by connecting a radio frequency (RF) source at point F with the inner and outer conductors of the cable. The phase delay of the inner conductor current I_{i1} is 180° at point C. Its end is connected to the outer conductor of the upper coaxial cable. The outer current I_{o2} flows in the reverse direction (see Figure 4.13), and the direction coincides with I_{o1}. This alternate connection produces in-phase excitation for

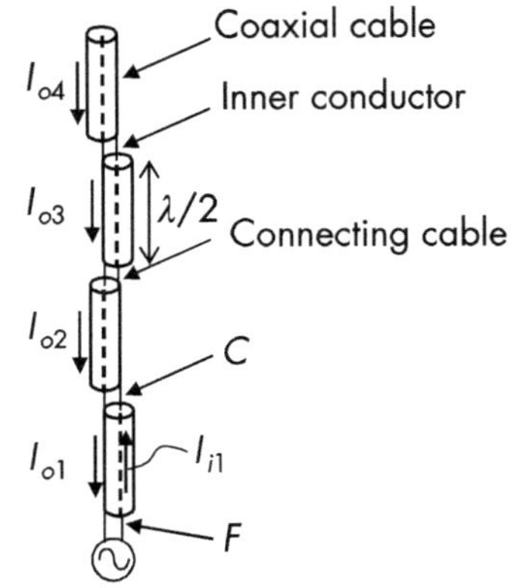

Figure 4.13 Colinear array antenna.

all the current on the outer conductor of the coaxial cables. The radiation pattern of the colinear array is omnidirectional in the horizontal plane, and the gain is increased by the number of cables in the vertical plane. A slight phase change at the connection of the cables creates variations in the beam tilt angle in this array antenna. The tilt angle adjustment and the impedance matching at the feed position need to be considered in the design of the colinear array antenna.

4.2.2 Linear Array

The radiation pattern in the vertical plane is controlled by the array antenna for cellular base-station antennas. Figure 4.14 shows a tournament feed network of a linear array antenna where the amplitude of each element is assumed to be identical. Each antenna element is excited through the phase shifter, and the beam tilt angle in the vertical plane is changed. The maximum tilt angle of the base stations is limited to approximately 10°. Then, the four-subarray feed network is used to reduce the number of phase shifters, as shown in Figure 4.15. A pair of elements excited in phase is connected to a phase shifter, which halves the number of phase shifters used in this example. If the subarray spacing $2d$ is double the length of the array shown in Figure 4.14, the grating lobes appear in the visible region, as described in Section 1.8. Figure 4.16 shows the antenna pattern, array factor, and subarray pattern of the eight-element array by four-element subarray, as shown in Figure 4.15. The grating lobes that appeared around

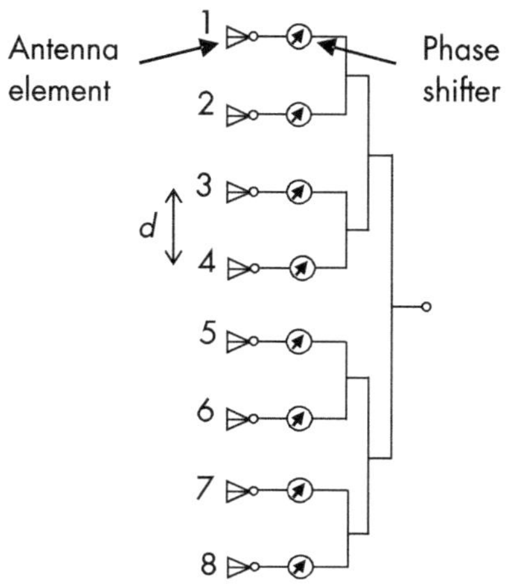

Figure 4.14 Feed network of array antenna.

$\theta = \pm 40°$ by the array factor is suppressed by the subarray pattern. Two-element in phase subarrays have null points around $\theta = \pm 40°$, which suppresses the grating lobes, as shown in Figure 4.16.

This undesired lobe suppression is effective only for the small tilt angle. The radiation patterns of the subarray are shown in Figure 4.17. Another grating lobe comes into the visible region by increasing

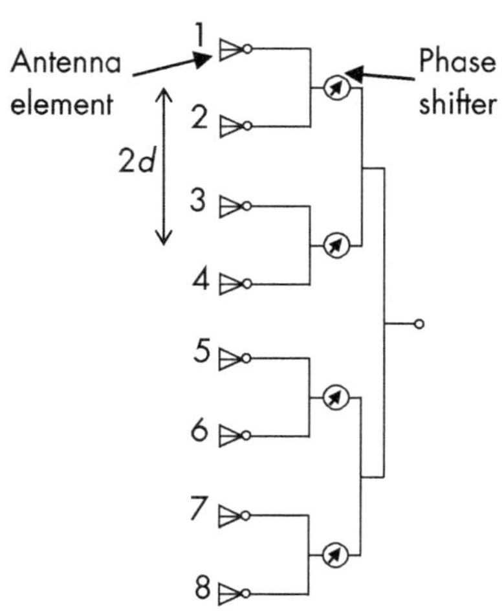

Figure 4.15 Feed network of subarray antenna.

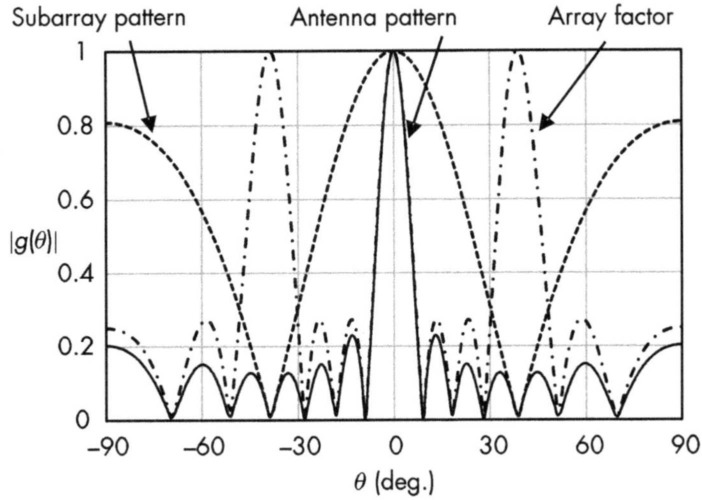

Figure 4.16 Radiation pattern of four-subarray antenna; element spacing $d = 0.8\lambda$.

the tilt angle θ_o, and the null positions of the subarray pattern are shifted from the position of the grating lobes, which increases the side-lobe level. The range of the beam tilt angle should be determined by the level of the side lobes.

Base stations use side-lobe suppression to reduce the interference to the outside of the covered area. The side-lobe level is controlled by changing the amplitude weight of each element. Analytical formulas that specify the uniform side-lobe level are given by Tschebycheff polynomials [1], and the peak side-lobe level is reduced by the Taylor weighting distribution to improve the antenna aperture efficiency [2]. In addition, the weights are obtained by a binomial array to achieve very low side-lobe levels [3]. Figure 4.18 shows the radiation patterns of the four-element array antenna with uniform amplitude distribution and a weighted amplitude; the pattern was obtained by using a binomial distribution. The weight of the binomial array is given as $a_1 = a_4 = 0.33$ and $a_2 = a_3 = 1$, where a_i is the amplitude weight of the antenna element. The binomial array provides very low side-lobe levels, as shown in Figure 4.18; however, the extended beamwidth reduces the antenna directivity gain.

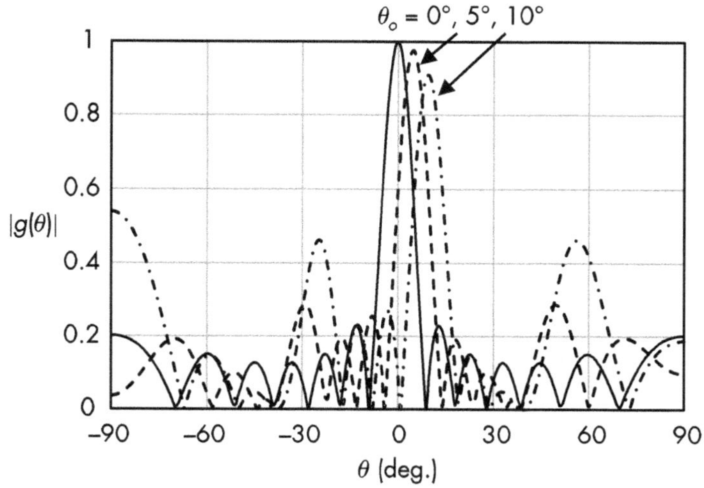

Figure 4.17 Radiation pattern of tilted four-subarray antenna; element spacing $d = 0.8\lambda$.

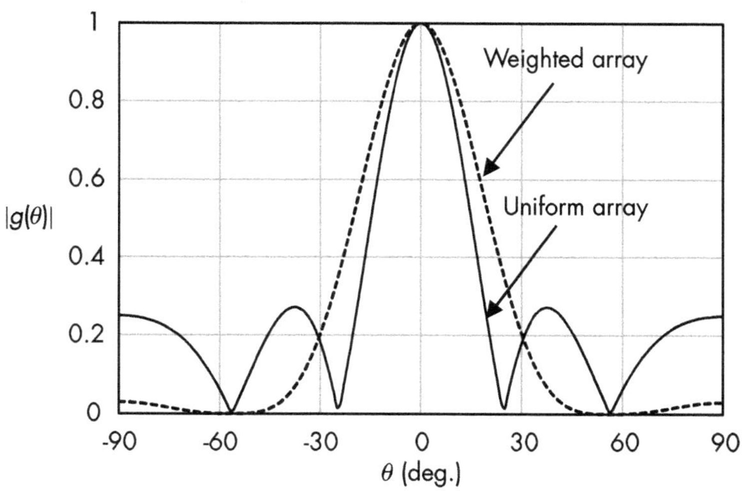

Figure 4.18 Radiation pattern of four-element array antenna; element spacing $d = 0.6\lambda$, amplitude weight = 0.33.

4.2.3 Antenna Pattern Synthesis

To reduce the interference between the base stations, the antenna patterns are synthesized to follow the required pattern. In the antenna pattern synthesis, the excitation condition of each antenna element is evaluated as described next.

To evaluate the amplitude and phase, other numerical methods for optimization have been proposed, such as the Fourier series expansion [4] and the Woodward–Lawson method [5]. As a design example, we present the least squares method for pattern synthesis. If the desired radiation pattern is $g(\theta)$, the pattern given by the actual feeding network of $g(\theta)$ is expressed by the amplitude I_n and phase ϕ_n of each element, as follows:

$$g(\theta) = f(\theta)\sum_{n=1}^{N} I_n \exp\{j(n-1)k_o d\cos\theta + j\phi_n\} \tag{4.7}$$

where n and d represent the number of elements and the element spacing, respectively; k_0 is the wave number in free space, and $f(\theta)$ is the element pattern.

To obtain the amplitude I_n and phase ϕ_n for the desired antenna pattern, the error value should be minimized in the synthesized pattern. An example of the desired antenna pattern mask is shown in Figure 4.19. The actual radiation pattern should fall in the hatting pattern. In the following vector, the discreet points of M express the desired pattern in the vertical plane, and the transposed matrix t is given as follows:

$$\boldsymbol{G_d}(\theta) = \begin{bmatrix} g_{d1}(\theta_1) & \cdots & g_{dM}(\theta_M) \end{bmatrix}^t \tag{4.8}$$

Using (4.7), $g(\theta)$ is rewritten in matrix form as follows:

$$\boldsymbol{G} = \begin{bmatrix} g(\theta_1) \\ \vdots \\ g(\theta_M) \end{bmatrix} = \begin{bmatrix} A_{11}(\theta_1) & \cdots & A_{1N}(\theta_1) \\ \vdots & \ddots & \vdots \\ A_{M1}(\theta_M) & \cdots & A_{MN}(\theta_M) \end{bmatrix} \begin{bmatrix} a_1 \\ \vdots \\ a_N \end{bmatrix} \tag{4.9}$$

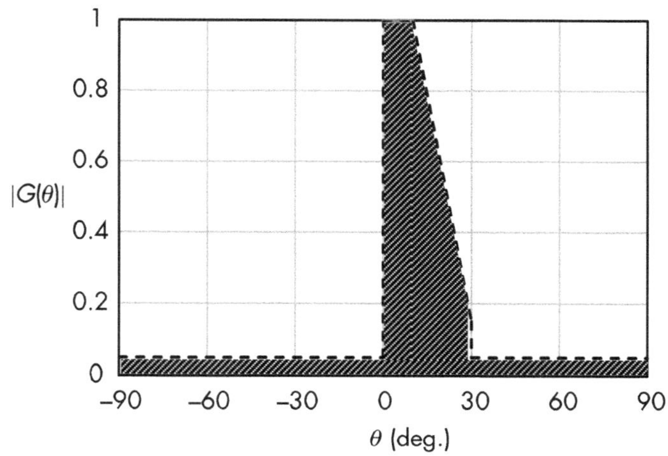

Figure 4.19 Desired pattern (antenna pattern mask).

$$A_{mn}(\theta_m) = f(\theta_m) I_n \exp\{j(n-1)k_o d\cos\theta + j\phi_n\} \tag{4.10}$$

The error function is defined by multiplying the difference between $\boldsymbol{G}$ and $\boldsymbol{G_d}$ and its Hermitian matrix as follows:

$$\varepsilon = \sum_{m=1}^{M}\left(g_{dm}(\theta_m) - \sum_{n=1}^{N} A_{mn}a_n(\theta_m)\right)\left(g_{dm}(\theta_m) - \sum_{n=1}^{N} A_{mn}a_n(\theta_m)\right)^H \tag{4.11}$$

Then, the condition for the excitation coefficient of a_n to minimize ε is given by the least square method as follows:

$$\frac{\partial\varepsilon}{\partial a_n^*} = \sum_{n=1}^{N}\left(g_{dm} - \sum_{n=1}^{N} A_{mn}a_n\right)A_{mn}^* = 0 \tag{4.12}$$

Note that an ideal synthesized pattern is given by the arbitral amplitude and phase distributions. A fixed amplitude or a limited phase range distribution is also provided by the optimization method under the constrained condition.

4.3 ANTENNA ELEMENTS OF THE MOBILE TERMINAL

The preceding sections describe the basics of antennas used in base stations. The antenna elements of mobile terminals are electrically small antennas to be built in the terminals. This section presents several antenna miniaturization methods and examples of small antennas. In addition, the fundamental limitations of electrically small antennas are presented.

4.3.1 Monopole on a Finite Ground Plane

According to image theory (see Figure 4.20), a $\lambda/4$ monopole antenna on the infinite ground plane has a radiation pattern identical with a $\lambda/2$ dipole antenna in the upper hemisphere space. Figure 4.21 shows the E-plane radiation pattern of the monopole antenna on the finite ground plane. The finite ground plane cannot produce the perfect image current under the ground plane, and the radiation pattern is different from the dipole antenna. The pattern is tilted upward, and the radiation level to the horizontal plane is 6 dB less than the radiation level to the $\lambda/2$ dipole. This is explained by the imperfect image current or the diffraction from the ground plane edge.

Figure 4.22 shows that the input impedance of the monopole antenna (Z_i = R + jX) is periodically changed by the size of the ground plane. A standing wave excited on the ground plane varies the voltage at the feeding point because a node is fixed at the ground plane edge. If the current reflected from the edge is decreased for a large ground plane, the input impedance converges at $Z_i = Z_d/2$

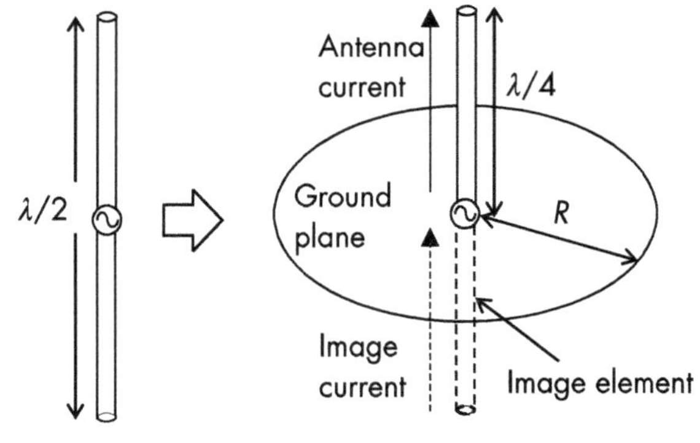

Figure 4.20 Monopole antenna on ground plane.

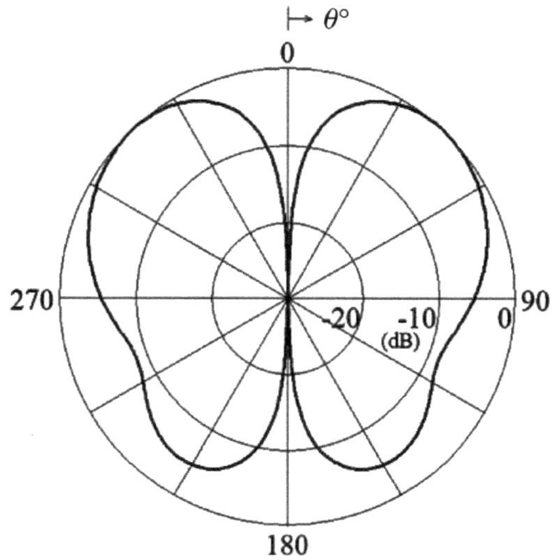

Figure 4.21 Radiation pattern of monopole antenna on ground plane; $R = 0.5\lambda$.

(see Section 3.1.4), where Z_d is the input impedance of $\lambda/2$ dipole antenna. As shown in Figures 4.21 and 4.22, the size of the ground plane changes the radiation pattern and the input impedance of the monopole antenna.

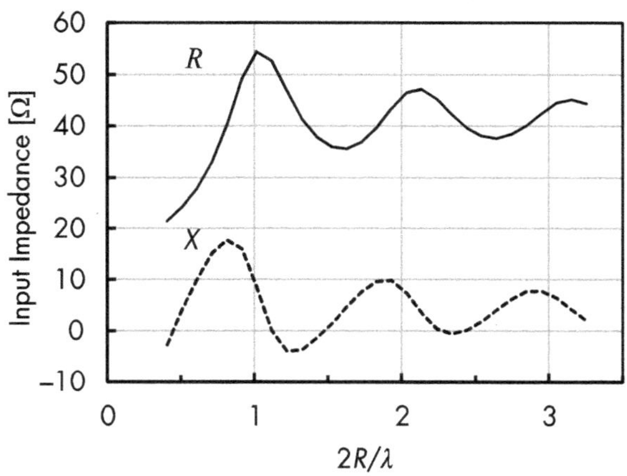

Figure 4.22 Input impedance of monopole antenna on ground plane.

A simple electrical model of a mobile terminal includes a rectangular conducting plate and a monopole antenna attached at the plate edge, as shown in Figure 4.23. The E-plane radiation pattern of the monopole antenna obtained by extending the plate length is shown in Figure 4.24. The pattern is a figure of eight for the short

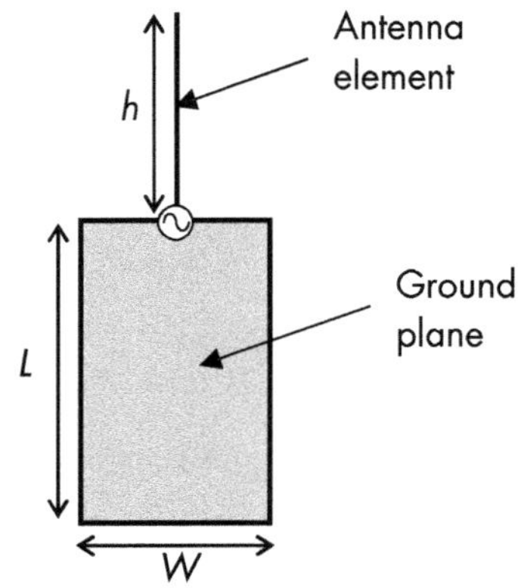

Figure 4.23 Monopole antenna on rectangular plate.

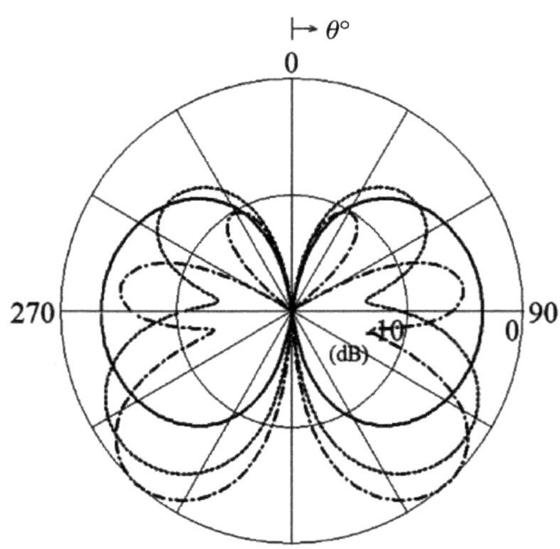

Figure 4.24 Radiation pattern of monopole antenna on rectangular plate; $h = \lambda/4$, solid line; $L = 0.1\lambda$, $W = 0.3\lambda$, dotted line; $L = 0.6\lambda$, $W = 0.3\lambda$, dash-dot line; $L = 1.0\lambda$, $W = 0.3\lambda$.

plate of $L = 0.1\lambda$; the sidelobes are increased by extending the plate length, for example, from $L = 0.1\lambda$ to 0.6λ. The radiation from the current on the plate is explained by strong standing waves on both sides of the plate, as shown in Figure 4.25.

Current distributions on the ground plane distort the radiation pattern in the E plane (see Figure 4.24). The ground plane size and the antenna location are design parameters of the mobile terminals.

4.3.2 Inverted-F Shaped Antenna

The height of the monopole antenna can be lowered by bending the top of the antenna element, as shown in Figure 4.26. This antenna, known as an inverted-L antenna, resonated at the length of $\lambda/4$. The image current in the vertical element flows in the same direction as that of the vertical element, which is known as the radiating element. When the current in the horizontal element flows in the direction opposite to the image current, as shown in Figure 4.27, the radiation

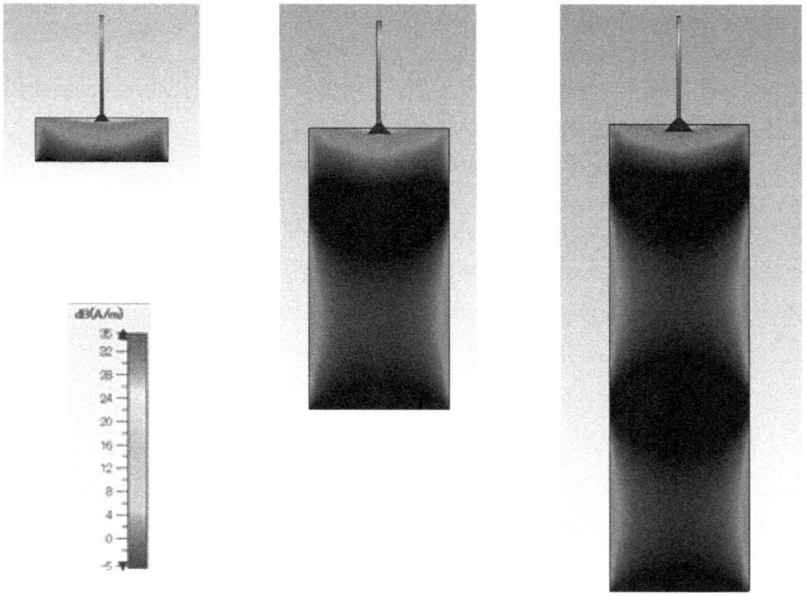

(a) (b) (c)

Figure 4.25 Current distributions of a monopole antenna on a rectangular plate; $h = \lambda/4$: (a) $L = 0.1\lambda$, $W = 0.3\lambda$, (b) $L = 0.6\lambda$, $W = 0.3\lambda$, and (c) $L = 1.0\lambda$, $W = 0.3\lambda$.

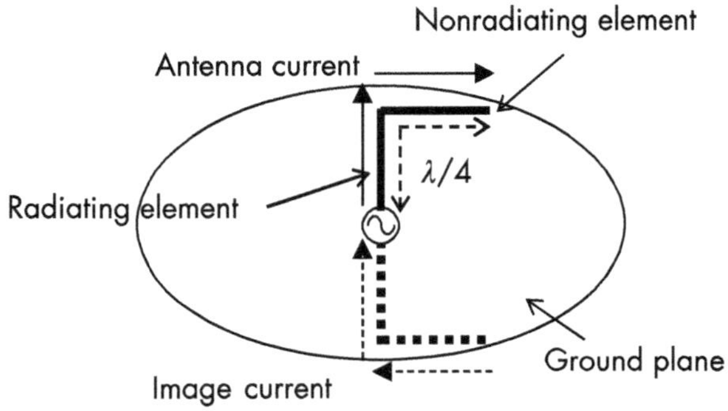

Figure 4.26 Inverted-L antenna.

from the horizontal element is canceled. This element does not mainly produce radiation and is known as a nonradiating element.

Radiation resistance is decreased by lowering the height of the antenna in the inverted-L antenna. To increase the radiation resistance, a short stub is inserted between the nonradiating element and the ground plane, as shown in Figure 4.27. This is an inverted-F antenna. The current on the short stub is in phase with the feeding current, which increases the radiation resistance by a factor of four due to the same principle as the folded dipole antenna. The resonance length of the antenna is $\lambda/4$ and is determined by the L-arm length $(h + w_1 + w_2)$, as shown in Figure 4.27. The main radiation from the two radiating elements provides a pattern that is similar to a monopole antenna; however, the radiation from the horizontal element cannot be ignored.

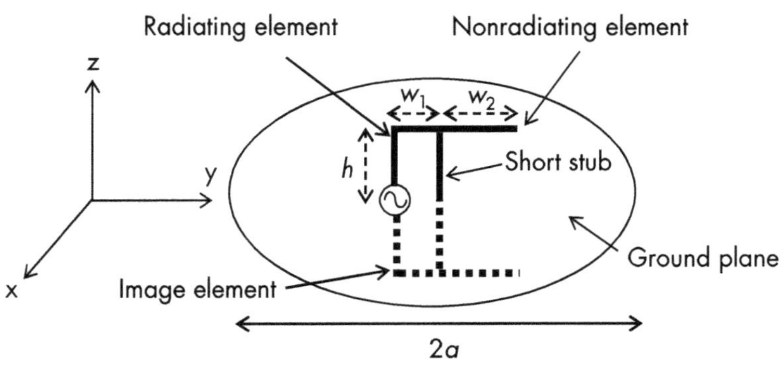

Figure 4.27 Inverted-F antenna.

Figure 4.28 shows the radiation pattern of the inverted-F antenna. The omnidirectional pattern is obtained in the *xy* plane, and the monopole-like pattern is obtained in the *zx* plane for the principal component E_θ; the omnidirectional and monopole-like patterns are given by the radiating and stub elements, respectively. The cross-polarization E_ϕ is increased by the horizontal antenna element in the *xy* and *zx* planes, and the null value in the zenith direction

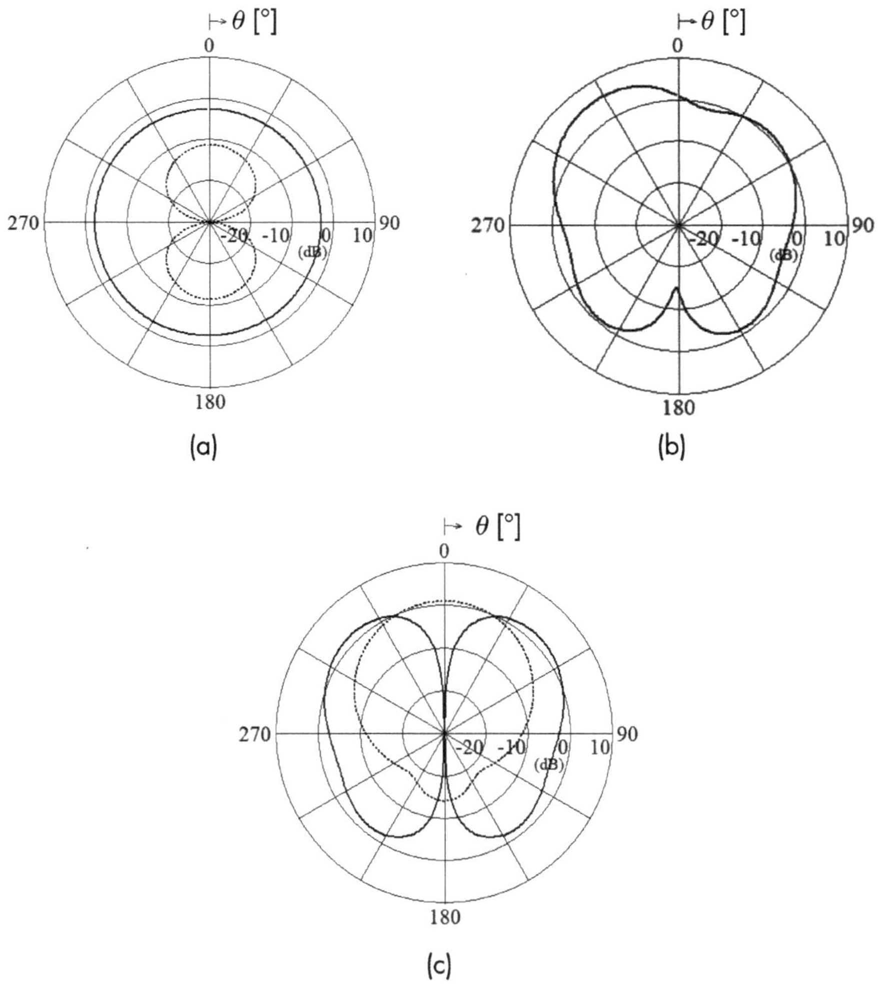

Figure 4.28 Radiation pattern of inverted-F antenna; solid line E_θ, dotted line E_ϕ, $2a = 122$, $h = 6$, $w_1 = 5$, $w_2 = 20.8$ (mm), $f_c = 2.45$ GHz. (a) *xy*-plane, (b) *yz*-plane, and (c) *zx*-plane.

disappeared in the *yz* plane. The increase in the E_ϕ radiation expands the frequency bandwidth of the inverted-F antenna.

4.3.3 Electrically Small Antenna

Small antennas are classified into (I) physically small antennas, (II) physically constrained antennas, and (III) electrically small antennas [6]. When we can put the antenna in our hand, it is called a physically small antenna. In the millimeter-wave band, if the antenna size is in millimeters, it is classified into category (I). The height of planar antennas, such as microstrip antennas, is much less than that of the operating wavelength and is very thin. This kind of antenna is classified into category (II). Similar to dipole antennas, the thickness of the wire antennas is also much less than the wavelength; however, wire antennas are not normally classified into physically constrained antennas. When the maximum size of an antenna is much less than the signal wavelength, it is known as an electrically small antenna (category (III)).

The inverted-L/F antenna described in Section 4.3.2 is not much smaller in size than the wavelength of operating frequency; however, the height of the antenna is much less than that of the monopole antenna. In general, the antennas with reduced sizes are also categorized as small antennas.

To reduce the antenna size, four miniaturization methods are used in practical applications. To produce resonance, the length of the current path on the antenna element should be $\lambda/2$, and this length is reduced to $\lambda/4$ using the infinite ground plane. The height of the monopole antenna can be reduced by winding the element spirally, as shown in Figure 4.29(a); this is known as a helix antenna. The resonant current of the microstrip antenna flows in parallel along the side of the antenna pattern, as shown in Figure 4.8(a), and the current path length is $\lambda_g/2$. If a slot cut in the antenna element, as shown in Figure 4.29(b), creates a detour in the current path, then the resonant side length is reduced. To reduce the antenna size, the current path on the antenna element is bent.

The second method of antenna miniaturization is the use of matching circuits. The input impedance of a short dipole antenna becomes capacitive when the element length is less than $\lambda/2$. We

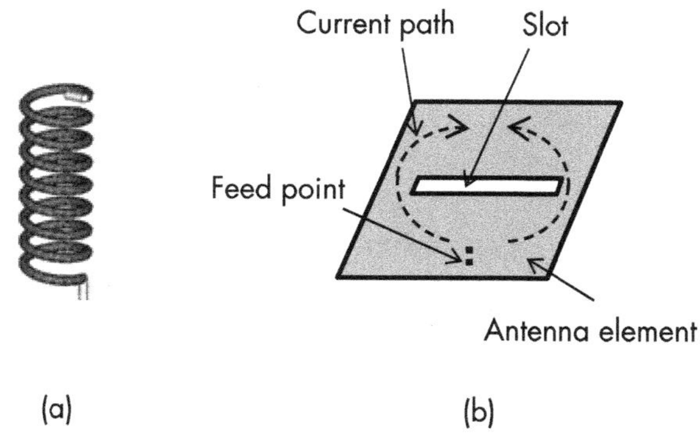

Figure 4.29 Bent current path: (a) helix element, and (b) slot cut in antenna element.

can compensate for this by inserting an inductor in series. A general matching circuit consists of the inductor L and capacitor C, as shown in Figure 4.30(a).

The matching circuit can shift the resonant frequency to the lower side and reduce the antenna size. The short stub of the inverted-L antenna was also the matching circuit. The part surrounded by a dotted line in Figure 4.30(b) was operated as a shorted transmission line. The inductive reactance is obtained for the height $h < \lambda/4$ by this stub, which provides the same effect as the matching circuit.

Materials having high dielectric constants reduce the antenna size because they use short guide wavelengths inside the dielectric. The loss tangent of the ceramic should be small enough for use in a microwave frequency band. Ceramic materials for the antenna are forsterite (Mg_2SiO_4, $\varepsilon_r = 7$), aluminum oxide (Al_2O_3, $\varepsilon_r = 9$), barium niobite magnesium acid ($Ba(Mg_{1/3}Nb_{2/3}O_3)$, $\varepsilon_r = 25$), and barium titanate

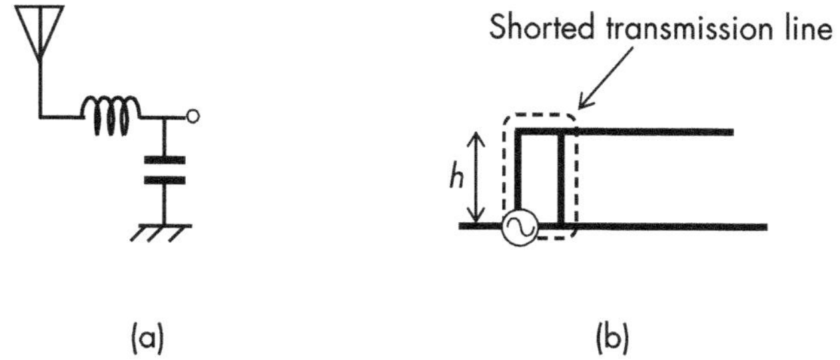

Figure 4.30 Matching circuit: (a) LC matching circuit, and (b) stub matching.

neodymium ($Ba_3Nd_{9.3}Ti_{18}O_{54}$), $\varepsilon_r = 85$). Antenna patterns are printed on the ceramic surface (see Figure 4.31) or inserted into multilayered ceramics such as low-temperature cofired ceramics. Ceramics are used for dielectric resonator antennas. A block of ceramic material mounted on the ground plane is fed by a probe. Dielectric resonator antennas do not have any conducting plates for the antenna element, which eliminates the conductor loss at millimeter-wave frequencies.

The ground plane mode excited by an antenna element can reduce the antenna size, but the frequency bandwidth is extended by the current flowing on the ground plane. This mechanism is almost the same for the monopole antenna at the plate edge, as shown in Figure 4.25. The antenna geometry is shown in Figure 4.32. The antenna is characterized by the size and shape of the ground plane, which is used for the built-in antennas of smartphones.

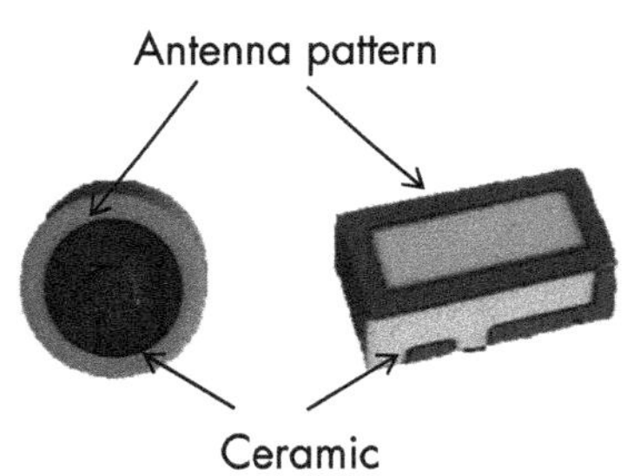

Figure 4.31 High dielectric constant material.

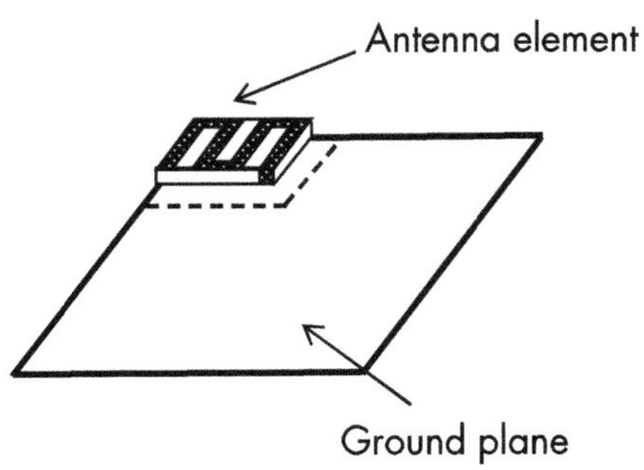

Figure 4.32 Ground plane mode.

4.3.4 Theory of Electrically Small Antennas

The reduction in the antenna size causes a decrease in the frequency bandwidth under a lossless condition. The bandwidth is inversely proportional to the unloaded Q factor of the antenna [7, 8].

Figure 4.33 shows a sphere of a radius (r_a) surrounding an antenna element; S and v denote the surface and internal volume of the sphere, respectively. Assuming that the sphere radius is much smaller than the wavelength, a short current element can be used for the antenna element inside the sphere. The element length is $2r_a$, and it is placed at the origin along the z axis. The element thickness $2a$ is not considered in this discussion. The radiating energy from the short current element is given by integrating the radiation field over the surface S as follows:

$$W_r = \frac{1}{2Z_o} \int_S \left| E_\theta^{\text{rad}} \right|^2 dS \tag{4.13}$$

where E_θ^{rad} is the radiation field component obtained by (1.16), and Z_o is the characteristic impedance in free space. The nonradiating energy becomes infinite in the vicinity of the origin [9] and is approximately calculated by integrating the nonradiating energy

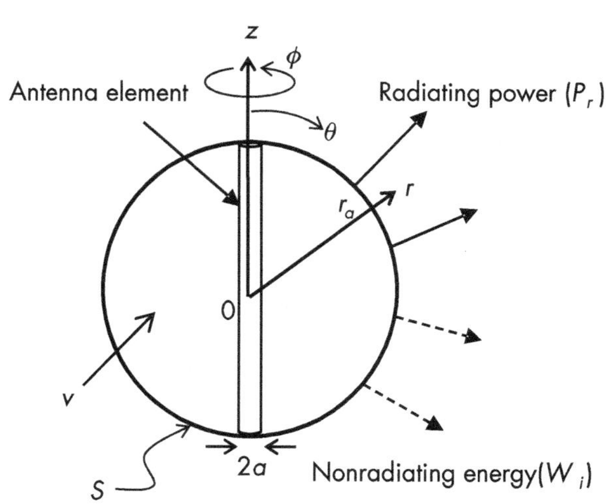

Figure 4.33 Sphere surrounding radiating element.

component of the short current element over the outside of the volume v as follows [10]:

$$W_e = \frac{1}{4}\int_0^{2\pi}\int_0^{\pi}\int_{r_a}^{\infty}\frac{\varepsilon_o}{4}\left(|E_\theta|^2 + |E_r|^2 - |E_\theta^{\text{rad}}|^2\right) r^2 \sin\theta \, dr \, d\theta \, d\phi \tag{4.14}$$

where the integral from r_a to infinity with respect to r eliminates the divergence of nonradiating energy. The internal sphere can be used for the antenna element to improve its performance. Finally, the Q factor of the short current element is defined as follows:

$$Q = \frac{2\omega W_e}{W_r} = \frac{1}{(k_o r_a)^3} + \frac{1}{(k_o r_a)} \tag{4.15}$$

Equation (4.15) provides the minimum Q factor for the antenna enclosed by the sphere of radius r_a, which means that the maximum bandwidth of the antenna is theoretically limited by 1/Q.

The nonradiating energy is approximately calculated in this formula by eliminating the divergence components of the integral in (4.14). Another minimum Q factor can be obtained by the input impedance of the short dipole antenna described in Section 1.7.2. The approximate formula for the input impedance of the dipole antenna is given by (1.42), which can be further approximated for the short dipole antenna as follows:

$$Z_i = R_r + jX = 20\pi^2\left(\frac{r_a}{\lambda}\right)^2 - j\frac{60}{k_o r_a}\left(\ln\frac{2r_a}{a} - 1\right) \tag{4.16}$$

where the short dipole length is $2r_a$, and the wire radius is a, as shown in Figure 4.33. Assuming that the ideal impedance matching condition cancels the reactance of the short dipole, only the radiation resistance R_r is considered at the matching frequency f_c. According to the circuit theory, the Q factor is given as follows:

$$Q' = \frac{f_c}{2R_r}\frac{\partial X}{\partial f_c} \tag{4.17}$$

Figure 4.34 shows the Q factors defined by (4.15) and (4.17) as functions of the sphere radius r_a. The Q factor of (4.17) is slightly higher than that of (4.15) because the nonpropagating energy in the vicinity of the antenna element is included. The Q factors of several small antennas are also plotted in Figure 4.34. The height of the disk-loaded dipole antenna in (a), (b), and (c) of Figure 4.34 is reduced by the top and bottom plates, and input impedance matching is performed by adding the short stub near the feeding element. The antenna (d) is the inverted-F antenna with the image element, which eliminates the infinite ground plane, as shown in Figure 4.27. The $\lambda/2$ dipole antenna is plotted as (e).

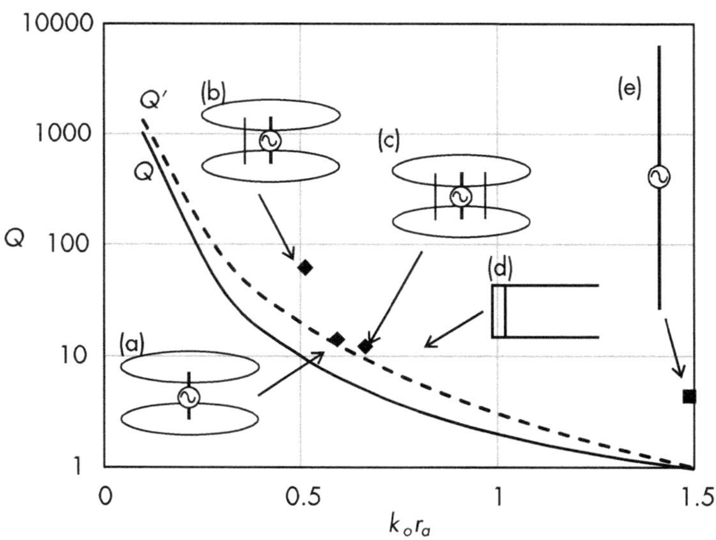

Figure 4.34 Minimum Q value and Q factors of small antennas.

References

[1] Dolph, C. L., "A Current Distribution for Broadside Arrays Which Optimizes the Relationship Between Beamwidth and Side-Lobe Level," *Proc. IRE and Electrons*, June 1946.

[2] Taylor, T. T., "Design of Line-Source Antennas for Narrow Beamwidth and Low Side-lobes," *IRE Trans. Antennas Propagat.*, Vol. AP-3, No. 1, January 1955, pp. 16–28.

[3] Stone, J. S., United States Patents No. 1,643,323 and No. 1,715,433.

[4] Chiba, T., "On a Pattern Synthesis Method for a Linear Array," *1958 IRE International Convention Record*, March 1966, pp. 172–179.

[5] Woodward, P. M., and J. D. Lawson, "The Theoretical Precision with Which an Arbitrary Radiation Pattern May Be Obtained from a Source of Finite Extent," *J.IEE*, Vol. 95, Pt. II, September 1966, pp. 363–370.

[6] Fujimoto, K., and H. Morishita, *Modern Small Antennas*, New York: Cambridge University Press, 2013, p. 8.

[7] Chu, L. J., "Physical Limitations of Omni-Directional Antennas," *J. Appl. Phys.*, Vol. 19, December 1948, pp. 1163–1175.

[8] Hansen, R. C., "Fundamental Limitations in Antennas," *Proc. IEEE*, Vol. 69, No. 2, February 1981, pp. 170–182.

[9] Collin, R. E., and S. Rothschild, "Evaluation of Antenna Q," *IEEE Trans. Antennas Propagat.*, Vol. AP-12, January 1964, pp. 23–27.

[10] McLean, J. S., "A Re-Examination of the Fundamental Limits on the Radiation Q of Electrical Small Antenna," *IEEE Trans. Antennas Propagat.*, Vol. AP-44, No, 5, May 1996, pp. 672–676.

5

DESIGN OF BASE STATION ANTENNAS

This chapter presents the design of base station (BS) antennas for cellular systems. An overview of mobile cellular systems is provided, and the categories of coverage areas used in cellular systems are summarized to determine the specifications of base stations. BS antennas consist of antenna elements, phase shifters, and feeding networks that satisfy the required frequency spectrum allocation, radiation pattern shape, and antenna gain. Beam tilt and null filling used for high-quality cell realization are also presented. Diversity schemes in BS antennas are explained through the combination of space and polarization as one of multiple antennas; herein, antenna pattern correlation is defined as an evaluation factor. Other applications of multiantenna systems are beam forming and multibeam antennas for enhancing the channel capacity of systems. Active integrated for 5G systems are described as examples of multiantennas.

5.1 CATEGORIES OF BASE STATIONS

A terrestrial mobile communication system consists of many coverage areas illuminated by electromagnetic waves by base station antennas. A group of coverage areas resembles a cellular tissue, which is the origin of the *cellular system* name. The shape of a coverage area is a rough circle affected by reflection and diffraction in a real propagation environment. This shape provided by an omnidirectional radiation pattern of a BS antenna is presented in this section. The circular area is divided into several sectors to increase the number of mobile terminals. The sector antenna is one of the most popular antennas in the cellular system, and it is detailed as the second topic. This chapter also presents other shapes of coverage areas and their service places.

5.1.1 Circular Coverage Antennas

An omnidirectional pattern is defined in the horizontal plane, and the vertical plane pattern varies with the design of the cell area. Service areas are categorized into large, macro-, micro-, and picocells. Their sizes are not clearly defined and are roughly explained as follows according to cell radii a: (a) $a \geq 10$ km for large cells, (b) 1 km $\leq a < 10$ km for macrocells, (c) 100m $\leq a < 1$ km for microcells, and (d) $a < 100$m for picocells. A high-gain antenna is provided by a narrow beam width in the vertical plane and is given by a colinear array antenna, as discussed in Section 4.2.1. The size of the service area is determined by the transmitting power, propagation loss estimation, and vertical plane pattern. The antenna gain is controlled by the number of colinear array elements N as:

$$G = 1.64N \tag{5.1}$$

where 1.64 is the gain of a single element given by $\lambda/2$ dipole antenna.

The cell size is restricted by tilting the beam in the vertical plane, as shown in Figure 5.1. The tilted beam angle θ_t toward the

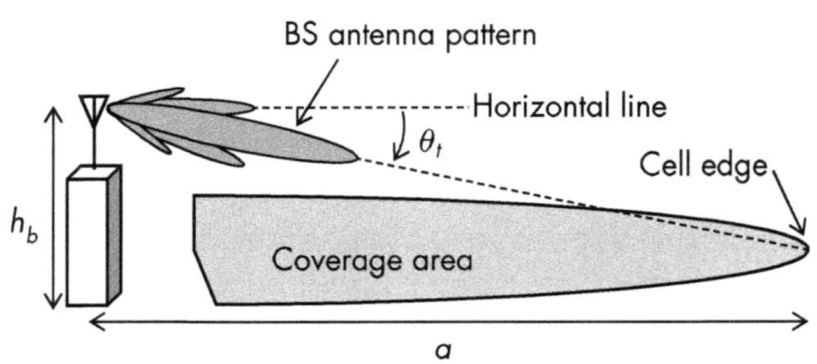

Figure 5.1 Beam tilt of BS antenna.

edge of the coverage area is obtained by the height of the BS antenna h_b and the radius of the cell a as:

$$\theta_t = \tan^{-1}\left(\frac{h_b}{a}\right) \tag{5.2}$$

The actual tilt angle used in the cellular system is larger than that set by (5.2) to reduce the interference in adjacent cells. The effect of beam tilt is verified by the propagation measurement in a semiflat land, as shown in Figure 5.2. The propagation loss is larger than 13 dB at $r = 2$ km and is effective in controlling the size of the coverage area [1]. The tilted beam is obtained by a linear array antenna, as explained in Section 1.8. Examples of the feeding network are described in Section 5.2.2.

5.1.2 Sector Antennas

A circular cell is divided into several parts to increase the number of user terminals inside the coverage area (Figure 5.3). A circular cell in Figure 5.3(a) can be separated into a small circle and a ring cell, as shown in Figure 5.3(b). The radiation pattern in the vertical plane is controlled to provide a tilted beam for the small circular cell and a shaped pattern for the ring one [2].

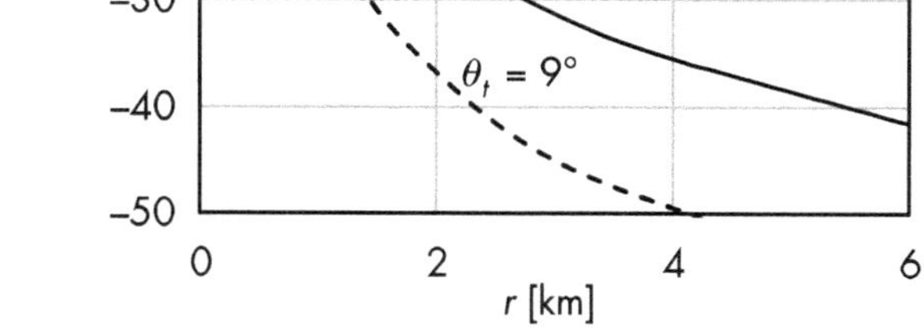

Figure 5.2 Propagation loss of tilted beam; $f = 920$ MHz, $G_t = 18$ dBi, $h_b = 34$m, $h_m = 1.5$m.

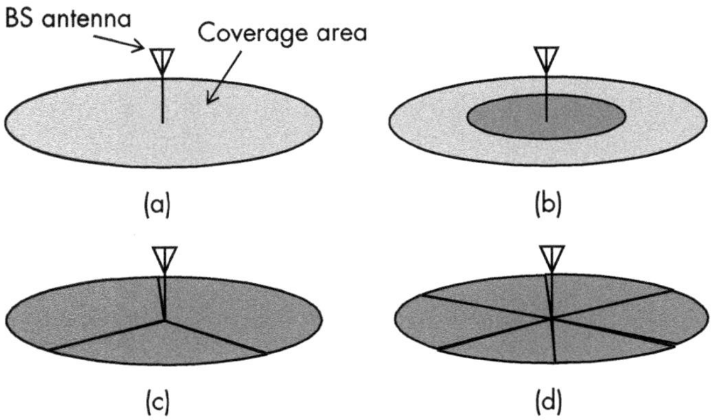

Figure 5.3 Circular and sector coverage area: (a) circular cell, (b) ring and circle, and (c) three-sector, (d) six-sector.

Another method is to divide the circular cell into several sectors, as shown in Figures 5.3(c) and (d). The sector coverage areas are given by shaping the horizontal radiation pattern of BS antennas. The sector angle is given by $\theta_h = 2\pi/N$, where N is the number of partitions of a circle. A sector pattern denoted by the beam width θ_h (Figure 5.4) is very difficult to synthesize by the small number of antenna elements. An ideal pattern presented by a gray sector is narrower than a real pattern of a half power beam width (HPBW) of θ_h, as depicted by the solid line in Figure 5.4. A correction factor α ($0.5 \le \alpha \le 0.75$) is multiplied with θ_h to simplify the design process of

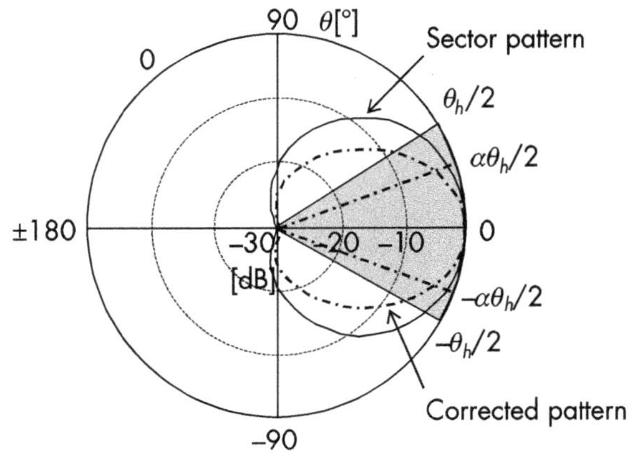

Figure 5.4 Sector pattern.

BS antennas. The value of α is selected to satisfy the system requirement. The corrected pattern by α is close to the original sector, as shown in Figure 5.4.

The antenna beam width is inversely proportional to the antenna size and is approximated by the following equation:

$$\theta_w = \sin^{-1}\left(\frac{\lambda}{D}\right) \tag{5.3}$$

where D is the antenna aperture width in the horizontal plane. The width D is determined by the number of antenna elements lined up in the horizontal plane and the reflector size to adjust the beam width of the sector antenna. The BS antenna width is estimated in the following examples. Assume a six-sector antenna ($N = 6$) and a correction factor $\alpha = 0.5$; the antenna width given by $\theta_w = \alpha\theta_h$ is 30°. Figure 5.5 provides $D = 2\lambda$, and the antenna width is 67 cm at 900 MHz and 33 cm at 2 GHz. The width of 67 cm is too wide for installation on top of buildings or towers. Six-sector antennas are used for frequencies over 2 GHz.

5.1.3 Other Pattern Shapes for Coverage Areas

Figure 5.6 shows a variety of coverage areas used in the current cellular system. A spot beam is used to illuminate train stations, public squares, and other places where people gather. The radiation

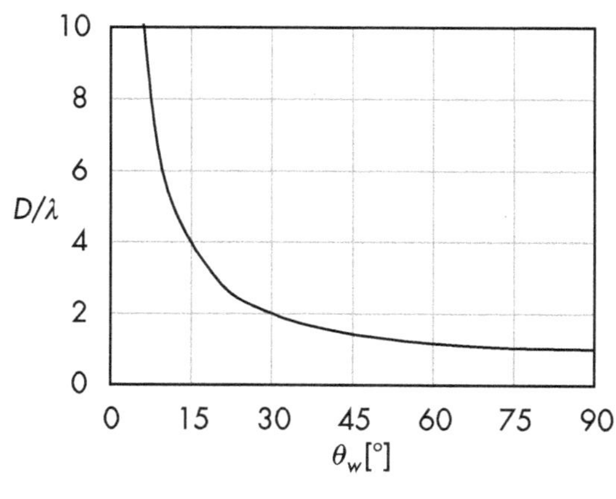

Figure 5.5 Antenna aperture width as a function of beam width.

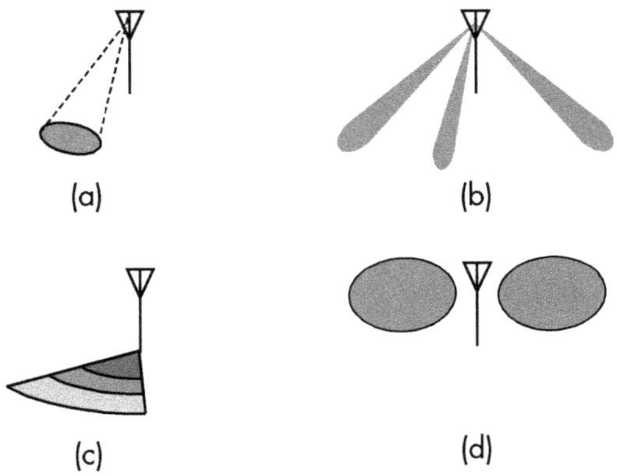

Figure 5.6 A variety of coverage areas in a cellular system: (a) spot beam, (b) multibeam, (c) multisector, and (d) bidirectional pattern.

direction of a high-gain flat BS antenna illuminates the target area. A part of the coverage area is overlaid by spot beams to increase the number of user terminals. Spot beams are given one by one by the BS antennas while a multibeam antenna provides several spot beams simultaneously. This beam configuration is introduced in 5G cellular systems by using the active integrated antenna described in Section 5.3. An advantage of the multibeam antenna is the change of beam direction according to the status of the user terminals in use.

The preceding section explains that the sector antenna size depends on the number of sectors and that the antenna width expands to increase the sectors. Another method to increase subareas is obtained by dividing a sector into several parts in the radial direction in a process known as multisector or vertical sectorization (Figure 5.6(c)) [3]. This coverage area is given by extending the vertical length to make a vertical array. The feeding network is the same as that of the multibeam antenna. Beam scanning or switching feeding is presented in Section 5.3.

A bidirectional radiation pattern is used to make the coverage areas inside tunnels and along streets. A longitudinal coverage area is provided by out-of-phase excited twin dipoles or a notch array [4]. A bidirectional antenna with low height excites a waveguide mode propagating along the streets sandwiched by buildings in big cities. The waveguide mode along streets can be obtained by an omnidirectional antenna installed at places lower than surrounding buildings.

Nevertheless, bidirectional antennas are effective in extending coverage areas.

The BS antennas described in this section are mainly for outdoor use. Indoor BS antennas are installed in places that cannot be reached by electromagnetic waves from outdoor BSs, such as subway stations, the inside of large buildings, and underground public spaces. Indoor BS antennas are installed on ceilings and walls. Ceiling antennas are monopole and sleeve antennas, and wall antennas are patch antennas that illuminate indoor coverage areas uniformly. Indoor antennas should be unobtrusive so as not to interfere with the appearance of rooms. Monopole antennas are replaced by disk-loaded monopole antennas to maintain a low profile similar to patch antennas.

5.2 DESIGN OF BS ANTENNAS

This section presents the array antenna configuration of outdoor BS antennas and their components. A subarray topology is a unique feature to obtain small beam tilt angles at a low cost. Dipole and patch antennas are basic BS antenna elements, and their modified geometries are described herein. The advanced design of BS antenna patterns is given by the array antenna, and a null filing technique and cosec^2 main lobe profile are presented as examples.

5.2.1 Antenna Configuration of BS Antennas

Figure 5.7 shows an overview of a 10-element BS antenna. The antenna is inserted into a radome made with a fiberglass reinforced plastic (FRP) cover to protect the antenna components from wind, rain, snow, and so on. A cover with a circular cross section offers better protection against strong wind pressure than a cover with a rectangular cross section. A microstrip antenna is used as a dual polarization element with two feeding points. As shown in Figure 5.7(b), double layered antenna substrates are used to achieve vertical and horizontal polarization diversity. The feeding circuits for both polarizations beneath the antenna substrates consist of thin coaxial cables, two phase shifters, and an adjustor, as shown in Figure 5.7(c). A dial at the antenna bottom is connected to the adjuster to change the phase of the antenna elements by using the phase shifter. The two connectors at the bottom are output ports for the two polarizations.

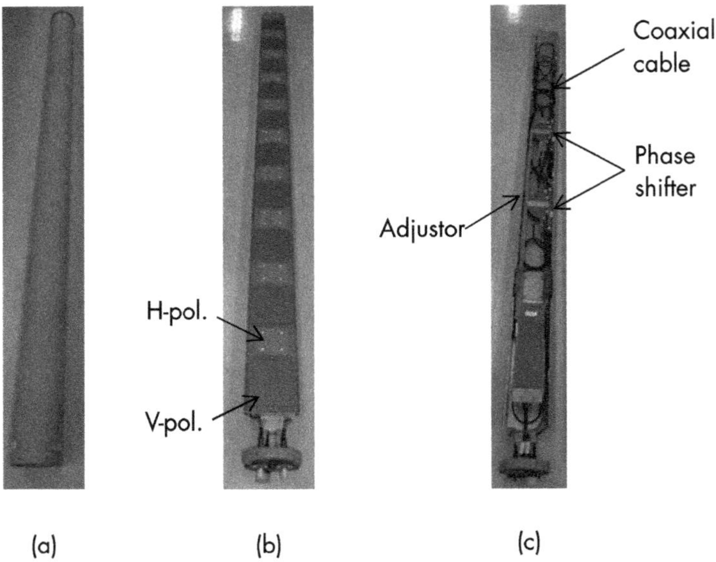

Figure 5.7 Overview of dual polarization antenna (courtesy of Nihon Dengyo Kosaku): (a) radome, (b) antenna element, and (c) feeding network.

Figure 5.8 shows the diagram of the feeding network of a 10-element antenna. Four subarrays are connected to a four-port phase shifter. Two subarrays, B and C, in the middle, are two-element subarrays, and the top and bottom are three-element subarrays that suppress the grating lobes for beam titling.

Many thin coaxial cables are used to obtain the feeding network of BS antennas. A simple and low-loss feeding network can be provided by a strip line, which has a conducting strip embedded in a dielectric substrate with top and bottom ground planes. The strip conductors are partly supported by plastic posts to remove the dielectric substrate and to obtain the air-filled strip lines, which are used for the feeding networks of BS antennas (Figure 5.9). A loop slot is excited by two probes for ±45° polarization, and a parasitic patch is added to extend the frequency bandwidth. Two independent strip lines consist of a tournament-feeding network and an arch-shaped phase shifter with four output ports. Tilt angle adjustors are placed at the top and bottom of the feeding network.

The adjustor for manual phase change is replaced with a built-in motor driven adjustor to realize remote phase control. The beam tilt angle is fixed at the initial installation of the BS antenna. However,

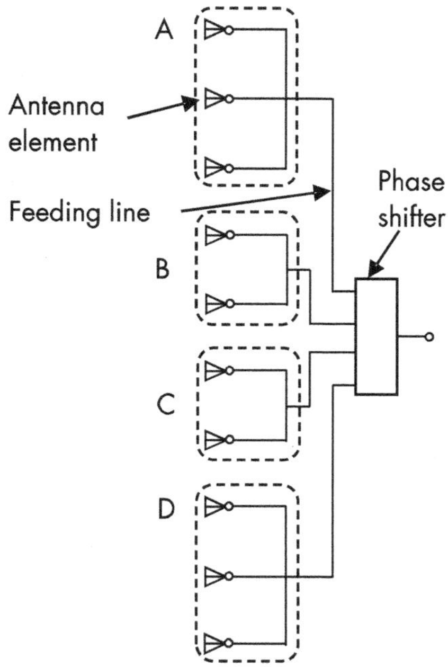

Figure 5.8 Feed network of subarray antenna.

the construction of a new BS near existing BSs requires coverage area adjustment and effective remote control of tilt angles. BS antennas are locked to the post, as shown in Figure 5.10. A mechanical beam tilt is obtained by changing the fixed angle θ_t; it provides a simple and low-cost beam-tilted antenna. This beam tilt does not change the antenna gain, whereas a mechanical tilt not only increases the wind pressure resistance but also contributes to the bad appearance of the structure.

A polarization diversity BS antenna employing both vertical and horizontal polarization uses a printed twin dipole with parasitic elements as shown in Figure 5.11(a). The parasitic element expands the frequency bandwidth while keeping the beam width at the same value in the E and H plane [5]. Figure 5.11(b) presents three-sector polarization diversity antenna built in one radome.

5.2.2 Components of BS Antennas

The basic BS antenna elements, namely, the dipole and microstrip antenna, were described in Section 4.1. In the current section, a

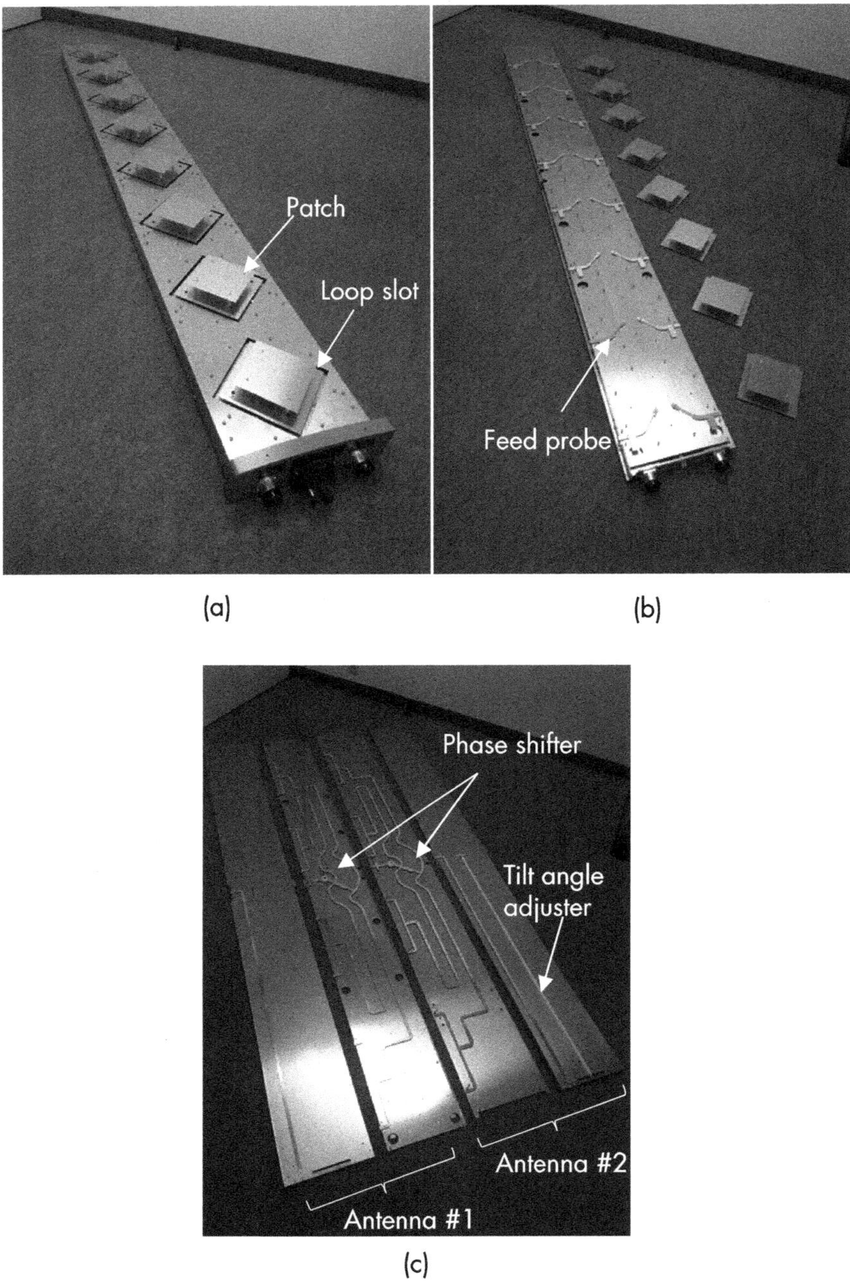

Figure 5.9 Structure of eight-element BS antenna: (a) antenna elements, (b) antenna feeding probes, and (c) feeding network. (Courtesy of Nazca Industrial Technology.)

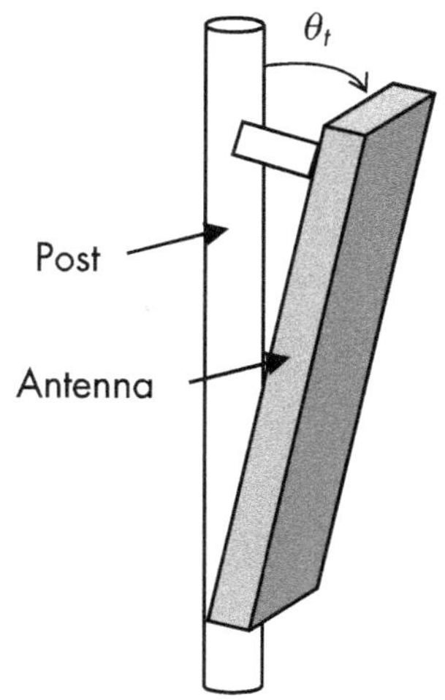

Figure 5.10 Mechanical tilt.

modified crossed dipole antenna for ±45°-polarization diversity for the sector antenna is presented. Figure 5.12 depicts the crossed dipole with feeding lines. Simple feeding is obtained by a Lecher line, and a pair of ports 1–1' or 2–2' excites the orthogonal dipole antennas. A compact and wide band crossed dipole antenna element is given by a square loop with a corner gap, as shown in Figure 5.13(a) [6]. Four square loop elements are closely placed and are coupled with one another. When the feeding port 1–1' is excited, the current on the elements flow in the direction of the solid arrow lines in Figure 5.13(b). Such current radiates a +45° polarization. The induced currents on the unexcited elements shown as dotted arrow lines make loop currents, which do not radiate to the front direction (z-axis) but radiate slightly in the xy plane. These coupled loop currents extend the effective antenna area and increase the frequency bandwidth. The other geometries of the crossed dipole antenna are given by a square loop without a gap [7], a bow tie-shaped element [8], and others.

The feeding circuits of a printed dipole antenna consist of a built-in balun and reflector, as shown in Figures 4.4 and 4.5. This printed antenna becomes complicated geometry when combining two elements as the crossed dipole. A simple feeding circuit is given by connecting the antenna elements with the inner and outer

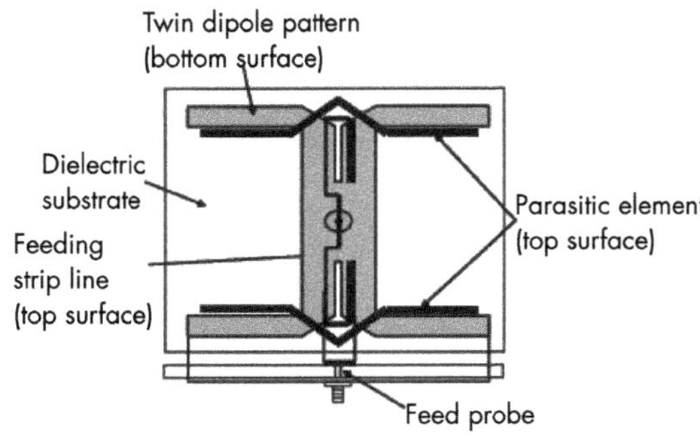

(a)

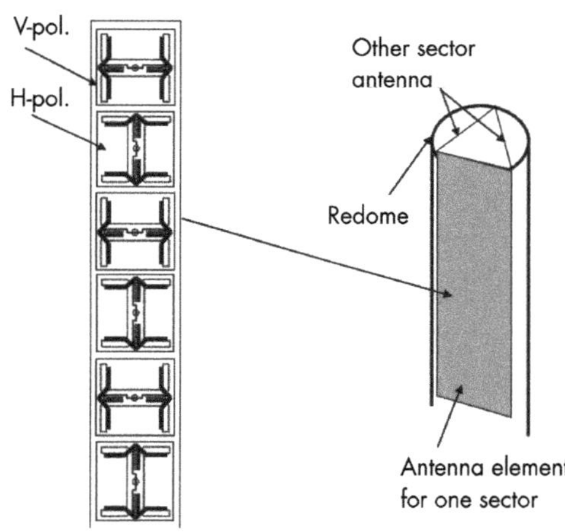

(b)

Figure 5.11 Polarization diversity antenna: (a) printed twin dipole element with parasitic element, and (b) three-sector polarization diversity antenna.

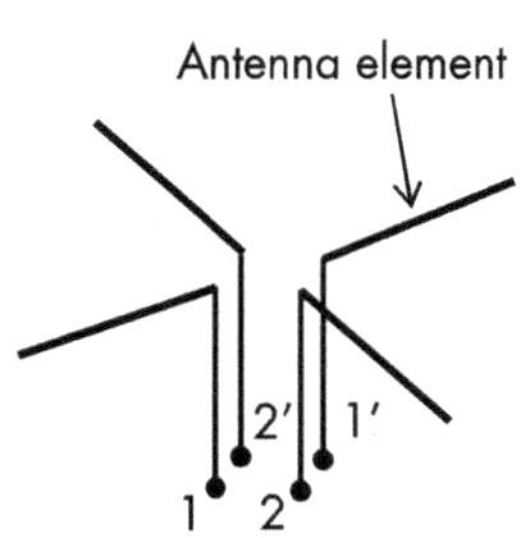

Figure 5.12 Crossed dipole and feeding.

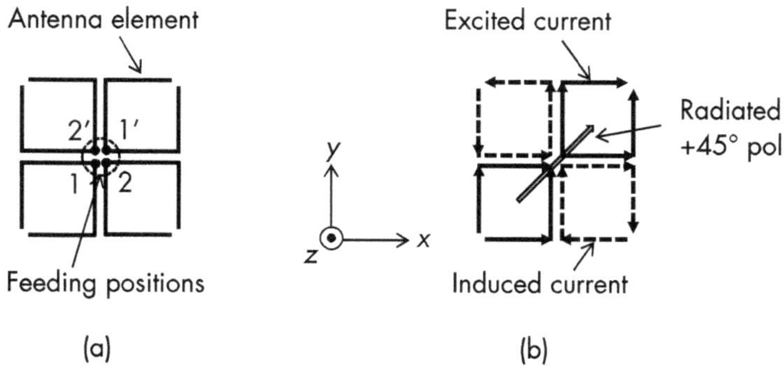

Figure 5.13 Modified crossed dipole and feeding: (a) antenna element, and (b) current flow on elements.

conductors of the coaxial cable (Figure 5.14). The feeding coaxial cable on the reflector measures $h = \lambda/4$ in height and is operated as a balun. A bent element is used to not touch the two orthogonal feeding lines at the antenna center. A small difference in the coaxial waveguide height denoted as g is introduced for the bent element.

The reflector size is determined by adjusting the beam width of the antenna element. Figure 5.15 shows the geometry of the dipole backed by the reflector. Let the ground plane seize being infinite ($W \to \infty$, $L \to \infty$); then, an image current at $z = -d$ flows in the reverse direction, and the radiation pattern in the zx plane is given as:

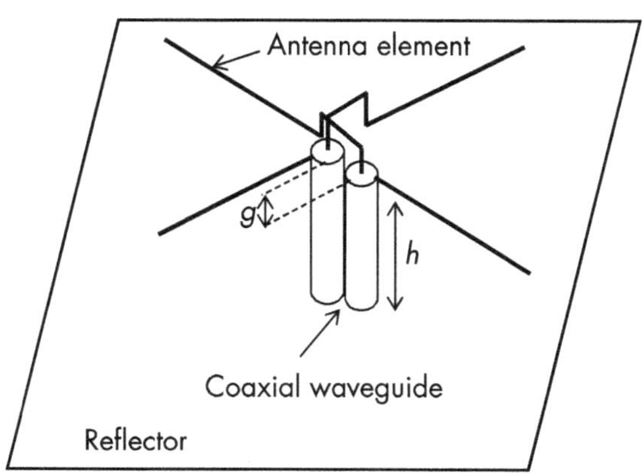

Figure 5.14 Dipole antenna and coaxial cable feeding.

$$D(\theta) = C\sin\left(\frac{k_o d}{2}\cos\theta\right) \tag{5.4}$$

where C is a constant and k_o is the wave number in the free space. The HPBW of 120° is obtained by $d = \lambda/4$ and is wide for the sector antenna element. The H plane pattern (zx plane) is adjusted by the reflector width W and flange height H. The E plane pattern is not changed because the null direction is along the reflector length L. Figure 5.16 presents the HPBW (B_W) by changing the reflector parameters. The beam width is determined to be 60° to 65° by using the reflector and is adjusted by a few degrees by the flange height.

An omnidirectional antenna used in outdoor cellular base stations is given by a colinear array antenna, as described in Section 5.1.1. The antenna consisting of coaxial cables arrayed in the vertical orientation is replaced with a loop slot antenna on the dielectric substrate, as shown in Figure 5.17(a) [9]. The substrate width W is 0.2λ, and the slot size of $a \times b = 0.1\lambda \times 0.3\lambda$ gives the 1λ circumferential length of the slot. A rectangular strip surrounded by the loop is fed by a microstrip line on the backside of the substrate. The current on the strip flows along the z direction and radiates the omnidirectional vertical polarization in the xy plane. The frequency bandwidth is extended by adding parasitic elements at both sides of the substrate, as shown in Figure 5.17(b) [10]. The antenna gain is increased by stacking the element along the z direction, and the feeding network is given by microstrip lines. The narrow substrate geometry facilitates the insertion of a cylindrical radome.

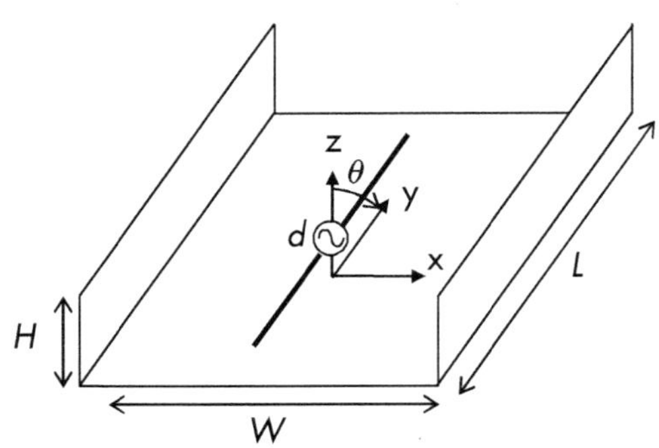

Figure 5.15 Dipole antenna backed by reflector.

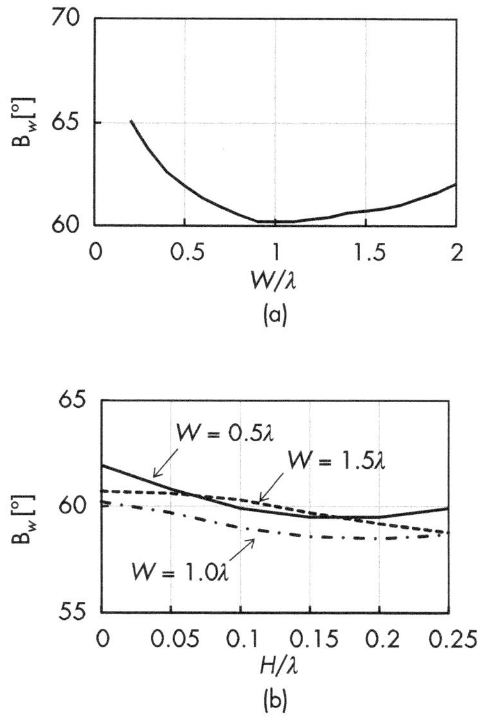

Figure 5.16 Beam width in *H* plane by reflector, $d = \lambda/4$, $L = \lambda$: (a) beam width versus W/λ; $H = 0$, (b) beam width versus H/λ.

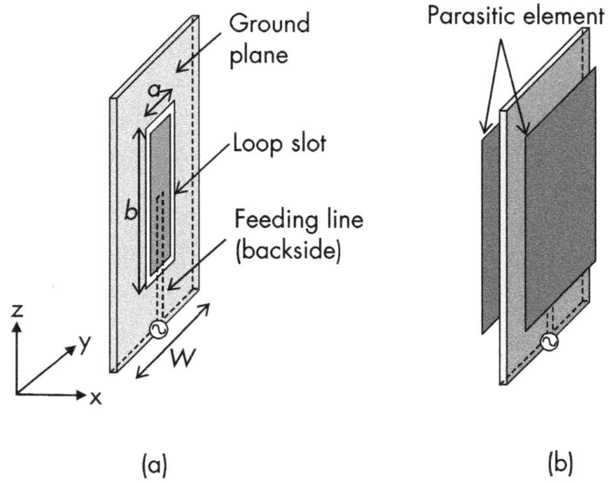

Figure 5.17 Vertical polarized loop slot antenna: (a) loop slot and feeding circuit, and (b) loop slot with parasitic elements.

The crossed dipole antenna (Figure 5.12) radiates an omnidirectional horizontal polarization by giving the phase difference of 90° between ports 1–1' and 2–2'. This crossed dipole is known as a turnstile antenna. The pattern ripple in the *xy* plane of ±0.5 dB is greater than the loop slot, and the diameter is double the loop slot. Figure 5.18 depicts a rounded folded dipole antenna, which is known as a halo antenna [11]. The resonant frequency is determined by the diameter *D*, and impedance matching is realized by adjusting the gap length *g* and element width *w*. A small diameter $D < 0.2\lambda$ and an omnidirectional pattern are provided by this antenna. A combination of the loop slot and halo antenna can be used as a vertical and horizontal diversity antenna, as shown in Figure 5.19. A parasitic element is added to extend the frequency bandwidth of the halo antenna [12].

The omnidirectional pattern that radiates a 45° slant polarization is given by four crossed dipoles [13] and a monopole antenna with slant parasitic elements on the ground plane [14]. A simple geometry for slant polarization can be obtained by a slotted conducting cylinder, as shown in Figure 5.20. The cylinder has four thin slots measuring 0.8λ in length and is inclined at a 45° angle. A $\lambda/2$ dipole antenna is placed at the cylinder center. The dipole antenna that excites the vertical polarization and is loaded with parasitic cylindrical slots inclined at 45° produces the omnidirectional radiation pattern. The feeding dipole is replaced with the loop slot antenna, and a ±45° polarization antenna is obtained by aligning two polarizations vertically (Figure 5.21). The element spacing is $d = 0.8\lambda$, and the array spacing is $s = 1\lambda$ for the cylinder radius $R = 0.15\lambda$. A slim ±45° polarization diversity antenna is provided by the slotted cylinder with a loop slot-feeding network.

A BS antenna consists of antenna elements, feeding cables, and phase shifters. Coaxial cables are used for the feeding network, and

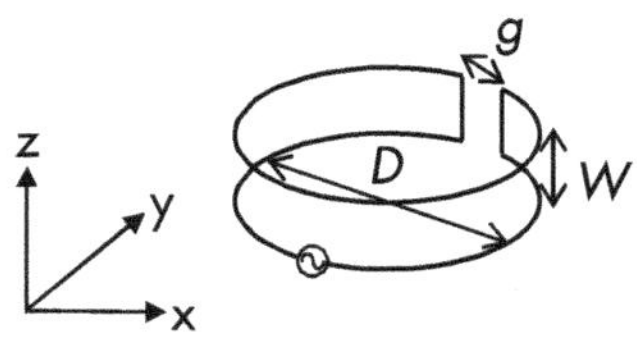

Figure 5.18 Halo antenna.

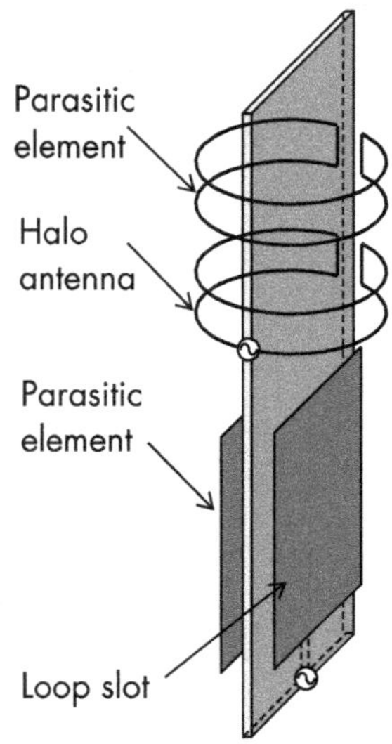

Figure 5.19 Omnidirectional V/H-polarization diversity antenna.

a phase shifter is inserted into them to obtain the beam tilt function. The adjustment of the tilt angle is mainly performed when the BS antenna is initially installed. The adjustment cycle is not frequent, and a semifixed phase shifter is used for the BS antenna. Several tens of watts of RF power are fed to the phase shifter of the BS antenna,

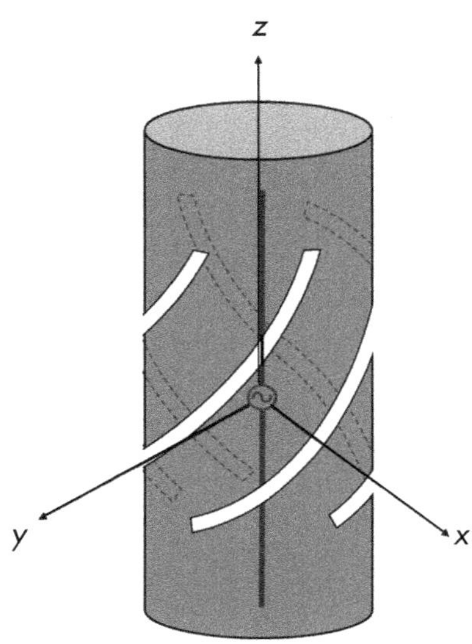

Figure 5.20 Omnidirectional slant-polarization diversity antenna.

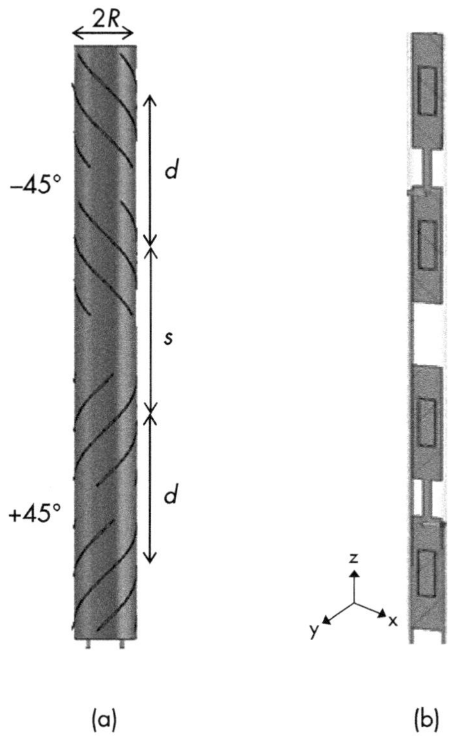

Figure 5.21 Omnidirectional ±45°-polarization diversity antenna: (a) antenna geometry, and (b) loop slot feeding circuit.

then interpassive modulation is avoided such that it does not interfere with the receiving frequency band. Because of the power handling capability and nonlinearity performance of semiconductors, PIN diode-based phase shifters are not used in cellular systems.

Figure 5.22 shows a microstrip line T-junction with a moveable feeding line connected to the input port. The middle position of the feeding line denoted as input port 0 provides equal amplitude and in-phase output at 1 and 2. The phase difference between two output ports is given by $(l_1 - l_2)/\lambda_g$ for the feeding line position at 0', where λ_g is the guide wavelength of the microstrip line. This description reflects the basic principle of the phase shifter. A multioutput port phase shifter is obtained by the same technique (Figure 5.23). Output ports are connected to an arched microstrip line, and the moveable

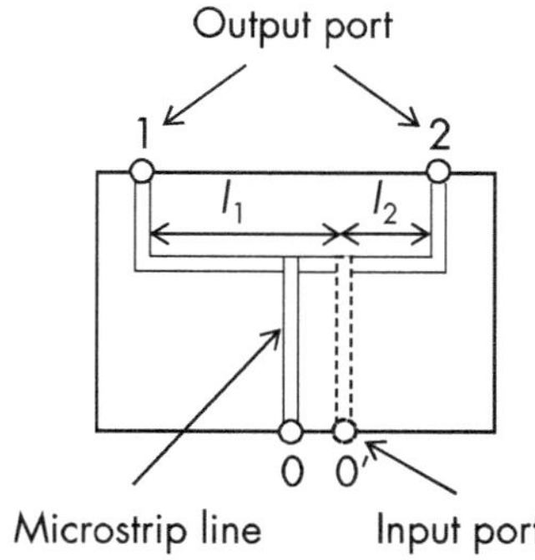

Figure 5.22 Semifixed phase shifter.

strip is rotated by the angle of ϕ. The phase difference at ports 2 and 3 is greater than that at ports 1 and 4 by a factor of $(r_b - r_a)\phi$ because of the different radii of the arches. This geometry is appropriate to use for a BS antenna with a subarray structure (Figure 5.8).

Another phase shifter consists of a meander microstrip line and a moveable dielectric plate (Figure 5.24) [15]. The guided wavelength of the dielectric covered line is shortened by the high dielectric constant of the plate. The electrical length of the microstrip line is changed by the position of the dielectric plate, and the phase delay between ports 1 and 2 is changed. This phase shifter has high power handling capability and a wide frequency band because of the absence of contact electrodes.

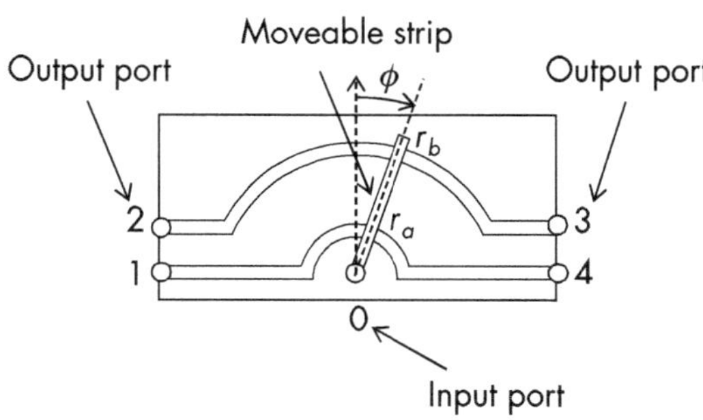

Figure 5.23 Four-port arch shaped phase shifter.

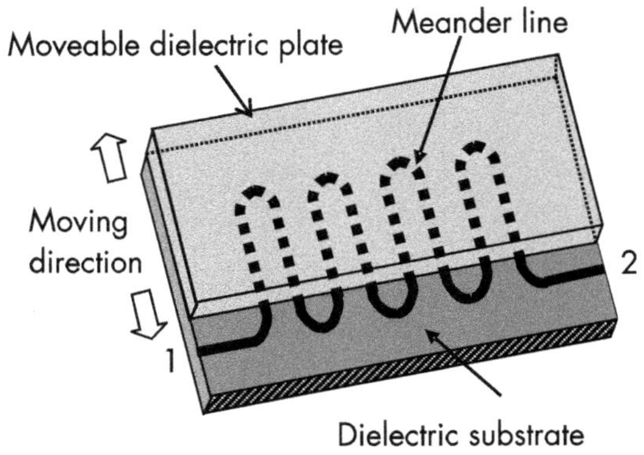

Figure 5.24 Phase shifter with moveable dielectric plate.

5.2.3 Design of Array Pattern

An advanced design of a vertical plane pattern for a base station antenna is shown in Figure 5.25. Given $\theta = 0°$ as the horizon, the lower region $0 < \theta < +90°$ is a pattern corresponding to $\text{cosec}^2\theta$, and the upper region $-90° < \theta < 0°$ is designed as a Taylor distribution, as described in Section 4.2.2. In particular, the level in the direction of another cell using the same frequency is often specified as very low ($-30° < \theta < 0°$). This design is based on antenna pattern synthesis (Section 4.2.3), and the excitation condition of each antenna element is obtained by adjusting the amplitude and phase distribution [16]. The $\text{cosec}^2\theta$ pattern can also be obtained by the phase control of one or two elements and a simple amplitude weight in the following procedure [17].

Consider an equispaced linear array of N elements along the z-axis. The array factor function $f(\theta)$ may be transformed into a product of polynomials as:

$$f(\theta) = \sum_{n=0}^{N} I_n e^{jn\left(k_o d \sin\theta + \phi_o\right)} = I_N \prod_{n=1}^{N}\left(w - w_n\right) \Rightarrow I_N \prod_{n=1}^{N}\left(w - (1+\varepsilon)w_n\right) \tag{5.5}$$

where k_0 is the free-space wavenumber, d is the spacing between elements, I_n is the magnitude of the nth current element, ϕ_0 is the phase

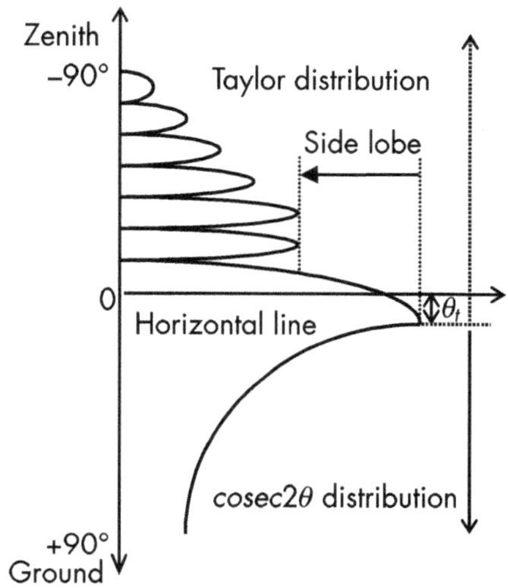

Figure 5.25 Vertical plane radiation pattern of BS antenna.

difference between the elements for the main beam tilting, w is $e^{j\psi} = e^{jk_0 d\sin\theta + \phi_o}$, and w_k denotes the roots of the array factor polynomial. The variable w moves in the unit circle according to the change of θ, and the nulls of $f(\theta)$ are filled by adding the constant ε, which relocates some of the roots inside or outside the unit circle [18].

Figure 5.26 presents an example of a null-filled pattern, in which the current amplitude is given by $I_n = (N - n)/N$, $(n = 0, 2, \ldots, N - 1)$.

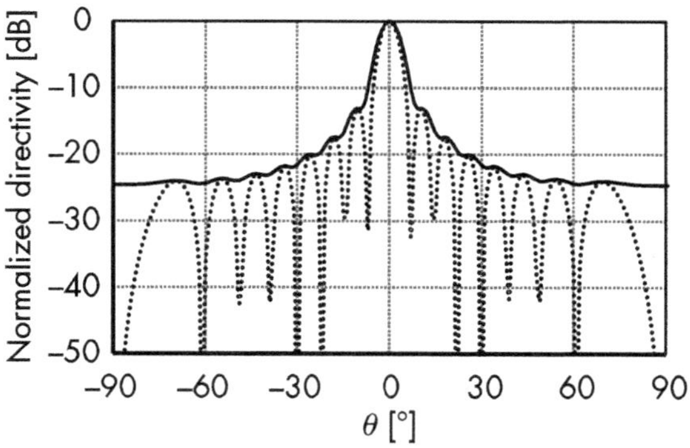

Figure 5.26 Null-filled radiation pattern; $N = 16$, $d = 0.5\lambda$, $\phi_o = 0$, $\varepsilon = 0.1$, solid line is amplitude slope of $1/N$ and dotted line is uniform amplitude.

All the nulls of the uniform amplitude weight disappear, and the null-filled pattern is obtained. This null-filled pattern is transformed into a cosec$^2\theta$ pattern by adding a phase difference of $\Delta\phi_1$ to the first element ($n = 1$), in addition to the beam tilt phase difference of ϕ_o [17]. Assuming the slope current amplitude distribution of $1/N$, the array factor $f(\theta)$ is divided into the first term and the other terms as:

$$f(\theta) = e^{j\Delta\phi} + \sum_{n=1}^{N-1} \frac{N-n}{N} e^{jn\psi} \tag{5.6}$$

The pattern of $\Delta\phi_1 = 0$ is symmetrical with respect to $\theta = 0$, and an asymmetrical term of $j\sin\Delta\phi$ appears because of the additional phase difference of $\Delta\phi_1$. This asymmetrical term lifts up the pattern in the region $\theta > 0$ and lowers the pattern in $\theta < 0$, which provides a quasi cosec$^2\theta$ pattern (Figure 5.27). The phase difference of $\Delta\phi_2$ is added to the second element to suppress the side lobes for $\theta < 0$. The conventional pattern synthesis methods shown in Figure 5.27 give a sharp cutoff around $\theta = 50°$ [19, 20]; in contrast, the simple procedure described in this section provides an almost similar radiation pattern and does not require iterative calculations to control the amplitude and phase distributions.

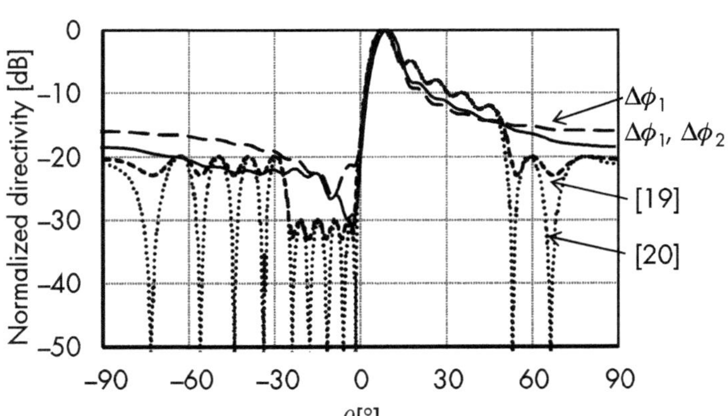

Figure 5.27 Cosec$^2\theta$ radiation pattern; $N = 16$, $d = 0.5\lambda$, $\phi_o = 8°$, $\varepsilon = 0.1$, $\Delta\phi_1 = 90°$, $\Delta\phi_2 = 25°$.

5.3 DIVERSITY AND MIMO ANTENNA

This section presents the BS antenna configuration for diversity and MIMO systems. The correlation coefficient used to evaluate the antenna performance is described by a complex radiation pattern (i.e., the amplitude and phase distribution of the far field pattern). A spot cell antenna is given by a high-gain antenna, and a multispot or vertical sectorization is provided by a multibeam antenna. The feeding circuit of a multibeam antenna consists of a Butler matrix and an active integrated antenna. This section describes these feeding circuits and their antenna configurations.

5.3.1 Antenna Pattern Correlation

Diversity and MIMO performance is characterized by the correlation coefficient, as described in Section 3.3.2. Figure 5.28 shows two-antenna space diversity with an antenna spacing of d. An incoming wave E_i direction is denoted as the unit vector $\hat{\boldsymbol{r}}$. The received electric fields at antennas #1 and #2 are expressed as follows:

$$e_1 = \sum_{i=1}^{N} E_i G_{1i} e^{j\phi_i}, \quad e_2 = \sum_{i=1}^{N} E_i G_{2i} e^{j(\phi_i - k_o \hat{\boldsymbol{r}} \cdot \boldsymbol{d})} \tag{5.7}$$

N incident waves arrive at the antenna with an amplitude E_i and a phase ϕ_i. G_{1i} and G_{2i} are the antenna gains in the direction of the i-th incident wave. Assuming that incoming waves change randomly with time and have the same amplitude, the mean values of e_1 and

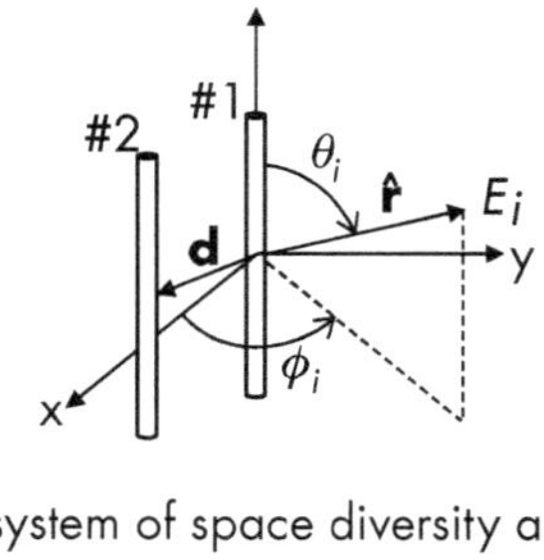

Figure 5.28 Coordinate system of space diversity antenna.

e_2 become zero, that is, $\langle e_1 \rangle = \langle e_2 \rangle = 0$, and the correlation coefficient defined by (3.33) is rewritten as:

$$\rho = \frac{\frac{1}{2}\left\langle \left(e_1 - \langle e_1 \rangle\right)^* \left(e_2 - \langle e_2 \rangle\right) \right\rangle}{\frac{1}{2}\sqrt{\left\langle \left(e_1 - \langle e_1 \rangle\right)^2 \left(e_2 - \langle e_2 \rangle\right)^2 \right\rangle}} = \frac{\sum_{i=1}^{N} E_i^* G_{1i}^* E_i G_{2i} \exp\left(-k_o \hat{\boldsymbol{r}} \cdot \boldsymbol{d}\right)}{\sqrt{\sum_{i=1}^{N} G_{1i}^2}\sqrt{\sum_{i=1}^{N} G_{1i}^2}} \quad (5.8)$$

The ith incoming wave from the (θ_i, ϕ_i) direction is expressed by the delta function $\delta(x)$ as:

$$E_i = \frac{\delta\left(\theta - \theta_i\right)\delta\left(\phi - \phi_i\right)}{\sin\theta} \quad (5.9)$$

where $\sin\theta$ is a metric coefficient in the spherical coordinates. Equation (5.8) may be converted into an integral form using the distribution function of the incoming waves in the θ and ϕ directions. The cross-polarization ratio defined by (3.32) is also included.

Uplink diversity reception is mandatory for base station antennas to mitigate the effect of multipath fading and to compensate for the power imbalance between the uplink and the downlink because the uplink signal is weaker than the downlink. For a space diversity system, the same antenna is installed with a spacing of more than 10λ. The correlation coefficient for this space diversity is calculated by assuming that the uplink signals have a uniform angular spread of ψ_m. The correlation coefficient of ρ_e is evaluated from the definition of (5.8) as:

$$\rho_e = \left(\frac{\sin X}{X}\right)^2, \quad X = 2\pi\left(\frac{\Psi_m d}{\lambda}\right)\sin\alpha \quad (5.10)$$

The calculated correlation is shown in Figure 5.29. For a design value of $\rho_e < 0.5$ for diversity reception, the antenna spacing of d should be more than 10λ because $\psi_m > 1.5°$ is obtained from propagation measurements in cities [21].

The same polarization components are used in the space diversity system, and the low correlation between antennas is given by the

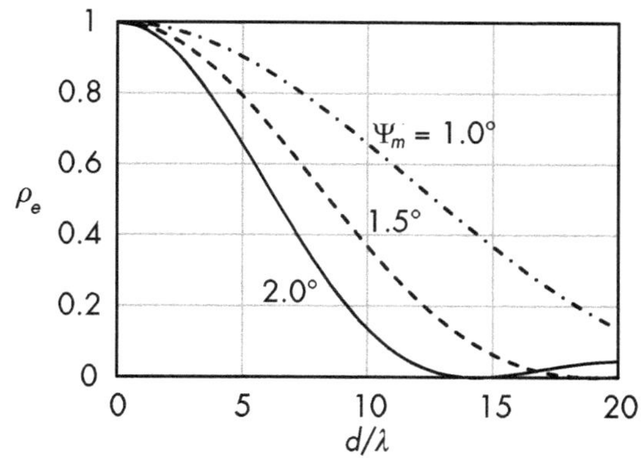

Figure 5.29 Correlation coefficient of space diversity antenna.

large distance, as explained previously. The polarization diversity by the orthogonal polarization components provides a very small correlation coefficient of less than 0.1 because of the small mutual coupling between two polarization components.

5.3.2 Butler Matrix and Multibeam

The cellular system is overlaid to enhance the channel capacity (Figure 5.30), and a multibeam cell is introduced, in addition to the micro- and spot cells. A multibeam is provided by an array antenna illuminating a part of a macrocell by dividing the part into several spots. Multibeam antennas are obtained by an analog feed circuit known as a Butler matrix or an active integrated antenna described in the next section.

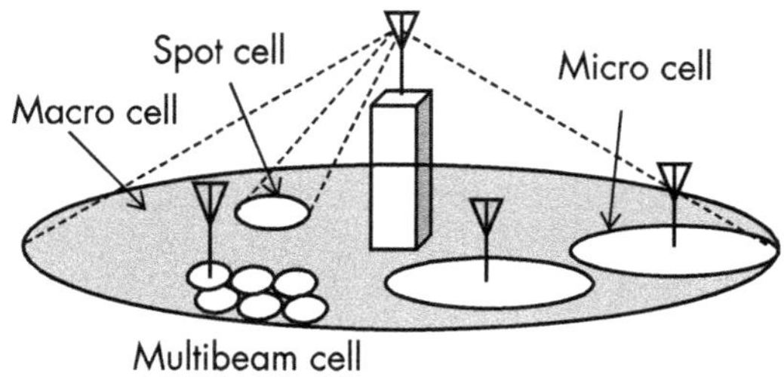

Figure 5.30 Overlaid multicell system.

The Butler matrix consists of a 3-dB hybrid coupler and delay lines. Figure 5.31 presents the operating principle of the 3-dB hybrid coupler. When a signal is inputted to port 1, equal amplitude with a phase difference of $\pi/2$ outputs is observed at ports 3 and 4. In this case, no signal is outputted to port 2. By switching the input port from 1 to 2, the phase delay at ports 3 and 4 is reversed, as shown in Figure 5.31(b). Figure 5.32 shows a diagram of the Butler matrix, where the phase delay of $\pi/4$ is given by the delay line and four output ports are terminated in the antenna elements. The output phase difference is shown for the port 1 and 2 inputs, and the beam direction is depicted by solid arrow lines. The output phase differences between the antenna elements are $-\pi/4$ by port 1 and $+3\pi/4$ by port 2. These differences change the beam direction, as shown in Figure 5.32. When ports 3 and 4 are used for the input, the beam directions are reversed, as depicted by the dotted arrow lines.

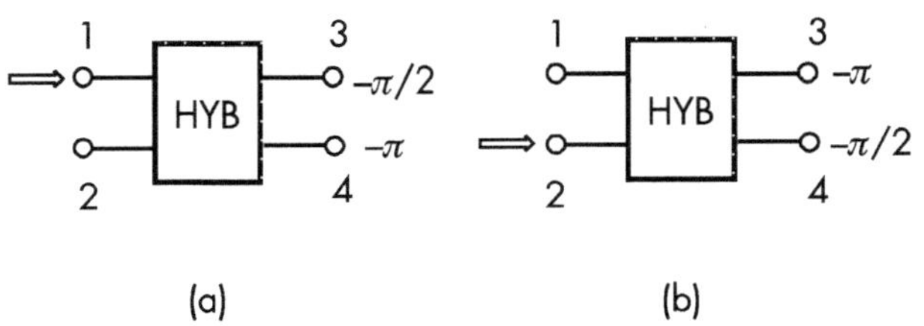

Figure 5.31 Output phase of a 3-dB hybrid coupler: (a) input at port 1, and (b) input at port 2.

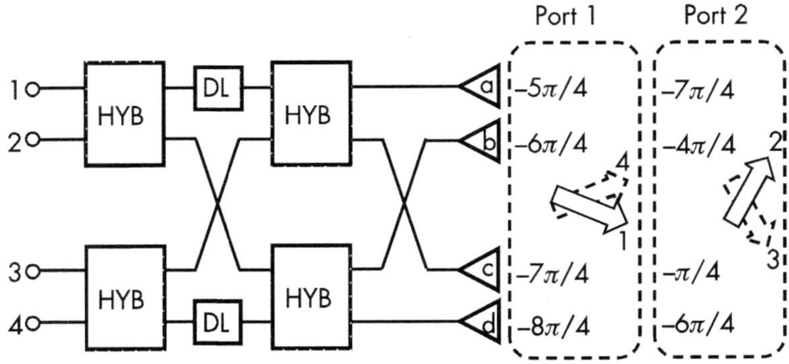

Figure 5.32 Butler matrix and output phase (HYB: 3-dB hybrid coupler, DL: $\pi/4$ delay line).

Figure 5.33 presents the radiation patterns by the Butler matrix, where the beam number corresponds to the input port number of the feeding circuit. For each beam direction, the maximum radiation angle matches the null point angle of the other beam. Hence, the correlation between the beams is very small at less than 0.1. An advantage of the Butler matrix is that it radiates several orthogonal beams simultaneously, but it does not radiate the front beam. All the radiated beams by the Butler matrix are tilted from the front direction.

The example in this section is an application of a multibeam by a linear array, which can be extended to a two-dimensional array to produce a multibeam on the plane surface.

5.3.3 Active Integrated Antenna

Figure 5.34 shows three types of BS antenna structure, where the antenna components inside a radome are depicted in the area surrounded by the dotted line. The feeding network of a conventional BS antenna, known as an analog feeding network, comprises analog phase shifters and power dividers, as shown in Figure 5.34(a). A power line is connected to the BS antenna to supply the electrical power with inside circuits, and several coaxial cables are pulled out to receive and transmit RF signals from a transceiver unit. The analog feeding network is used to tilt the main beam and to synthesize the radiation pattern. The beam tilt is used at a semifixed angle,

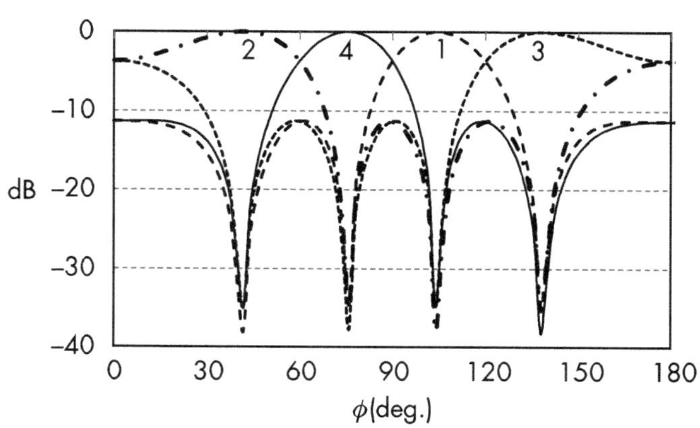

Figure 5.33 Patterns by four-port Butler matrix.

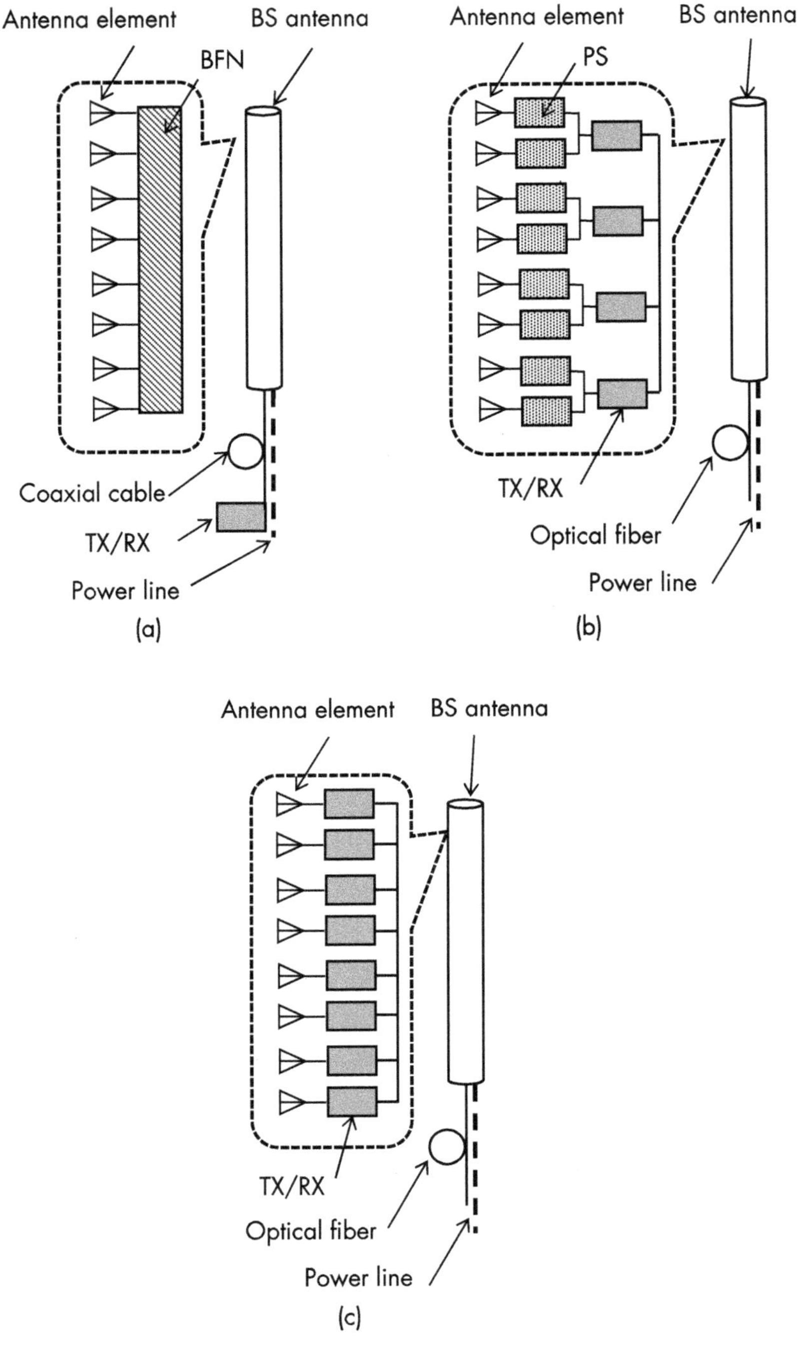

Figure 5.34 Beam forming network (BFN), phase shifter (PS): (a) analog beam forming antenna, (b) hybrid beam forming antenna, and (c) digital beam forming antenna.

and the synthesized pattern cannot be changed after the BS antenna installation.

An advanced BS antenna system requires dynamic pattern scanning to change the coverage area according to the status of terminal users. A hybrid beam-forming network gives the beam scanning, and its antenna structure is shown in Figure 5.34(b). A built-in transceiver unit is connected to a subarray, and the beam forming is obtained by the combination of analog phase shifters and transceiver units [22]. This structure does not need the coaxial cable output, and the signal is provided by an optical fiber to remove the transmission loss in high frequency. The beam forming is given by two stages. The beam angle change is controlled by the analog phase shifters in the vertical plane and is given by transceiver units in the horizontal plane.

Figure 5.35 shows the coordinate system of the beam coverage area. Beam scanning is obtained by changing the tilt angle θ from the horizontal line and the azimuth angle ϕ. The maximum tilt angle range of less than 20° is smaller than the azimuth one and is given by the subarray with analog phase shifters. The azimuth scanning range is around $\phi \leq |60°|$ to cover a sector area as a quick response to track the user movement. This beam scanning is provided by a digital beam forming technology to remove the analog phase shifters. The amplitude and phase of the output RF signal of the built-in transceiver can be dynamically changed and are controlled by the baseband signal generated by the digital processing unit. The baseband signal information is transmitted through the optical fiber to

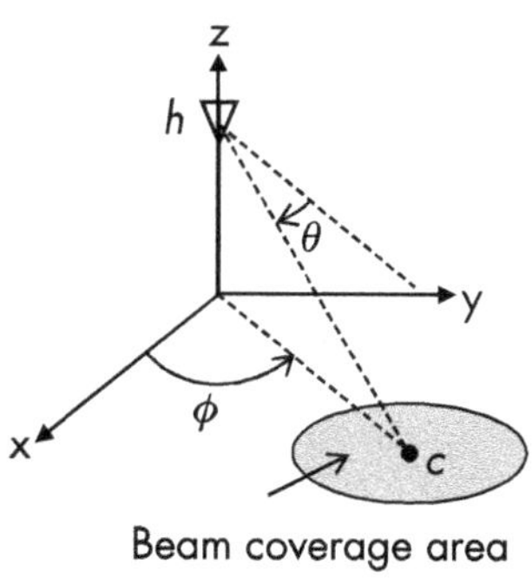

Figure 5.35 Beam coverage area.

the built-in transceiver, and thus no RF cables are used in the analog phase shifter.

A digital beam-forming antenna is given by removing the analog phase shifters from the hybrid beam-forming network, and the transceiver units are connected to all the antenna elements, as shown in Figure 5.34(c). The beam scanning can be obtained by changing the tilt angle θ and azimuth angle ϕ simultaneously.

References

[1] Ebine, Y., and S. Kozono, "Beam Tilting Effects of Base Station Antenna for Land Mobile Communications," *IEICE Technical Report*, AP82-23, June 1982 (in Japanese).

[2] Nakano, M., et al., "Evaluation of Ring-Omni Cells for Cellular System Using Antenna Patterns on Vertical Plane," *IEICE Trans. Commun. B*, Vol. E96-B, No. 9, September 2013, pp. 916–925 (in Japanese).

[3] Inoue, Y., et al., "Field Experimental Evaluation of Null Control Performance of MU-MIMO Considering Smart Vertical MIMO in LTE-Advanced Downlink Under LOS Dominant Conditions," *IEICE Trans. Commun.*, Vol. E97-B, No. 10, October 2014, pp. 2136–2144.

[4] Arai, H., and K. Kohzu, "A Bi-Directional Notch Antenna," *Conf. Dig., 1996 IEEE AP-S Int. Symp.*, Baltimore, MD, July 1996, pp. 42–45.

[5] Arai, H., and K. Cho, "Cellular and PHS Base Station Antenna Systems," *IEICE Trans. Commun.*, Vol. E86-B, No. 3, March 2003, pp. 980–992.

[6] "Dual-Polarized Dipole Antenna," AU2000018647, Australia Patent.

[7] Bao, Z., et al., "A Novel Broadband Dual-Polarization Antenna Utilizing Strong Mutual Coupling," *IEEE Trans. Antennas and Propagation*, Vol. 62, No. 1, January 2014, pp. 450–454.

[8] Cui, Y., et al., "A Broadband Dual-Polarized Planar Antenna for 2G/3G/LTE Base Stations," *IEEE Trans. Antennas and Propagation*, Vol. 62, No. 9, June 2014, pp. 4836–4840.

[9] Karikomi, M., and T. Matsuoka, "An Omnidirectional Board Bandwidth Microstrip Antenna Using a Parasitic Cylinder," *IEICE Trans. Commun.*, Vol. E76-B, No. 12, December 1993, pp. 1514–1517.

[10] Yokoo, Y., and H. Arai, "Dual-Frequency Loop Slot Antenna with Parasitic Elements," *2004 International Symposium on Antennas and Propagation*, No.4D3-1, Sendai, Japan, August 2004.

[11] Harrison, C. W., Jr., and R. W. P. King, "Folded Dipoles and Loops," *IEEE Trans. Antennas and Propagation*, Vol. 9, No. 2, March 1960.

[12] Matsuno, H., et al., "An Omni-Directional Slim Polarization Diversity Antenna for Cellular Base Station," *2010 IEEE International Workshop on Antenna Technology*, PS2.12, Lisbon, Portugal, March 1–3, 2010.

[13] Quan, X. L., and R. L. Li, "Analysis and Design of a 45° Slant-Polarized Omnidirectional Antenna," *IEEE Trans. AP.*, Vol. 13, No. 1, January 2014.

[14] Amin, M., et al., "Single Feed Low Profile Omnidirectional Antenna with Slant 45° Linear Polarization," *IEEE Trans. AP.*, Vol. 55, No. 11, 2007.

[15] Asaka, S., et al., "The Structure and Characteristics of a Revolving Variable Phase Shifter Using Slot Couplings," *ISAP2000*, Fukuoka, Japan, August 2000, pp. 305–308.

[16] Rodriguez, J. A., "Synthesis of Shaped Beam Antenna Patterns with Nullfilling in the Sidelobe Region," *IET Electronics Lett.*, Vol. 33, No. 24, November 1997, pp. 2004–2005.

[17] Kimura, Y., et al., "Relations Between Vertical Radiation Pattern and Downlink Capacity for Base Station Antenna of IMT-2000," *IEICE Technical Report*, AP2006-14, 2006 (in Japanese).

[18] Yamamoto, M., et al., "Simple Design Method for Cosec Square with Null-fill in Equispaced Linear Array," *2017 IEEE International Workshop on Electromagnetics*, PS-23, London, U.K., May 2017.

[19] Buckley, M. J., "Synthesis of Shaped Beam Antenna Patterns Using Implicitly Constrained Current Elements," *IEEE Transactions on Antennas and Propagation*, Vol. 44, No. 2, February 1996.

[20] Kim, Y. U., and R. S. Elliott, "Shaped-Pattern Synthesis Using Pure Real Distributions," *IEEE Transactions on Antennas and Propagation*, Vol. 36, November 1988, pp. 1645–1649.

[21] Katagiri, Y., et al., "Analysis on the Reflection Method for the Radiation Efficiency Measurement Using the Transmission Line Model," *Proc. ISAP 2006*, Singapore, November 2006.

[22] Kuriyama, T., et al., "Antenna Pattern Evaluation of 28GHz Band Massive MIMO Antenna RF Frontend Module for 5G," *Proc. IEEE Antennas Propag. Soc. Int. Symp.*, July 2019, p. 609.

6

DESIGN OF MOBILE TERMINAL ANTENNAS

Built-in antennas for smartphones and mobile terminals are presented in this chapter. Antenna space is very limited in terms of integration with other electric devices in smartphones. The location of the antenna element as an antenna design parameter is described to include the effect of a hand holding the terminal. Antenna miniaturization is provided using the effect of the chassis to mount the antenna element. To remove or make use of the effect of the chassis, an antenna design using chassis eigenmodes is presented. The chassis mode requires matching of the circuits to excite a desired mode effectively, and the design procedure of the matching circuits is described. Multiple antennas for diversity or multiple-input and multiple-output (MIMO) applications used in mobile terminals are explained by an evaluation factor of the correlation coefficient. Decoupling methods to suppress the mutual coupling of closely spaced antenna elements are explained to integrate multiple antennas into the limited space of mobile terminals. Finally, numerical and real phantoms to emulate the human body are presented for stable evaluation of antennas under the effects of human bodies. The effects of the hand and body are also discussed in terms of antenna design.

6.1 BUILT-IN ANTENNA

This section describes the process of mounting antenna elements into smartphones and the effect of the chassis on antenna characteristics. First, the space for the built-in antenna is depicted, and several examples of antenna location are presented. The mounting of the antenna elements in the limited space in smartphones is described for the microwave frequency band. The design parameters of built-in antennas are limited, and the input impedance is adjusted by matching circuits, which is presented as the second topic in this section.

6.1.1 Antenna Location and the Effect of the Chassis

Figure 6.1 shows the location of the built-in antenna in a smartphone, where the antennas can be placed at the four edges denoted by A, B, C, and D. Upper part E is used for the near field communication (NFC) antennas. The NFC system uses the frequency band of 13.6 MHz, and a spiral inductor is placed on the back side of the smartphone so the display is not disturbed. The top and bottom antenna positions (A and B) are appropriate when the smartphone is held by a single hand in browsing mode. Bottom position B is effective in decreasing the exposure of the head to electromagnetic waves in talk mode. The side positions (C and D) are used for the antennas in landscape mode because the top and bottom parts (A and B) of the smartphone are being held in this case.

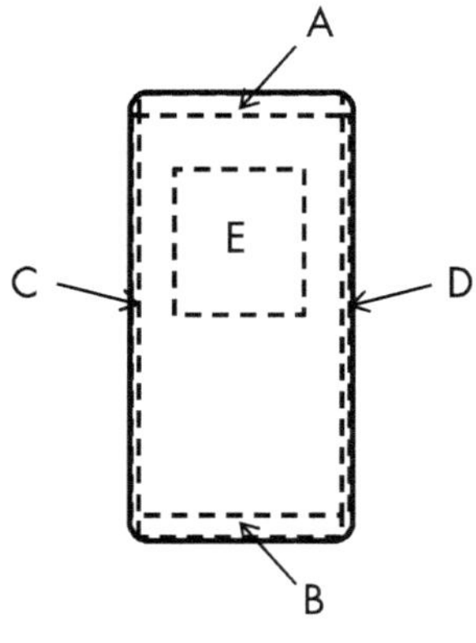

Figure 6.1 Antenna location in a smartphone.

Figure 6.2 shows examples of an antenna location in cellular smartphones. The frequency spectrum used in a smartphone is extended from around 700 MHz to 6 GHz for the use of the cellular system, wireless local area network (WLAN), and global positioning system (GPS). In example (a) in Figure 6.2, the main antenna is placed at position B, and another antenna for the use of WLAN and GPS is placed at position A. The main antenna and subantennas are used in example (b) for diversity reception or a MIMO system. These two examples are assumed to be mainly used in the talk and viewer modes. Example (c) allows for use in the landscape mode, and the main and sub-WLAN antennas are separated from the GPS antenna.

Figure 6.3 presents three examples of smartphone antennas. Printed circuit board antennas are etched on the substrate edge of the radio frequency (RF) circuits, as shown in Figure 6.3(a), where a bent monopole antenna is used. The substrate of the RF circuit is a part of the ground plane of the monopole antenna, which changes the antenna radiation pattern. The layout design is flexible, and the antenna is inexpensive [1]. A film antenna printed on a polyethylene terephthalate (PET) substrate can be easily mounted on a conformal shape because of its flexibility, as shown in Figure 6.3(b) [2]. The direct laser structuring process enables a 3-D antenna design and provides embedded antennas for smartphones, as shown in Figure

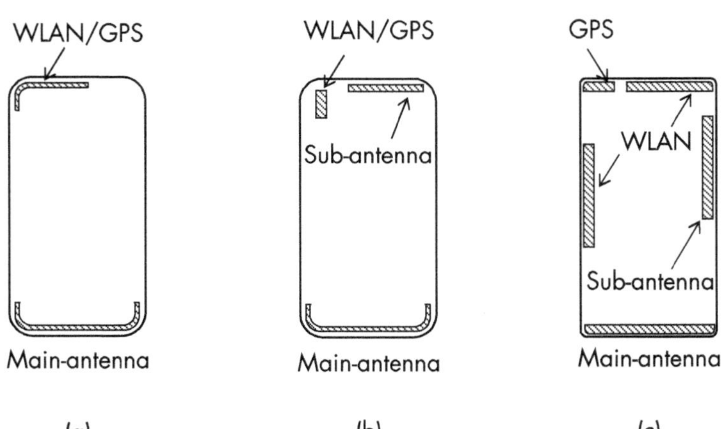

Figure 6.2 Antenna location examples of a smartphone: (a) a single cellular antenna, (b) main and subcellular antenna, and (c) multiantenna.

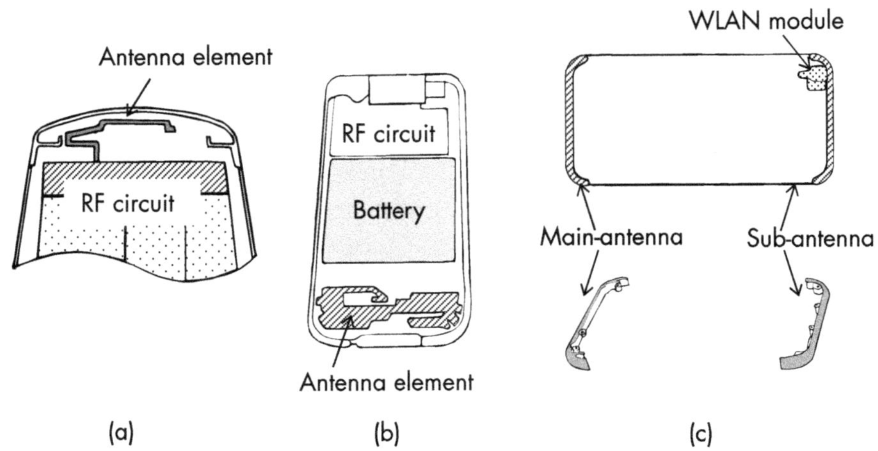

Figure 6.3 Antenna examples used in smartphone: (a) PCB antenna, (b) thin film antenna, and (c) molded interconnect antenna.

6.3(c). An antenna element is metalized on the part of the plastic made by injection molding [3], which is easily embedded inside the smartphone.

6.1.2 Antenna Elements

Figure 6.4 shows a basic design model for a built-in antenna. The hatched area at the top is an antenna space in this example, and L and inverted-F antenna elements are shown in Figures 6.5(a) and

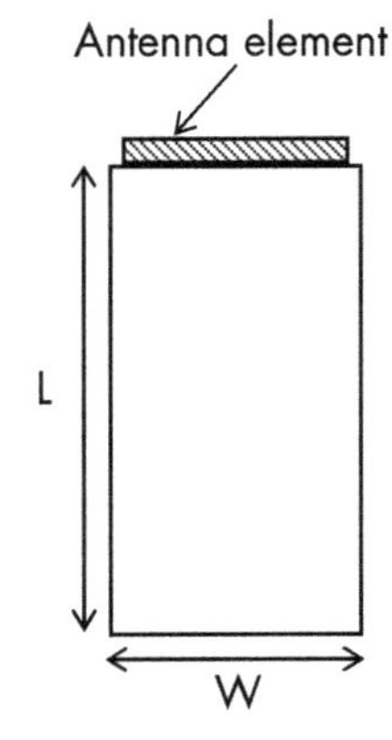

Figure 6.4 Geometry of built-in antenna.

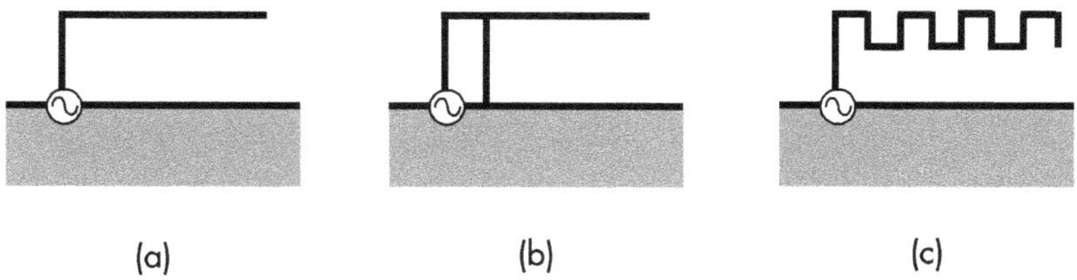

Figure 6.5 Built-in antenna element: (a) L antenna, (b) inverted F antenna, and (c) meander line antenna.

(b), respectively, and are used for the built-in antenna. The electrical length of the antenna is extended by a meander line (Figure 6.5(c)) to lower the resonant frequency. Dual resonance is obtained by adding a branch element (l_2), as shown in Figure 6.6(a), and the branch element can also be extended to the outer side, as shown in Figure 6.6(b). A parasitic element (p_2) near a feeding element (l_1) also provides dual resonance, as shown in Figures 6.6(c) and (d). Multiresonance can be obtained by increasing the number of branch or parasitic elements [4, 5].

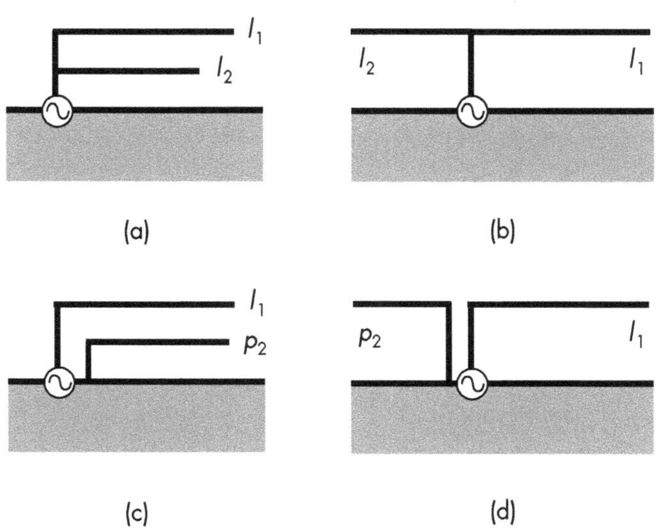

Figure 6.6 Multiresonance antenna element: (a) inner blanch element, (b) outer blanch element, (c) inner parasitic element, and (d) outer parasitic element.

The electrical length of the antenna elements described in this section is $\lambda/4$, which requires the ground plane to resonate at the designed frequency. A prefect image current is given by the infinite ground plane perpendicular to the antenna element, and the smartphone chassis denoted by $L \times W$ in Figure 6.4 is then operated as a part of an asymmetrical dipole antenna rather than the ground plane. Figure 6.7 shows the asymmetrical dipole antenna, characterizing the geometry in Figure 6.5 as a simple model in which the antenna is described as an element of $h = \lambda/4$ and the chassis is approximated by an element of length h'.

Figure 6.8 presents the normalized input impedance of the asymmetrical dipole $Z_a = Z/R_o = r_a + jx_a$, where R_o is the input resistance at $h' = h = \lambda/4$. The normalized resistance increases by a factor of $(h'/h)^n$, where $n = 1$ for $h'/h < 1$ and $n = 2$–3 for $h'/h > 1$. The normalized reactance is capacitive for $h'/h < 1$ and inductive for $h'/h > 1$. The electrical length of the chassis should be adjusted to obtain resonance at the designed frequency. However, this cannot be carried out in practical applications. Several antenna elements must be used to cover the wide band frequency spectrum assigned for the cellular system, forcing antenna designers to apply matching circuits for each antenna resonance.

6.1.3 Matching Circuit

The input impedance of the antenna Z_a is matched to the characteristic impedance of the transmission line Z_o connected to the terminal 1–1' by matching circuits, as shown in Figure 6.9, where the reactances of x_1 and x_2 are assumed to be purely imaginary. Two matching circuits are selected depending on the value of the antenna resistance normalized by Z_o, which is usually 50Ω. Circuit A is used for the case of normalized resistance $r_a < 1$, and circuit B is used for $r_a > 1$.

The combined impedances at terminal 1–1' are given as:

$$z_A = \frac{1}{\dfrac{1}{jx_2} + \dfrac{1}{jx_1 + r_a + jx_a}}, \qquad z_B = jx_1 + \frac{1}{\dfrac{1}{jx_2} + \dfrac{1}{r_a + jx_a}} \tag{6.1}$$

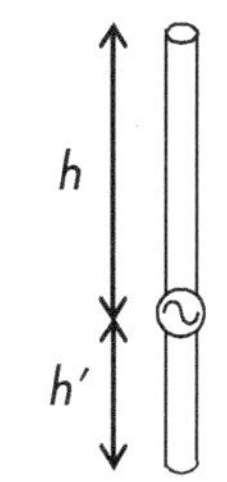

Figure 6.7 Asymmetrical dipole antenna element.

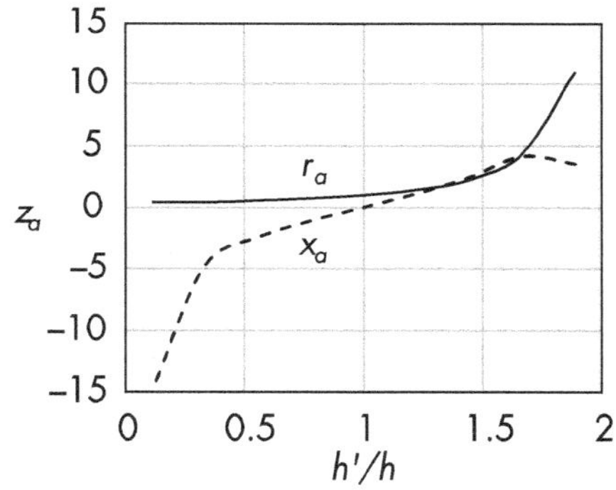

Figure 6.8 Normalized input impedance of asymmetrical dipole; $h = \lambda/4$.

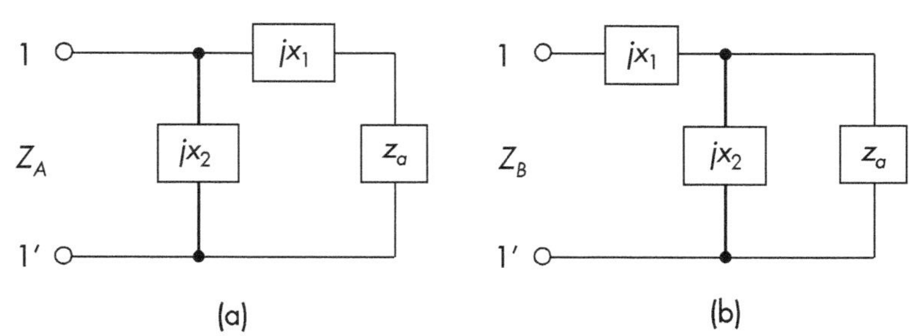

Figure 6.9 Circuit topologies of matching circuit; z_a is normalized antenna input impedance: (a) circuit A, and (b) circuit B.

The impedance matching condition is obtained by solving the following equations:

$$\mathrm{Re}(z_i) = 1, \quad \mathrm{Im}(z_i) = 0, \quad i = A, B \tag{6.2}$$

where *Re* and *Im* represent the real and imaginary parts of the impedance z_i, respectively. The solutions of (6.2) are given as follows:

$$x_1 = \pm\sqrt{r_a(1 - r_a)} - x_a, \quad x_2 = \mp\frac{r_a}{\sqrt{r_a(1 - r_a)}}, \qquad r_a < 1 \tag{6.3}$$

$$x_1 = -\frac{x_2(r_a^2 + x_2^2 + x_a x_2)}{r_a^2 + (x_a + x_2)^2}, \quad x_2 = \frac{2x_a \pm 2\sqrt{r_a^2(r_a - 1) + r_a x_a^2}}{2(r_a - 1)}, \quad r_a > 1 \tag{6.4}$$

In each case, two sets of solutions are obtained. It should be noted that x_2 is determined at first in (6.4) to find the value of x_1.

Figure 6.10 presents the reactance values of both matching circuits as a function of r_a. The solutions chosen in Figure 6.10 correspond to the upper signs in (6.3) and (6.4). When circuit A is used for $r_a < 1$, a series inductance x_1 is the dominant factor in the matching circuit and the range of shunt reactance x_2 is –0.3 to –0.5. The

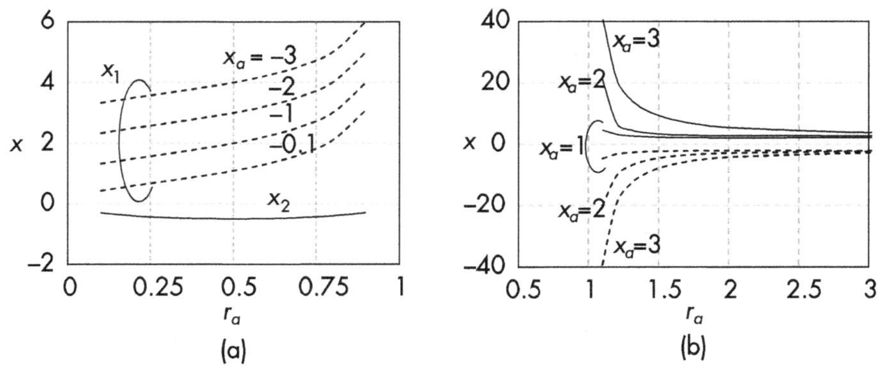

Figure 6.10 Reactance of matching circuit; solid line is x_2 and dotted line is x_1: (a) circuit A, (b) circuit B.

reactance values of circuit B converge to almost the same absolute value for a large r_a.

It should be noted that the matching circuit loss is mainly determined by the reactance elements. The loss of the chip inductors and capacitors used in the matching circuits is expressed by the factor Q as follows:

$$Q = \frac{X}{R_l} \tag{6.5}$$

where X and R_l are the reactance value and its resistance component, respectively. Radiation efficiency depends on $R_l/R_a = X/QR_a$, where R_a is the radiation resistance of the antenna. The loss of the matching circuit should be minimized to improve the radiation efficiency of the antenna. Then, it is necessary to satisfy the equation of $X/QR_a << 1$. This condition can be rewritten as $X << QR_a$. The Q value of the chip inductors and capacitors is greater than 10 above 1 GHz, which suggests that the reactance value should be $X << 10R_a$ so as not to degrade radiation efficiency. The matching circuit parameters should also be designed to satisfy this criterion.

6.2 MULTIANTENNA SYSTEM

A multiantenna system is introduced into mobile terminals to mitigate the fading effect under multipath propagation environments and enhance the channel capacity for high-speed data transmission. The former application is known as diversity reception, and the latter is based on a MIMO antenna system. Multiantenna systems require more space in mobile terminals. However, the space available for cellular systems is limited because of the increase in the number of antennas used to cover a wide frequency spectrum. The number of chassis operating modes can be increased by adding feeding probes to the chassis of mobile terminals. First, the theory and feeding methods of the chassis mode are presented here. This section also describes how to install multiple antennas in a mobile terminal and presents the effect of mutual coupling to measure antenna performance. Several antennas are arranged close to one another in the limited space of the terminal, which increases the mutual coupling between antenna elements. The third topic is decoupling methods used to decrease the coupling effect for multiple antennas in MIMO applications.

6.2.1 Chassis Mode

The antennas described in Section 6.1.1 use the chassis as a part of the radiating element, and resonance is given by the combination of the antenna element and chassis. The chassis has self-resonance when its side length is greater than $\lambda/2$. This section presents the theory of chassis resonance by introducing the characteristic mode [6, 7].

Figure 6.11 shows a conducting body surrounded by S_o, the radiated electric field $\boldsymbol{E}^s$, a current $\boldsymbol{J}$ on the body surface, an incident electric field $\boldsymbol{E}^i$, and an observation point p on surface S. The electric field generated by current $\boldsymbol{J}$ is given by the summation of an infinitesimal current element, as described in Section 1.2:

$$\boldsymbol{E}^s(\boldsymbol{r}) = -j\omega\mu_0 \int_{S_o} \left(\boldsymbol{I} + \frac{1}{k_o^2}\nabla\nabla \right) \cdot \boldsymbol{J}(\boldsymbol{r}_o) \frac{e^{-jk_o|\boldsymbol{r}-\boldsymbol{r}_o|}}{4\pi|\boldsymbol{r}-\boldsymbol{r}_o|} dS_o \tag{6.6}$$

where $\boldsymbol{r}$ and $\boldsymbol{r}_o$ are observation and source position vectors, and ω, μ_o, and k_o are the angular frequency, permittivity, and wave number in free space, respectively. The boundary value condition on the conductor surface is used to eliminate the tangential electric field on surface S as follows:

$$\boldsymbol{n} \times \left(\boldsymbol{E}^s + \boldsymbol{E}^i \right) = 0 \tag{6.7}$$

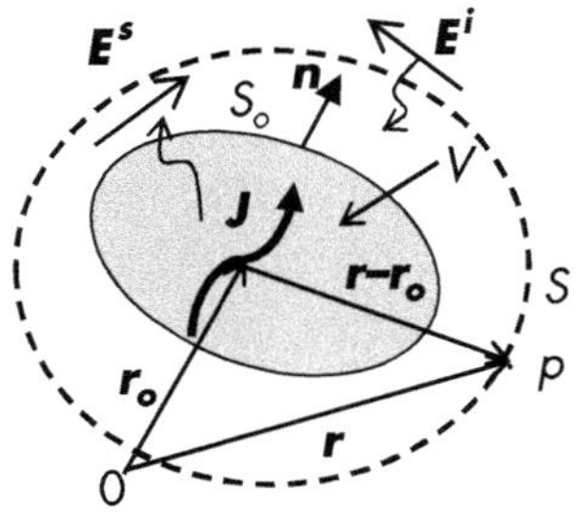

Figure 6.11 Geometry of conducting body and parameters.

where $\boldsymbol{n}$ is the normal vector on surface S. Assuming no incident field $\boldsymbol{E}^i = 0$, the boundary value condition on S is expressed as follows:

$$\boldsymbol{n} \times \boldsymbol{E}^s(\boldsymbol{r}) = \boldsymbol{R}(\boldsymbol{J}) + j\boldsymbol{X}(\boldsymbol{J}) = 0 \tag{6.8}$$

The solutions of (6.8) are given by the complex angular frequency as $\omega_n = \omega_{nr} + j\omega_{ni}$. The mode number is represented by n, and the corresponding current distribution is denoted as $\boldsymbol{J}_n$. This solution, known as the natural mode, is rigorous. However, it is cumbersome to determine the numerical solutions. Another solution can be obtained via the analogy of the resonance of a series RLC circuit as $\boldsymbol{X}(\boldsymbol{J}) = 0$. This approximate solution is derived from the eigenvalue problem and is easily calculated numerically. This solution is known as the characteristic mode.

Figure 6.12 shows the geometry of a rectangular conducting plate as a simple model of the chassis. The characteristic mode analysis provides eigenvalues to determine the corresponding frequency and current distribution. The current distributions of the dominant three modes of the conducting plate ($b/a = 3$) are shown in Figure 6.13, where J_i ($i = 1, 2, 3$) denotes the mode number and the arrow represents the rough current flow.

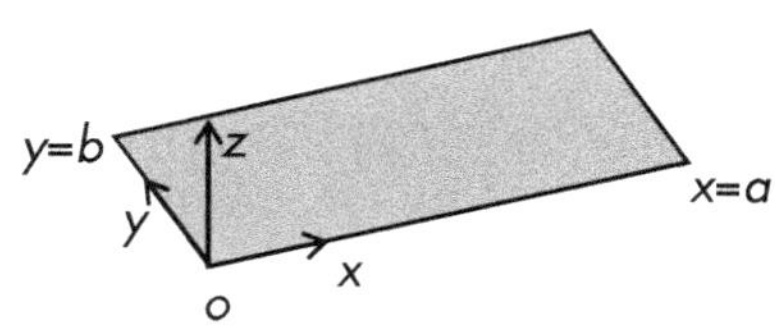

Figure 6.12 Geometry of rectangular conducting plate.

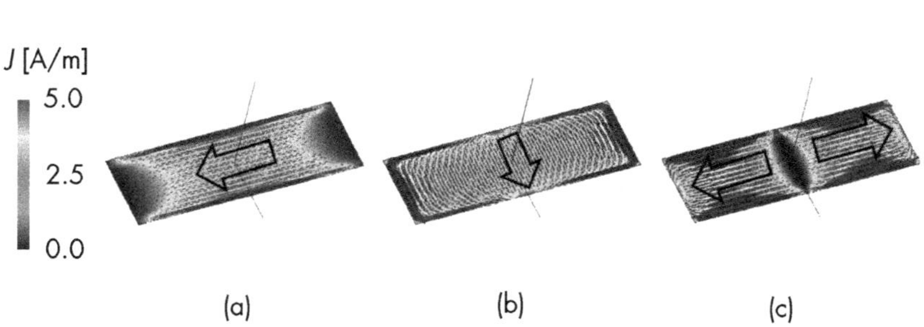

Figure 6.13 Current distribution of characteristics mode; $b/a = 3$: (a) J_1, (b) J_2, and (c) J_3.

Mode J_1 is known as the dipole mode due to a side length of $b \cong \lambda/2$, and J_2 is given by $a \cong \lambda/2$. Mode J_3 is the second harmonic one of mode J_1. These modes are almost the same as those of the microstrip antenna described in Section 4.1.2. The chassis mode does not require any radiating elements, but it does require the probe to excite the mode efficiently.

The probe should be nonresonant so as not to disturb the chassis mode [8]. Figure 6.14(a) presents a nonresonant T-bar probe to excite mode J_1. Two T-bar probes are mounted at the middle of both sides of the plate [9], and the excited current distribution shown in Figure 6.14(b) is almost the same as J_1 in Figure 6.13(a). A T-bar probe should be connected with the feeding line through the matching circuit to create resonance at the designed frequency. The two T-bars in Figure 6.14(a) are excited in phase to maintain the symmetrical current distribution along the y-axis.

6.2.2 Antenna Position and Mutual Coupling

More than two antennas are used in diversity reception, and multiple antennas are mounted at the chassis edge, as shown in Figure 6.15. The antenna spacing d, a dominant factor in mutual coupling, should be suppressed to lower than −10 dB to obtain the required diversity gain. Mutual coupling is estimated by a simple model of two $\lambda/2$ dipole antennas, as shown in Figure 6.16. This model provides the lower boundary of the mutual coupling because the antenna shape and chassis increase the coupling in actual mobile terminals.

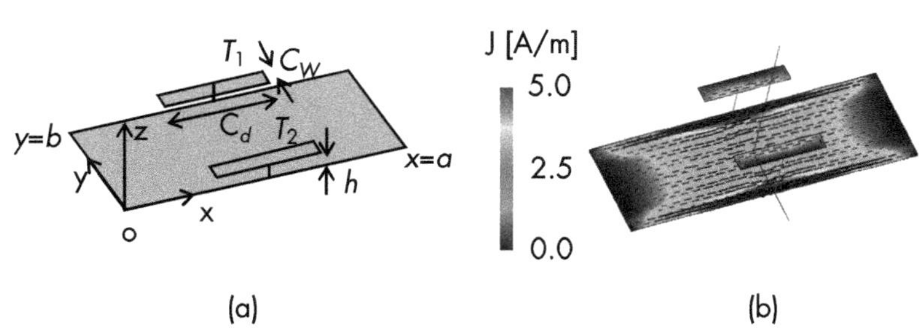

Figure 6.14 T-bar probe on plate; $b/a = 3$, $C_w/a = 0.17$, $C_d/a = 0.9$, $h/a = 0.17$: (a) geometry of T-bar probe, and (b) current distribution.

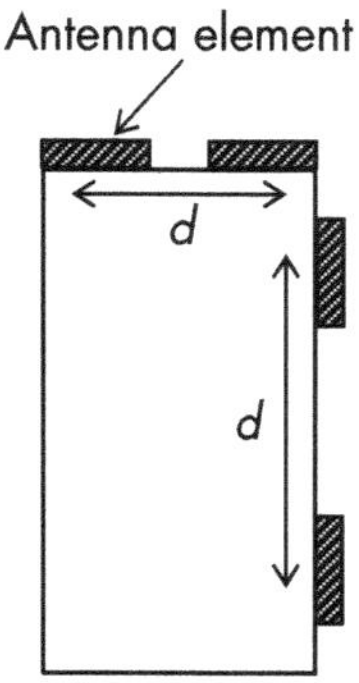

Figure 6.15 Antenna position on chassis.

Figure 6.17 shows mutual coupling as a function of frequency by changing antenna spacing. Mutual coupling was suppressed by less than –10 dB for $d > 40$ mm and $f > 2$ GHz, which is acceptable for the installation of the antenna elements on the narrow side of the terminal. A long side can be used for low frequencies less than 1 GHz, and the top and bottom edges represent another antenna position. To verify these criteria, mutual coupling is demonstrated by two modified inverted-F antennas mounted on the narrow side of the chassis, as shown in Figure 6.18. A nonradiating element of the inverted-F antenna is folded to reduce its size, and the position of the matching stub is adjusted to obtain impedance matching at 2 GHz. The spacing d is defined as the distance between feeding positions. The mutual coupling S_{21} is approximately 3 dB larger than that in the two-dipole model, as shown in Figures 6.18(a) and (b), which is caused by the coupling between nonradiating elements. The

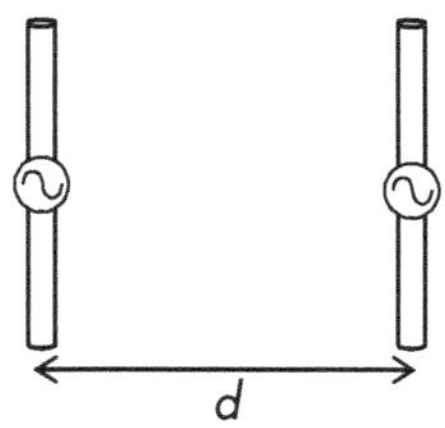

Figure 6.16 $\lambda/2$ dipole array.

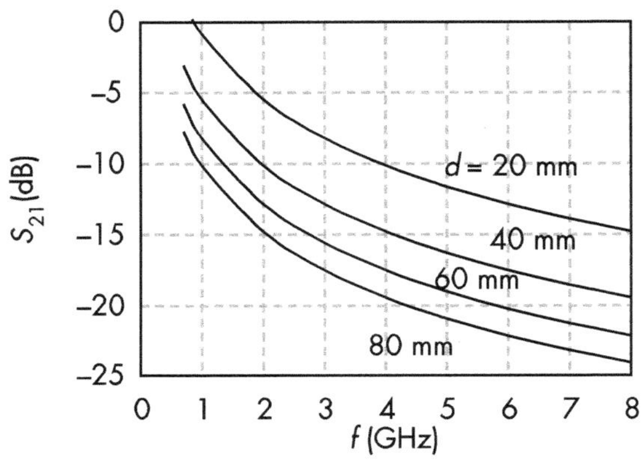

Figure 6.17 Mutual coupling of $\lambda/2$ dipole array.

feeding elements in Figure 6.18(c) are orthogonal to each other, and S_{21} is decreased compared with (a) despite the small distance d. The coupling of orthogonal antenna elements can theoretically be eliminated, and the coupling of (c) depends on the coupling between a feeding probe and nonradiating element of another antenna, with these components being parallel to each other. As described in this section, the dominant factors of mutual coupling are the distance between feeding elements and their relative orientation.

6.2.3 Decoupling Method

This section presents decoupling methods used to suppress the mutual coupling between closely spaced antenna elements. Figure 6.19(a) shows a ground plane notched to decrease the current flow between antenna elements along the ground plane edge. The notch depth l_d of $\lambda/4$ creates an open end at point P, which suppresses the current flow between antenna feeding points. The notch is able to stop the current at the ground plane edge. However, it cannot suppress the current induced through the space. A decoupling bridge connecting two antenna elements, as shown in Figure 6.19(b), cancels the induced current I_{BA} from antenna A to B by creating another current path, denoted as a dotted loop. This loop current is given by the image of the ground plane. An upward current I_{BA} on antenna element B induced by the electric field radiated from element A is

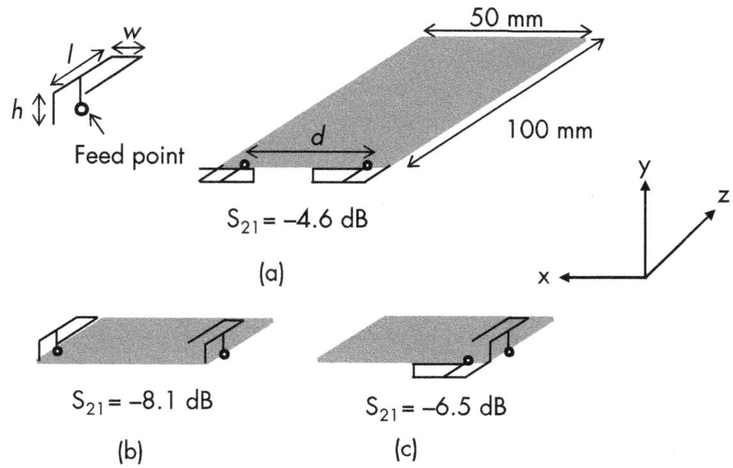

Figure 6.18 Inverted-F antennas on chassis: (a) d = 22 mm, (b) d = 50 mm, and (c) d = 12.7 mm.

canceled by the downward current I'_{BA} at the feed point when the loop length is λ.

These decoupling methods suppress mutual coupling via a simple modification of the antenna, and they heavily depend on the currents flowing on the ground plane. The center frequency is shifted by the decoupling parameters, and the antenna should then be redesigned to adjust the center frequency. The radiation pattern in the vertical plane is significantly altered by the decoupling notch as the current on the ground plane is quite different from that of the antenna without the notch. The decoupling bridge maintains the same radiation pattern due to the image current I'_{BA} canceling the radiation from the bridge current I_{BA}. The change in radiation pattern should be considered in the design of decoupling methods.

Mutual coupling is suppressed by adding a decoupling network at the feeding points, as shown in Figure 6.19(c). The additional circuit consists of delay lines and a shunt reactance [10]. The mutual coupling between antennas A and B is determined by an impedance matrix of order 2 × 2, where the element is denoted as Z_{ij} (i, j = A, B). The coupling factor Z_{AB} is adjusted by connecting a transmission line to make the real part of Z_{AB} zero, which is equivalent to having zero energy transfer between the two elements. As shown in Figure

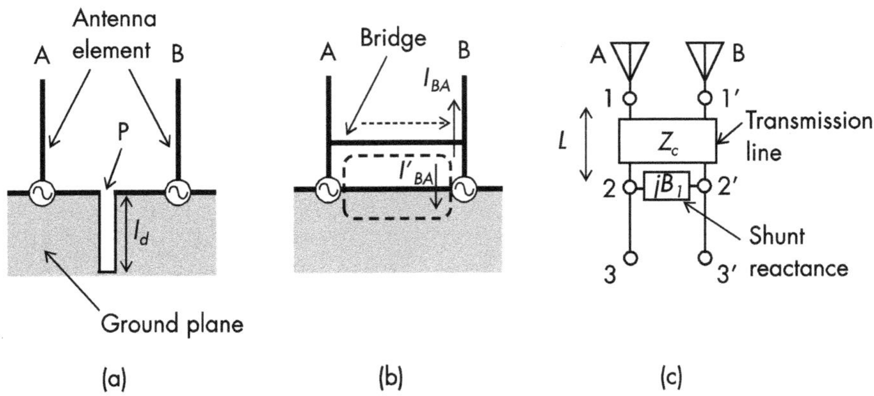

Figure 6.19 Decoupling method: (a) decoupling notch, (b) decoupling bridge, and (c) decoupling network.

6.19(c) and Section 3.1.1, Z_{AB} at 1–1' is transformed into $Z(L)$ at 2–2' as follows:

$$Z(L) = Z_r + jZ_i = \frac{Z_{AB} + jZ_c \tan(\beta L)}{Z_c + jZ_{AB} \tan(\beta L)} \tag{6.9}$$

where Z_C, β, and L are the characteristic impedance, propagation constant, and length of the transmission line, respectively. After setting the length L to $Z_r = 0$, the imaginary component Z_i is canceled at 2–2' by the shunt reactance of $jB = -Z_i$. The nondiagonal elements of the impedance matrix are set to zero in the above procedure, and the diagonal elements are not matched with the impedance of the feeding point. To use this decoupling network, the matching circuit should be connected with ports 3 and 3'. Figure 6.20(a) shows the input characteristics of this decoupling method and mutual coupling suppression in a wide frequency range.

The parameter setting $Z_r = 0$ leads to the absence of induced current on the antenna, which is equivalent to having one antenna element as a reflector of another element. The radiation pattern of the monopole antenna becomes unidirectional because of the reflector, as shown in Figure 6.20(b), which is a drawback of this decoupling method.

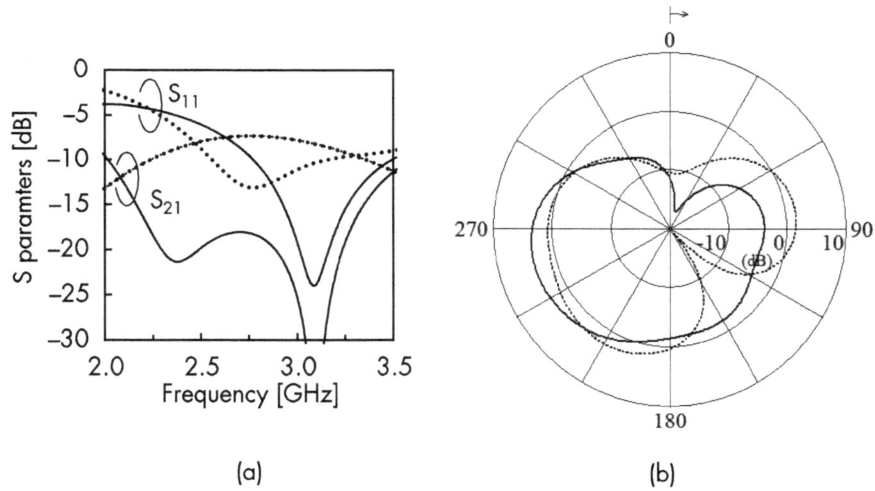

Figure 6.20 Decoupling network; solid line is with decoupling network, dotted line is without decoupling network: (a) input characteristics, and (b) radiation pattern.

6.3 BODY EFFECT

This section describes a special type of mannequin, called a phantom, used to evaluate the influence of the human body on antenna characteristics. It is not easy to evaluate antenna performance in the presence of a real human body, then an electrical equivalent of the body is required to emulate it. Several types of phantoms used in the design of mobile terminal antennas are described here. The effect from the body can be presented by a simple sphere phantom in terms of the radiation pattern of the antenna to replicate the effect of the human body. An upper body and a hand phantom are also described to evaluate these effects on the antenna. Finally, these are also compared with real human body results.

6.3.1 Numerical Phantom

A European work group called CO-operation in the field of Scientific and Technical Research 244 (COST 244) has proposed two simplified models of the head for numerical evaluations [11]. These two simplified models are a cube with a side length of 20 cm and a sphere with a radius of 10 cm. Both the cube and sphere are modeled on the basis that they are composed of a homogeneous dielectric material. One example of typically assumed material constants is given in Table 6.1.

To model the skin, an outer shell of a dielectric material with a thickness of 5 mm and a dielectric constant of 3 ($\varepsilon_r = 3$) is added. Such a simple form for the phantom is useful as a standard to check the validity of the computational codes and measurement setups.

The example of a sphere using the layered structure model shown in Figure 6.21 provides a precise analytical model, and the corresponding parameters are given in Table 6.2. Using the layered-sphere model, it is possible to compute the absorption of the electromagnetic wave by a thin layer, such as the skin [12]. Table 6.2 shows the assumed material characteristics of the various layers in the human head model.

An analysis object can be divided into very small volumes or cells by the finite difference and time domain method. Once a real object is discretized into cells, the characteristics of the model can be calculated numerically, thereby giving a close approximation to the actual characteristics. The phantom used in numerical simulations is based on an anatomy chart of the human body. A phantom with millimeter-resolution accuracy can be constructed automatically using magnetic resonance imaging data from a human body. The phantom used for numerical simulations must be specifically

Table 6.1
Electrical Parameters of the COST 244 Phantom

***f* (MHz)**	ε_r	σ
900	43	0.83
1800	41	1.14

Table 6.2
Parameters of Layered-Sphere Model

P	**Layer**	**R_p (cm)**	ε_r	**σ (S/m)**
6	skin	a	47.5	1.33
5	fat	$a - 0.15$	6	0.10
4	bone	$a - 0.27$	5	0.20
3	dura	$a - 0.70$	47.5	1.33
2	CSF	$a - 0.80$	83.2	1.33
1	brain	$a - 1.10$	59.4	1.00

(f = 2 GHz), dura = cranial dura matter; CFS = cerebrospinal fluid.

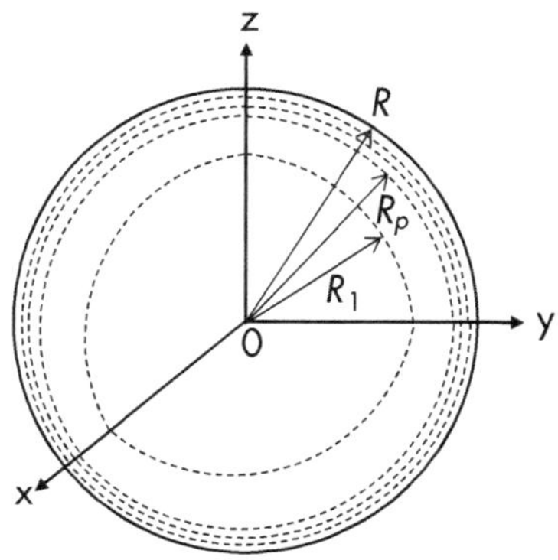

Figure 6.21 Layered-sphere phantom model; R_i: radius of each layer.

tailored to the purpose of the calculations. The Visible Human Project contains data on the electrical characteristics of human body tissues in relation to frequency [13].

6.3.2 Phantom in Measurement

Creating an exact replica of the human body is not cost-effective, and phantoms composed of one of two types of homogeneous dielectric material are normally used to emulate its shape: a dry phantom made of a ceramic material or a wet phantom consisting of a water solution or a jelly-like substance made into a thin plastic shell. One issue is that the dry phantom tends to be very heavy, although the hard material acts as a realistic imitation of the head, hand, and upper half of the body. The wet phantom offers the advantage that additives can be used to change its material characteristics. However, one major disadvantage is that it is difficult to preserve wet phantoms for more than one month due to the type of material used.

The human body consists of low-water content structures, such as the skin, fat, and bone, as well as high-water content structures, such as the brain, muscles, and internal organs. However, the electrical characteristics of the tissues vary greatly at frequencies of 10 MHz and more. For mobile communications operating at a frequency band of 800 MHz–2 GHz, the dielectric-loss tangent is 0.1–3.0 and the dielectric constant ranges from 20–70 [14]. For materials used in the construction of phantoms, it is not easy to obtain a material with a loss tangent (tan δ) between 1 and 10. However, ceramic materials can be used to obtain a dielectric constant (ε_r) between 10 and several 10,000s in the microwave frequency band.

A phantom with a loss tangent similar to that of the human body can be constructed by adding conducting powder to the ceramic [15]. The composition of the plastic shell used to form the phantom shape includes Ba, Ca, Ti, Sn, and carbon powder, which then mimics the electrical characteristics of a living body. The electrical characteristics of the plastic and oxide powder are $\varepsilon_r = 3.1$ and $\tan\delta = 0.1$ at 1 MHz and $\varepsilon_r = 20$ and $\tan\delta = 0.02$ at 1 MHz, respectively. The dry phantom shown in Figure 6.22 was proven to be particularly useful when the influence of the hand on the radiation characteristics of the mobile terminal needed to be considered.

The wet phantom (Figure 6.23) was composed of a salt solution and agar chosen for use in the microwave frequency band [16], and the loss tangent could be controlled using NaCl. Because the material of wet phantoms is equivalent to biological material [17], it also suffers from the problem that it can decompose over time. If the food preservative dehydroacetic acid sodium salt (DASS) is added to the phantom material, a preservation period of one month or more may be achievable at normal temperatures. The phantom should also be covered by a thin film for further protection.

The electrical characteristics of muscle tissues may also be imitated in the frequency range from 200 MHz to 2.5 GHz. One example of the types of material used in a wet phantom is shown in Table 6.3.

In the 900-MHz frequency band, the dielectric constant may be controlled (from 35 to 65) using polyethylene powder and conductivity can be controlled by introducing small quantities of NaCl

Table 6.3
The Composition of the Phantom

Ingredients	Muscle (g)	Brain (g)
De-ionized water	3375	3375
Agar	104.6	104.6
Sodium Chloride	39.2	23.1
DASS	2.0	2.0
TX-151	84.4	57.1
Polyethylene powder	337.5	548.1

Phantom volume is about 3,500 cm^3; agar for solidification; sodium chloride for conductivity control; dehydroacetic acid sodium salt (DASS) for preservation; TX-151 for gelling polyethylene powder for relative dielectric control.

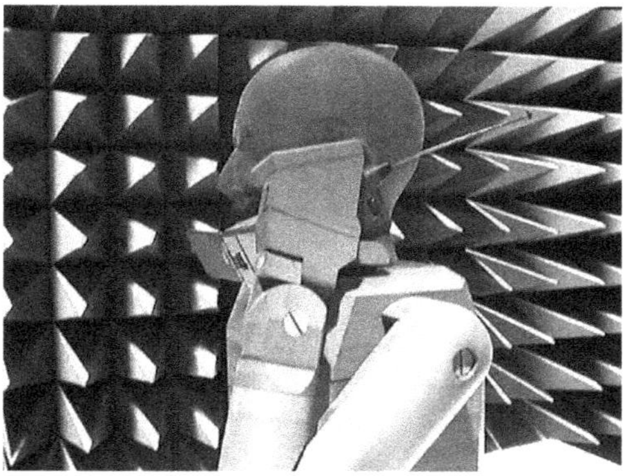

Figure 6.22 Ceramic dry phantom.

Figure 6.23 Wet phantom. (Courtesy of Prof. Ito, Chiba University.)

(0.3–2.5). Figure 6.24 gives an example of the dielectric properties of a wet phantom.

The dry and wet phantoms are designed to have the same electrical parameters as a real human body. Another design procedure is given for an electromagnetic-equivalent phantom with the same characteristic impedance on the phantom surface [18]. The input impedance and radiation pattern of the antennas near the phantom are affected by the reflection from the phantom surface, which means that the surface impedance is the dominant factor in determining the phantom's electrical performance. The equivalent phantom shown in Figure 6.25 consists of a head, an upper body, and a hand with an arm.

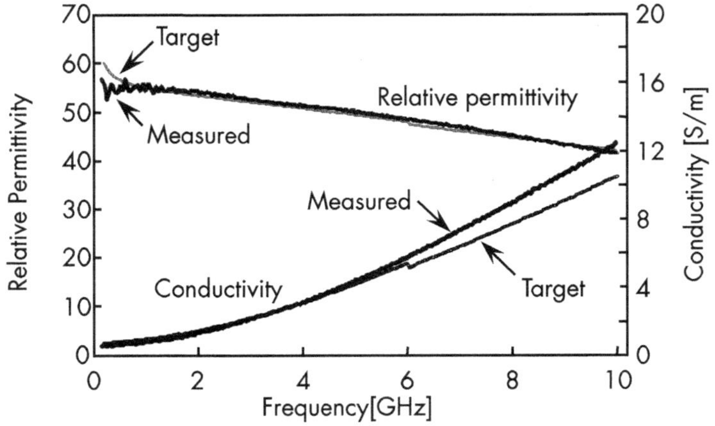

Figure 6.24 Dielectric properties of body tissues at RF and microwave frequencies; 20MHz–6 GHz (http://www.fcc.gov/fcc-bin/dielec.sh, 6–10 GHz; http://niremf.ifac.cnr.it/tissprop/#cred).

The shell of the phantom is made of a low-dielectric resin. An electromagnetic wave-absorbing sheet is placed on the inside of the shell to provide the same surface impedance with the human body. Although the weight of the head and upper body of the specific anthropomorphic mannequin phantoms is greater than 20 kg, that of the equivalent phantom is 3.6 kg, and it can be easily installed on a simple 3-D rotary positioner with azimuth and roll axes when using the great circle cut measurement system.

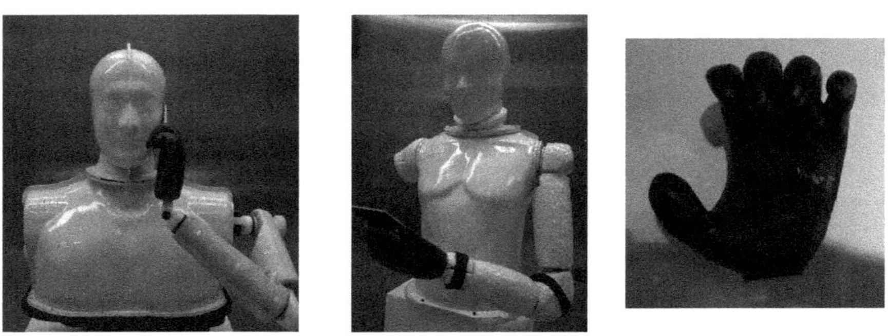

(a) (b) (c)

Figure 6.25 Electrical equivalent phantom (courtesy of Microwave Factory): (a) talk position, (b) viewer position, and (c) hand phantom.

6.3.3 Body Effect

This section presents the effects of the body on antenna performance using a sphere, an upper body, and a hand phantom in the design of mobile terminal antennas. First, basic measurement examples are presented to characterize the effect of the human body on antenna radiation pattern using a simple phantom emulating the human head, which, in this case, is a spherical wet phantom with a 10-cm radius and a dielectric constant of $\varepsilon_r = 52 + j19$. In the construction of wet phantoms, salt solution and agar have been the choice for those used in the microwave frequency band. The loss tangent may be controlled using NaCl. As stated earlier, because the material from which a wet phantom is made is equivalent to biological material, it also suffers the issue that it can decompose over time. For experimental use, it is ideal to have wet phantoms that can be preserved for a long period while maintaining suitable material characteristics. If the food preservative DASS is added to the phantom material, a preservation period of one month or more may be possible at a normal temperature. The phantom should also be covered by a thin film for further protection.

For the measurements, a standard dipole antenna was used, and the two specific measurements included (a) the position of the null of the radiation pattern toward the phantom ($\alpha = 0°$) and (b) the position of the maximum radiation from the antenna toward the phantom ($\alpha = 90°$), as shown in Figure 6.26, where the measurement frequency was 2.5 GHz. The pattern is depicted only for the *E*-plane due to the axial symmetry about the *y*-axis. The measurement pattern in Figure 6.27(a) shows a smaller change than that in Figure 6.27(b) because the null of the pattern faced toward the phantom. The small side lobes around $\theta = \pm 170°$ were caused by a creeping wave excited at around $\theta = 90°$ that propagated along the phantom surface and then radiated backward. In Figure 6.27(b), it can be clearly observed that the radiation pattern on the phantom side ($180° \leq \theta \leq 360°$) was suppressed, and the effects of the phantom were much larger than those in (a) in this case.

The reflection coefficient S_{11} of the dipole antenna is shown in Figure 6.28 for varying distances between the antenna and phantom and for two antenna orientations ($\alpha = 0°$ and $90°$). Figure 6.28(a) ($\alpha = 0°$) shows that the antenna matching condition at the feed point (S_{11}

< −20 dB) was not deteriorated, except for the antenna edge attached to the phantom ($d = 0$), because of the antenna current flowing to the phantom. The input characteristics were not affected by the small distance d in this orientation because the null in the radiation pattern faced toward the phantom. Figure 6.28(b) ($\alpha = 90°$) indicates that the distance d should be greater than 50 mm (0.41λ) to achieve stable S_{11} performance because the antenna radiating toward the phantom causes a large amount of reflection from the phantom.

The antenna input characteristics were not seriously affected by the distance (0.41λ in the previous example). However, the change in the radiation pattern due to blocking by the phantom cannot be neglected, as shown in Figure 6.27(b).

The second example is presented using a dipole antenna placed near the upper body phantom and human body (Figure 6.29). The change in radiation pattern should be carefully considered in the design of mobile terminal antennas. The radiation pattern presented in Figure 6.30 shows the experimental setup used to examine the horizontal and vertical antenna orientations. An electromagnetic-equivalent phantom (Figure 6.25) was used in the measurement. The radiation patterns using the phantom agreed well with the patterns in the presence of a human body in the measurement at 940 MHz. A small difference in the pattern suppression level by the phantom side ($180° \leq \theta \leq 360°$) was observed at 1,900 MHz due to the shape difference between the phantom human head and shoulder.

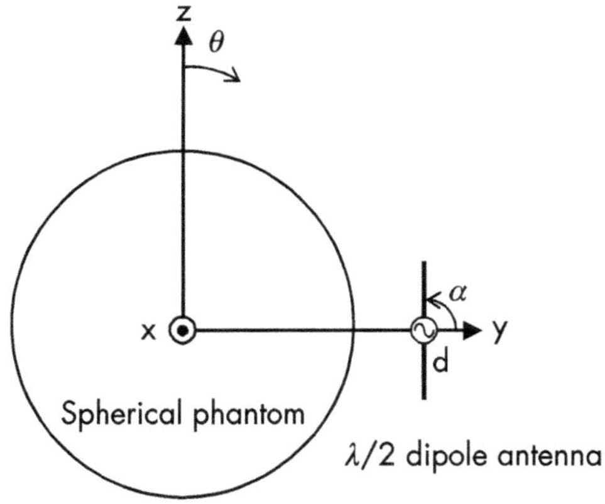

Figure 6.26 Spherical phantom and dipole antenna.

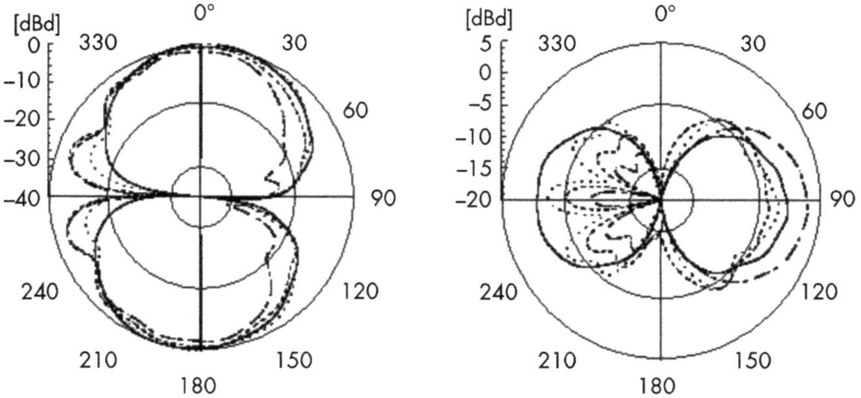

(a) (b)

Figure 6.27 E-plane radiation pattern: (a) $\alpha = 0°$, — $d = \infty$, — • — $d = 0$, — — — $d = 0.08\lambda$, - - - - $d = 0.96\lambda$ and (b) $\alpha = 90°$, — $d = \infty$, — • — $d = 0.17\lambda$, — — — $d = 0.42\lambda$, - - - - $d = 1.0\lambda$.

The third example is the effect of a user hand. The presence of the user hand also changes the performance of mobile terminals, especially the input characteristics [19]. A $\lambda/2$ dipole antenna was

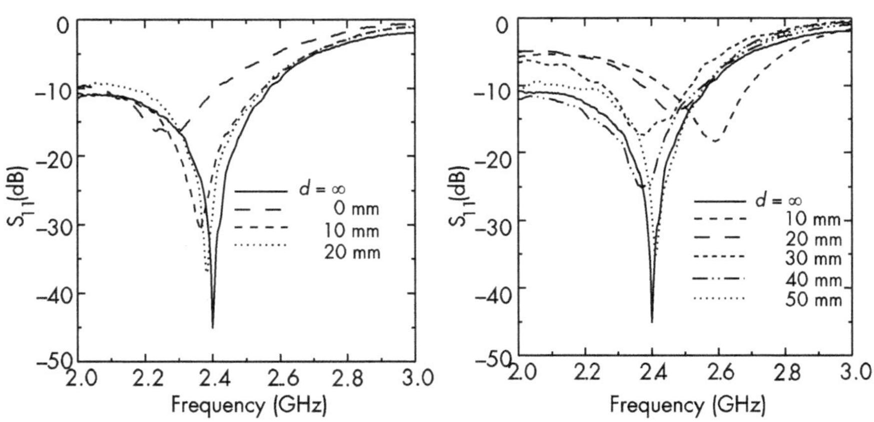

(a) (b)

Figure 6.28 Antenna input characteristics on the relative position of antenna and phantom for varying antenna positions: (a) antenna orientation, $\alpha = 0°$, and (b) antenna orientation, $\alpha = 90°$.

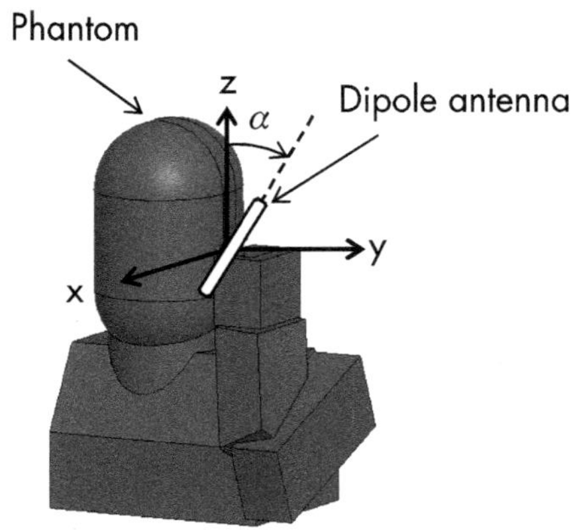

Figure 6.29 Upper body and dipole antenna.

placed in front of a hand as a basic evaluation model, as presented in Figure 6.31. The dipole was fixed using a polystyrene form spacer with a 5-mm thickness.

To ensure clear evaluations, the human test subject positioned his/her hand to align the knuckle of the middle finger with the center of the antenna. A measure of voltage standing wave ratio (VSWR) was used instead of S_{11} so that small changes around the center frequency could be observed. The VSWR performance of the antenna within two bands is shown in Figure 6.32, with three human test subjects and the hand phantom (Figure 6.25(c)) used [20].

Compared with the VSWR performance without test subjects, the matching frequencies in the 800-MHz and 2-GHz bands with human test subjects were shifted by more than 40 and 80 MHz, respectively. The VSWR performances with the phantom were virtually equivalent to those with human test subjects in these frequency bands. The VSWR losses at 840 MHz were greater than those at 1,970 MHz because the electrical distance between the antenna and hand at 840 MHz was shorter. In the design of mobile terminal antennas, the center frequency shift and loss should be considered.

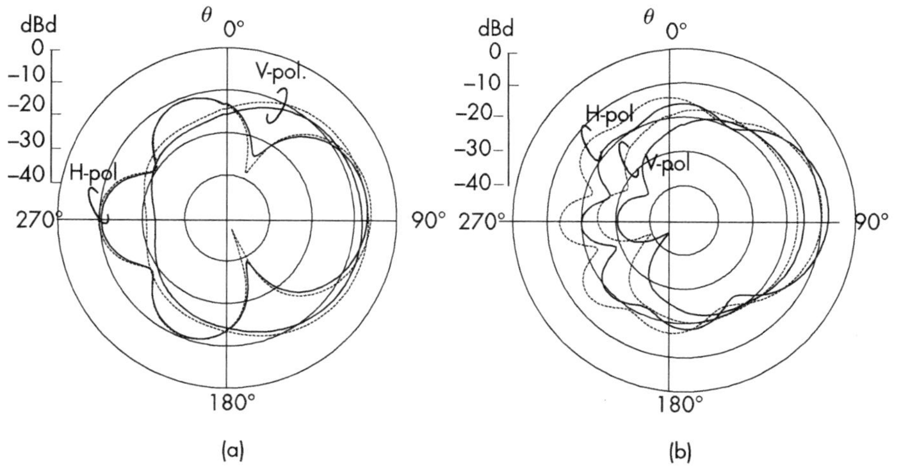

Figure 6.30 Radiation patterns from the antenna in the presence of typical human body and phantom; solid line is the pattern of human body, dashed line is the pattern of phantom, V-polarization is $\alpha = 0°$, H-polarization is $\alpha = 90°$, (a) 940 MHz, and (b) 1900 MHz.

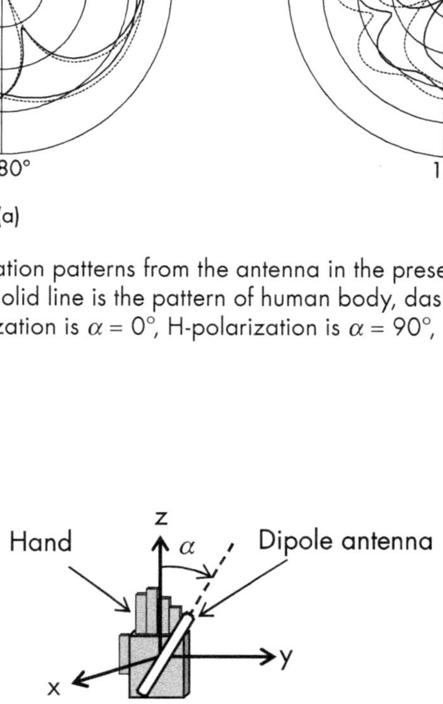

Figure 6.31 Hand and dipole antenna.

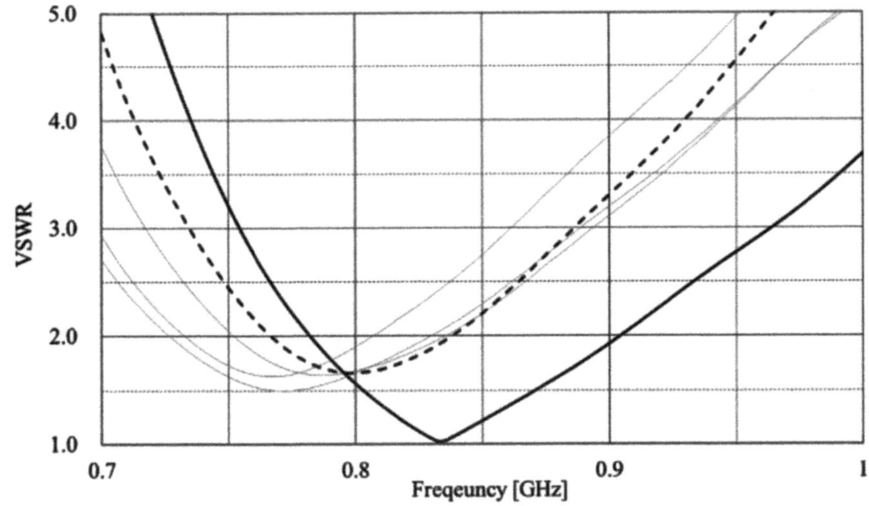

(a)

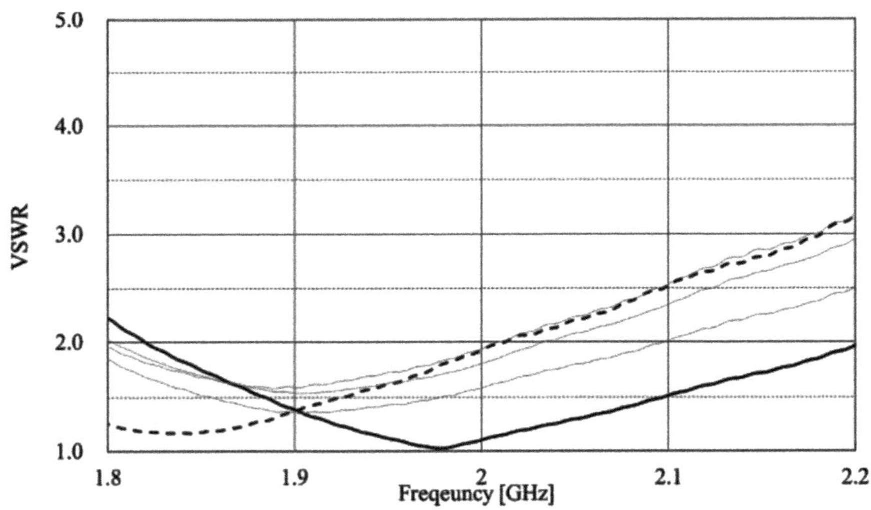

(b)

Figure 6.32 Input characteristics on effect of hand (solid line is antenna without human hand; dashed line is antenna with phantom; fine lines are typical VSWRs of test subject). (a) 940 MHz, and (b) 1900 MHz.

References

[1] Andersen, A., "Small Size 2.4 GHz PCB Antenna," Application Note AN043, https://www.ti.com/lit/an/swra117d/swra117d.pdf.

[2] Wang, L., "A Flexible Modified Dipole Antenna Printed on PET Film," *Proc. 2012 IEEE Asia-Pacific Conference on Antennas and Propagation*, Singapore, October 2012, pp. 230–240.

[3] TE Connectivity, https://www.te.com/usa-en/products/antennas/custom-antennas.html (accessed February 13, 2022).

[4] Ban, Y. L., et al., "Small-Size Multiresonant Octaband Antenna for LTE/WWAN Smartphone Applications," *IEEE Antennas and Wireless Propagation Letters*, Vol. 13, March 2014, pp. 619–622.

[5] Wong K. H., "Printed λ/8-PIFA for Internal Penta-Band Mobile Phone Antenna," *Proc. Third European Conference on Antennas and Propagation*, March 2009, pp. 539–537.

[6] Harrington, R., and J. Mautz, "Theory of Characteristic Modes for Conducting Bodies," *IEEE Trans. Antennas Propagat.*, Vol. AP-19, No. 5, September 1971, pp. 622–628.

[7] Harrington, R., and J. Mautz, "Computation of Characteristic Modes for Conducting Bodies," *IEEE Trans. Antennas Propagat.*, Vol. AP-19, No. 5, September 1971, pp. 629–639.

[8] Martens, R., E. Safin, and D. Manteuffel, "Inductive and Capacitive Excitation of the Characteristic Modes of Small Terminals," *Proc., Loughborough Antennas Propag. Conf.*, November 2011, pp. 1–4.

[9] Krishna, K. K., and S. V. Hum, "A Two-Port Chassis-Mode MIMO Antenna," *IEEE Antennas and Wireless Propagation Letters*, Vol. 12, 2013, pp. 690–693.

[10] Chen, S.-C., et al., "A Decoupling Technique for Increasing Port Isolation Between Two Strongly Coupled Antennas," *IEEE Transactions on Antennas and Propagation*, Vol. 56, No. 12, December 2008.

[11] d'Inzeo, G., "Proposal for Numerical Canonical Models in Mobile Communications," *Proc. COST 244 Meeting on Reference Models for Bioelectromagnetic Test of Mobile Communication Systems*, November 1994, pp. 1–7.

[12] Amemiya, Y., and S. Uebayashi, "The Distribution of Absorbed Power Inside a Sphere Simulating Human Head in the Near Field of a λ/2 Dipole Antenna," *Trans. IECE Japan*, J66-B, 9, 1982, pp. 1115–1122.

[13] National Library of Medicine, "The Visible Human Project," http://www.nlm.nih.gov/research/visible/visible_human.html (accessed February 13, 2022).

[14] Cook, H. F. "The Dielectric Behavior of Some Types of Human Tissues at Microwave Frequencies," *British Journal of Applied Physics*, 2, pp. 292–300, 1951.

[15] Itoh, K., "Human Models for the Research of Antennas and Propagation," *Journal of the IEICE*, Vol. J 82, No. 9, 1999, pp. 956–966.

[16] Kobayashi, T., et al, "Dry Phantom Composed of Ceramics and Its Application to SAR Estimation," *IEEE Trans. Microwave Theory and Tech.*, Vol. MTT-41, No. 1, 1993, pp. 136–140.

[17] Chan, K. W., "Microwave Phantoms," *Proc. COST 244 Meeting on Reference Models for Bio-Electromagnetic Test of Mobile Communication System*, November 1994, pp. 27–31.

[18] Yamaguchi, H., et al., "Lightweight Tissue-Equivalent Phantom for Evaluation of Antenna Performances," *Proc. Asia-Pacific Microwave Conference*, G2-3, Hong Kong, December 2008.

[19] Amano, Y., et al., "Ultra Light Carbon Phantom in RF Performance Test System for Mobile Terminals," *Proc. International Symposium on Antennas and Propagation*, Bangkok, October 2009, pp. 1131–1134.

[20] Amano, Y., et al., "Ultra Light Carbon Phantom Hand for RF Performance Evaluation of Mobile Terminals," *Proc. Asia-Pacific Microwave Conference*, TU4H-2 (#1485), Singapore, December 2009.

7

MEASUREMENT OF BASE STATION AND MOBILE TERMINAL ANTENNAS

This chapter describes practical measurement methods for BS and mobile terminal antennas. Open site and near-field measurements for the BS antennas are described as low power tests. Passive intermodulation (PIM) should be avoided in high-power BS antenna operations to avoid disturbing the frequency spectrum, and the theory and measurement of PIM are explained. In mobile terminal measurements, the over the air (OTA) test is used to measure evaluation parameters of built-in antennas that do not have a test port to separate the antenna from the transceiver. Mobile terminals are requested to avoid the interference of electromagnetic (EM) waves and are tested in terms of EM compatibility (EMC). The exposure to EM wave radiation should be maintained below the level defined by the specific absorption ratio (SAR) or the power density (PD) so that it does not harm the human body while using the mobile terminal. Finally, EMC and SAR/PD measurements are described.

7.1 BASE STATION ANTENNA MEASUREMENTS

BS antenna measurements are classified by the level of handing power. In low power tests, the antenna radiation pattern and input impedance are measured. BS antenna pattern measurement is performed in an open site or near-field measurement when the antenna has a high gain. The beam width and the position of the null point should be verified for the designed coverage area of the cell; then, this section presents the measurement method for high-gain antennas that cannot be measured in an anechoic chamber. PIM causes serious problems, especially when the BS antenna is operated at high-power levels. This section describes the definition of PIM and its measurement method.

7.1.1 Antenna Pattern Measurements in Open Sites

The distance r between the transmitting and receiving antennas should satisfy the condition $r \geq 2D^2/\lambda$, defined by (3.18), to measure the far-field radiation, as presented in Chapter 3, where the maximum dimension of the antenna is denoted as D. For a cell radius of less than 3 km, the vertical radiation pattern of the BS antenna can be controlled by electronically adjusting the phase of the array elements. Since the length of the BS antenna becomes greater than 10λ, and the measurement distance r requires more than 200λ as a far-field condition, and in the 900-MHz band, a distance greater than 67m is necessary, which is difficult to obtain in an indoor facility.

To satisfy this far-field condition, the transmitting antenna on a steel tower or building should be installed, as shown in Figure 7.1. For radiation pattern measurement, it is necessary to prevent interference from the radio link that is in use and to suppress reflections from surrounding buildings. To avoid receiving waves reflected from the ground, the antenna size D should satisfy the following equation, where h_r represents the height of the test antenna:

$$D \geq \frac{1}{2} \times 0.886 \frac{\lambda r}{2h_r} \tag{7.1}$$

This equation is derived so that the angle of ground reflection θ_t is larger than the half-power beamwidth θ_h ($\theta_t > \theta_h$), where the angle θ_t is approximated by $\theta_t = 2h_r/r$. If the test antenna is significantly greater than the wavelength, the half-power beamwidth $2\theta_h$ of the radiation pattern can be approximated by the following equation:

$$2\theta_h \cong 0.886 \frac{\lambda}{D} \tag{7.2}$$

Equation (7.1) is derived by $\theta_t > \theta_h$ and (7.2).

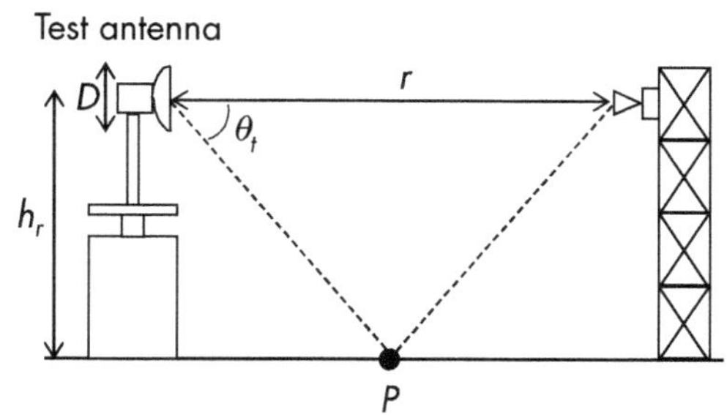

Figure 7.1 Far-field measurement using a tall tower.

When the required distance is unavailable for measurements, a diffraction fence may be installed on the reflection position P in Figure 7.1 to suppress the influence of the reflected waves. The diffraction fence should be mounted so that it does not interrupt the main beam between the transmitting and receiving antennas, and a saw-shape fence is used to provide effective diffraction.

Figure 7.2 shows a measurement method using ground reflection. The height h_t of the receiving antenna near the ground is adjusted to have a phase difference of $\lambda/2$ between the direct and reflection paths to maximize the received electric field strength because the reflection coefficient at the ground is assumed to be $\Gamma = -1$. The height h_t is given as:

$$h_t \cong \frac{\lambda r}{4h_r} \tag{7.3}$$

Figure 7.3 shows the slant range measurement to remove the effect from the ground reflection. The test antenna is installed on

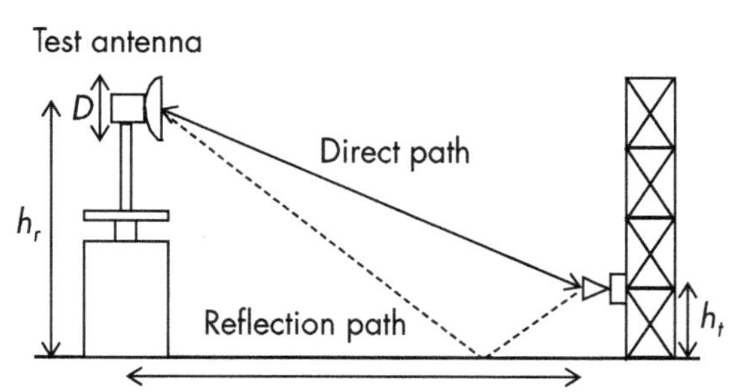

Figure 7.2 Far-field measurement using ground reflection.

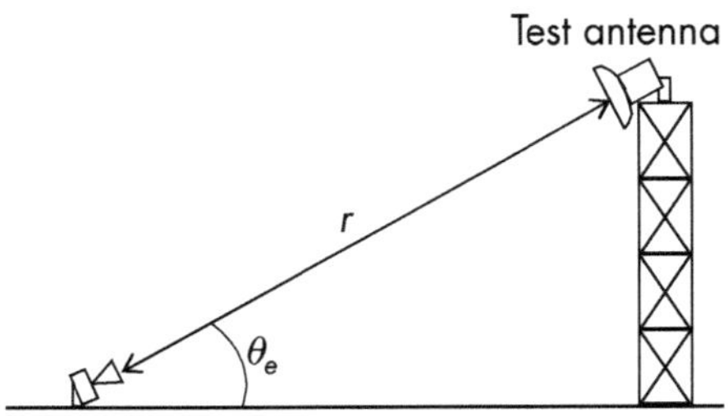

Figure 7.3 Far-field measurement using slant range.

the top of the tower, and the measurement antenna on the ground is pointed toward it with a large elevation angle θ_e. This is an effective method for determining the direct path component in the measurement.

7.1.2 Near-Field and Very-Near-Field Measurements

The near-field measurement is an effective method for characterizing the far-field of BS antennas in an anechoic chamber. The cylindrical near-field measurement is the most commonly adopted technique for a BS antenna with a fan beam pattern. The theory of near-field measurement was described in Section 3.2.3, and the distance r from an antenna under test (AUT) to a probe (Figure 7.4(a)) should be placed in the Fresnel region defined in Section 3.2.1. The probe scan range along the z axis often takes very long to realize sufficient dynamic range, which requires a large measurement facility inside an anechoic chamber.

Multiprobe measurement is one of the simple techniques for resolving this problem by reducing measurement and calculation times [1]. Another technique is the single cut near-field to far-field transformation [2]. This section presents near-field and very-near-field measurements for quickly obtaining the far-field of an AUT in a compact anechoic chamber.

The very-near-field measurement is based on an equivalent line electric current using a small loop probe to minimize disturbance of the EM field in the vicinity of the AUT. The equivalent currents are measured on a line in front of and in a circle around the AUT. The far-field pattern of each plane is estimated from the linear and circular distributions by applying the one-dimensional integral

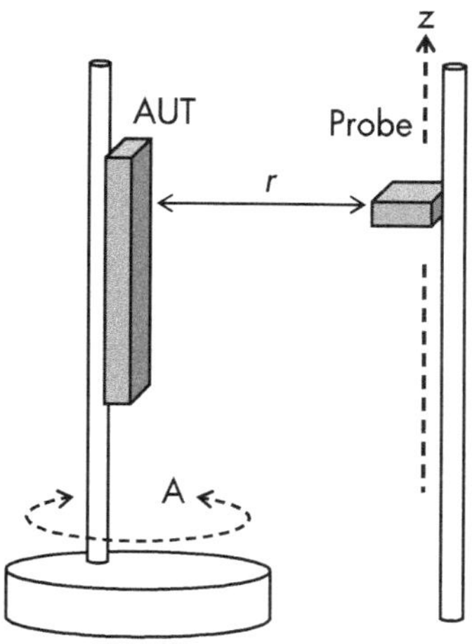

(a)

(b)

Figure 7.4 Cylindrical near field measurement: (a) geometry of measurement, and (b) photo of measurement setup.

equation, and the measurement time and the equipment can be kept minimal. The spherical far field is approximated from the product of the far field of each cut plane, and its maximum directivity can be estimated.

The equivalent current method regards the equivalent current on each measurement point as a small electric dipole, and the radiation field from the dipole is calculated [3]. The far field is calculated by integrating the radiation field from each dipole. Although each dipole is considered an infinitesimal dipole of zero length, the far field in the vertical plane is calculated similarly as a linear array antenna as follows:

$$E_\theta(\theta) = Z_o \sin\theta \sum_{m=1}^{M} J_z(m) e^{jk_o z_m \cos\theta} \tag{7.4}$$

$$E_\phi(\theta) = Z_o \sum_{m=1}^{M} J_y(m) e^{jk_o z_m \cos\theta} \tag{7.5}$$

where Z_0 represents the characteristic impedance in free space, subscription means a polarization, $J_z(m)$ and $J_y(m)$ is the acquired complex current, z_m represents its position on the z direction, and k_0 represents the wavenumber in free space. To measure in as compact an area as possible, the measurement length of the linear current distribution is assumed to be equal to the length of the AUT.

The far field in the horizontal plane is calculated from the circular current distribution around the AUT. Assuming that the distance from the AUT center to the circular scan line is farther than the wavelength, which is sufficient, and compared with the far-field criterion of an infinitesimal dipole, the horizontal plane far field is given as follows:

$$E_\theta(\phi) = Z_o \sum_{n=1}^{N} \left(1 + \hat{\phi} \cdot \hat{\phi}_n\right) J_z(n) e^{jk_o d_\phi} \tag{7.6}$$

$$E_\phi(\phi) = Z_o \sum_{n=1}^{N} \left(1 + \hat{\phi} \cdot \hat{\phi}_n\right) J_\phi(n) e^{jk_o d_\phi} \tag{7.7}$$

$$d = x_n \cos\phi + y_n \sin\phi \tag{7.8}$$

where N represents the total number of equivalent currents on the circle; x_n and y_n represent the coordinate points on the x and y axes, respectively; and, ϕ_n represents the azimuth angle of each equivalent current. $\hat{\phi}$ represents the unit vector of ϕ direction, and $\hat{\phi}_n$ represents the position vector of each sampling point [4].

By using the orthogonal cut planes obtained as described earlier, the spherical far-field pattern is approximated as:

$$E_p(\theta,\phi) = E_p(\theta,\phi = 0°)E_p(\theta = 0°,\phi), \quad p = \theta,\phi \tag{7.9}$$

where $E_p(\theta,\phi)$ is the virtually calculated full-sphere far-field, $E_p(\theta,\phi = 0°)$ is the far-field in the vertical plane according to (7.4) and (7.5), and $E_p(\theta = 0°,\phi)$ is far-field in the horizontal plane by (7.6) and (7.7). The horizontal far-field pattern of the electrically long antenna does not significantly change, depending on the change of the zenith angle.

The horizontal cut surface pattern for arbitrary θ can be approximated using a single horizontal cut plane pattern of $\phi = 90°$.

Figure 7.5 shows a measurement overview of a BS antenna in an anechoic chamber. The outer diameter of a small loop probe is 12 mm and is electrically sufficiently small to neglect the correction of the probe radiation pattern. The probe slides in the z direction, and the AUT rotates in the xy plane. The AUT is configured by eight radiation elements placed on a co-linear line, and the polarization is slanted at 45°. The scanning probe is tilted at 45° from the vertical axis of the AUT. Table 7.1 shows the measurement specifications of the current distribution.

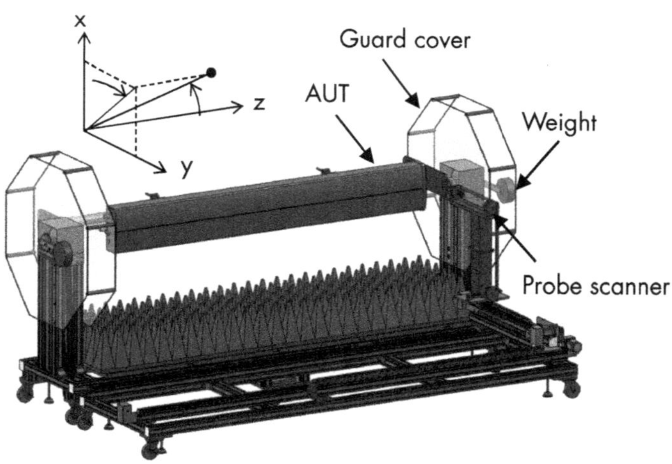

Figure 7.5 Cylindrical very-near-field measurement. (Courtesy of Nazca Industrial Technology.)

Table 7.1
Measurement Specifications

1-D measurement	Length	2155 mm (6.90λ)
	Probe distance	100 mm (0.32λ)
	Sampling interval	33 mm 0.11λ)
Circular measurement	Radius	660 mm (2.11λ)
	Sampling step	2° (0.07λ)
Fresnel region	> 3500 mm	
Far-field region	> 29700 mm	

Figure 7.6 shows the estimation results from the measured data, where the measurement frequency is at 0.79 GHz. The reference far-field patterns by near field to far field transformation using cylindrical measurement agree well with those by very-near-field measurement, and the directivity error is less than 0.4 dB [5].

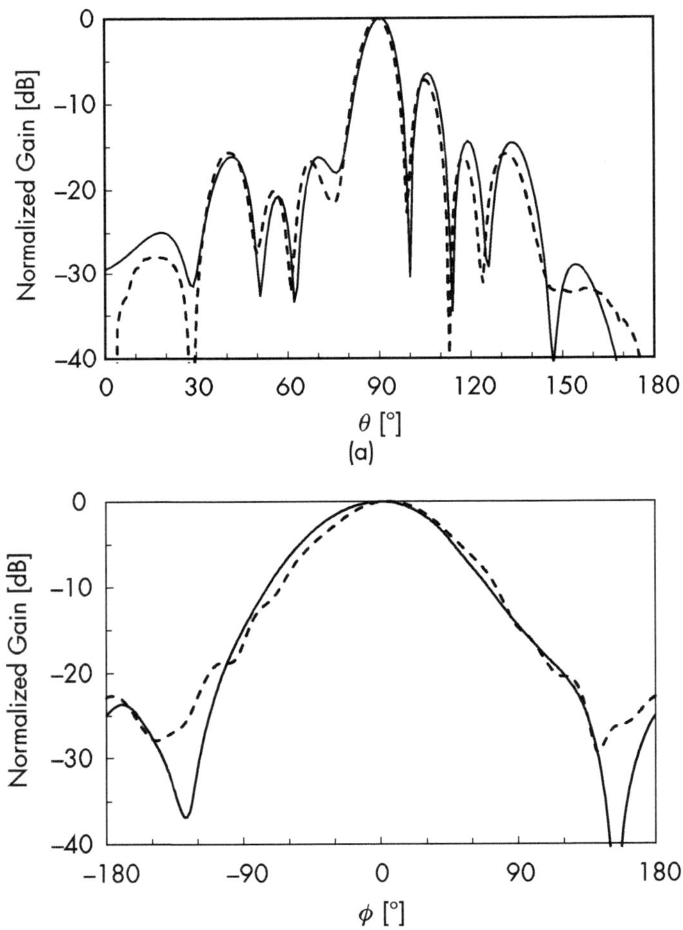

Figure 7.6 Measured radiation pattern (solid line: near field; dotted line: very near field): (a) vertical plane, and (b) horizontal plane.

7.1.3 Passive Intermodulation

PIM is a nonlinear phenomenon in which an undesired frequency signal is generated by combining several different frequency inputs in a circuit with nonlinear input-output characteristics [6]. Although an antenna is a passive device, very weak electric discharge phenomena occur at the junctions of the antenna parts during high-power operation. Such discharges often appear at the surface of the junctions made of different metals.

The BS antenna is used for both transmission and reception, and the transmit power level is several watts per channel. The uplink received electric field strength is between 10 and 60 dBμV, which is equivalent to −103 to −53 dBm in the 50Ω transmission line system. This power imbalance becomes greater than 140 dB with multichannel input to BS antennas and causes the PIM in the passive device. Most current cellular systems are duplex, with the transmitting and receiving frequency bands allocated at a certain interval, and the system uses both frequency bands simultaneously. When the PIM occurs at the transmitting antenna, it appears on the receive frequency band, and the receive channel cannot be used because of interference.

The PIM phenomenon causes the antenna to have a nonlinear input-output response that may be regarded as a two-port circuit with transmitter input and free space output.

$$Y = a_0 + a_1X + a_2X^2 + a_3X^3 + \cdots = \sum_{k=0}^{\infty} a_kX^k \tag{7.10}$$

If the two input signals have different angular frequencies, $\omega_1 = 2\pi f_1$ and $\omega_2 = 2\pi f_2$ as:

$$X_1 = A_1 \cos\omega_1 t, \ X_2 = A_2 \cos\left(\omega_2 t + \phi\right) \tag{7.11}$$

where A_1 and A_2 are the respective amplitudes of the two signals and the phase angle ϕ of X_2 is normalized with respect to that of X_1. Substituting (7.11) into (7.10), the components of the nonlinear output signals may be obtained. The angular frequency components appearing at the output port is expressed as follows: for terms up to X_i^3, for example, (a) ω_1 and ω_2 from the term for X^1, (b) $2\omega_1$, $2\omega_2$ from the term for X^2, and (c) ω_1, ω_2, $3\omega_1$, and $3\omega_2$ from the term for X^3. Taking the M-signal input for X, each power series from the output of the circuit with the input-output response of (7.10) is expressed as (7.12) by denoting each input signal as $X_i = A_i \cos\omega_i t$:

$$X^k = \left(\sum_{m=1}^{M} X_m\right)^k = \sum_{n_1,\cdots M} \frac{k!}{n_1!\cdots n_M!} X_1^{n_1} X_2^{n_2} \cdots X_M^{n_M} \tag{7.12}$$

where n_i represents an integer given by the expression of $k = n_1 + n_2 + \cdots + n_M$, in which the sum of all combinations of n_i is included. Then, the angular frequency ω_k of the PIM signal corresponding to this nonlinear response is given as:

$$\omega_k = n_1\omega_1 \pm n_2\omega_2 \pm \cdots n_k\omega_k \tag{7.13}$$

The parameter k is the order of PIM. The PIM signals for odd orders higher than the third cause severe problems in current cellular systems. The interference condition for the PIM can be rewritten by denoting p as a natural number as:

$$\omega_{PIM} = \omega_2 + p(\omega_2 - \omega_1), \quad \omega_{PIM} = \omega_1 + p(\omega_1 - \omega_2) \tag{7.14}$$

The order of $2p \pm 1$ frequency causes the most severe problems in downlink frequency because the odd order PIM frequency is close to the input frequencies.

Figure 7.7 shows the allocation of uplink and downlink frequency bands for cellular systems with upper and lower limit edge frequencies of f_1 and f_2, respectively. The frequency interval should be allocated so that it is unaffected by the third- and fifth-order PIM, as shown in Figure 7.7.

PIM signals are expressed as an absolute value in dBm and a relative value in dBc, normalized by the input signal. For example, for a PIM level of −100 dBm, the relative PIM value is −143 dBc, normalized by two input signals of 43 dBm.

Figure 7.8 shows the block diagram of the PIM measurement setup. The system comprises three units connected to three ports of the duplexer operating: (1) the transmit port, (2) the reception port,

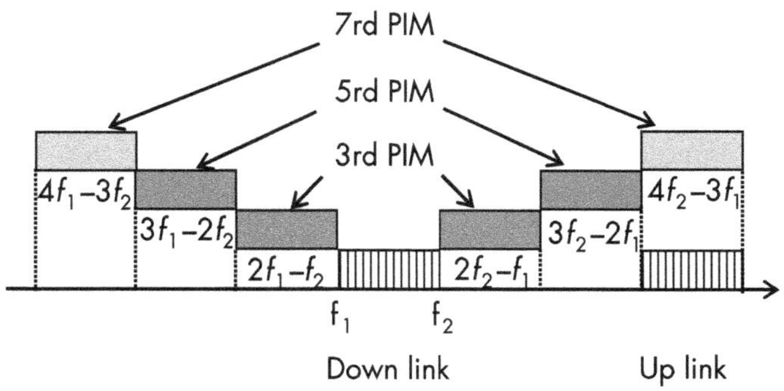

Figure 7.7 Frequency allocation for higher order PIM.

and (3) the common port for both transmission and reception. After amplification to the required power level, the two continuous signals P_1 and P_2, at different frequencies, are input to the transmit port of the duplexer through the power combiner. The device under test (DUT) is connected to the common port for both transmission and reception. When the DUT is used as a connector and a transmission cable, power is dissipated in a dummy load. When the DUT is an antenna, the AUT is set up inside an anechoic chamber.

With this measurement setup, the PIM signal detected at the reception port is measured by a spectrum analyzer. If the PIM signal level is less than the noise level of the spectrum analyzer, a low noise amplifier (LNA) is inserted to adjust the receiving level. By inserting a band-pass filter before the LNA, direct coupling from the transmitting port to the receiving port is eliminated. The band-pass filter

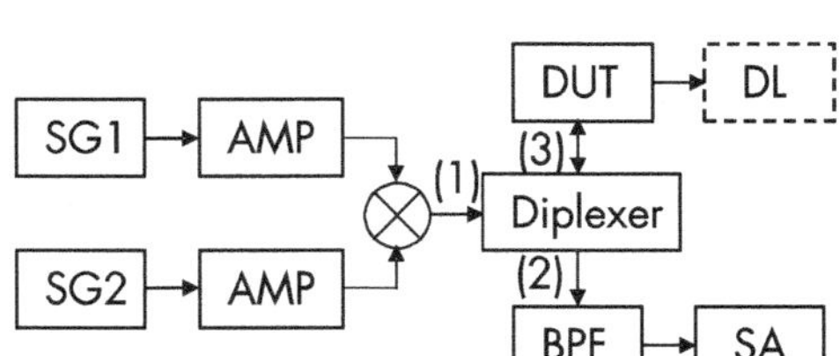

Figure 7.8 Block diagram of setup for antenna PIM measurement. SG: signal generator, AMP: amplifier, DUT: device under test, DL: dummy load, BPF: band-path filter, SA: spectrum analyzer.

suppresses the undesired PIM occurring in the LNA and spectrum analyzer. When the DUT is a connector and a transmission cable, a terminating load is used, as shown in Figure 7.8. To avoid the PIM in a dummy load, a long coaxial cable is frequently used instead of a resistor. A coaxial cable with a length of 50–100m is typically sufficiently long to serve as the dummy load. This long cable then dissipates most of the power. After confirming that there is no PIM excitation at the interconnection between the duplexer and DUT, the PIM is measured using the setup described earlier.

During PIM measurements on antennas, metallic objects should not be located near the AUT to avoid the occurrence of PIM by induced current excitation. The PIM for antennas should be measured inside an anechoic chamber even though measurements can also be made in open sites, where the main beam of the antenna is directed toward the sky. PIM phenomena may occur at several points, but PIM signals may be difficult to observe because of phase cancelation by each other. To avoid this phase cancelation, the PIM should be measured at several different frequencies.

7.2 MOBILE TERMINAL MEASUREMENT

This section presents measurement methods for evaluating mobile terminals. Transmission and reception tests are conducted using a test connector to separate a transceiver from an antenna in the conventional measurement. Mobile terminals with no test connector must be evaluated by the OTA tests. The principle of OTA measurement and the definition of measurement parameters are described in this section. OTA measurements provide a static performance of mobile terminals, whereas MIMO performance is tested in multipath environments. This section presents a spatial fading emulator to reproduce artificial multipath fading.

7.2.1 Measurement Parameters

Radiation power level is characterized by total radiation power (TRP) and effective isotropic radiated power (EIRP) in the transmit mode of the DUT. Figure 7.9 shows the coordinate system of the DUT measurement. An observation point P should be located in the far-field region determined by the DUT size, as defined in Section 3.2.1. Denoting electric field components at point P as $E_{f\theta}$, $E_{f\phi}$, TRP is given by integrating all radiated power as:

$$TRP = \frac{1}{2Z_o}\int_0^{2\pi}\int_0^{\pi}\left\{\left|E_{f\theta}(\theta,\phi)\right|^2 + \left|E_{f\phi}(\theta,\phi)\right|^2\right\} r^2 \sin\theta \, d\theta \, d\phi$$
$$= \frac{1}{2Z_o}\sum_{m=1}^{M}\sum_{n=1}^{N}\left\{\left|E_{f\theta}\left(\theta_{mn},\phi_{mn}\right)\right|^2 + \left|E_{f\phi}\left(\theta_{mn},\phi_{mn}\right)\right|^2\right\} r^2 \sin\theta_{mn} \frac{\theta}{N}\frac{\phi}{M} \tag{7.15}$$

where Z_o represents the characteristic impedance in free space. In the measurement, observation points are discretized, and M and N represent the number of sampling data points in the ϕ and θ directions, respectively. To suppress the error in TRP less than 0.1 dB, it is recommended that θ/N and ϕ/M should be smaller than 15° [7].

The maximum radiation level is determined by EIRP in the regulation. EIRP is defined as the power radiated by an isotropic antenna to achieve an identical power level in the direction of maximum radiation (θ_o, ϕ_o) of a DUT as:

$$EIRP = \frac{4\pi r^2}{2Z_o}\left\{\left|E_{f\theta}\left(\theta_o,\phi_o\right)\right|^2 + \left|E_{f\phi}\left(\theta_o,\phi_o\right)\right|^2\right\} \tag{7.16}$$

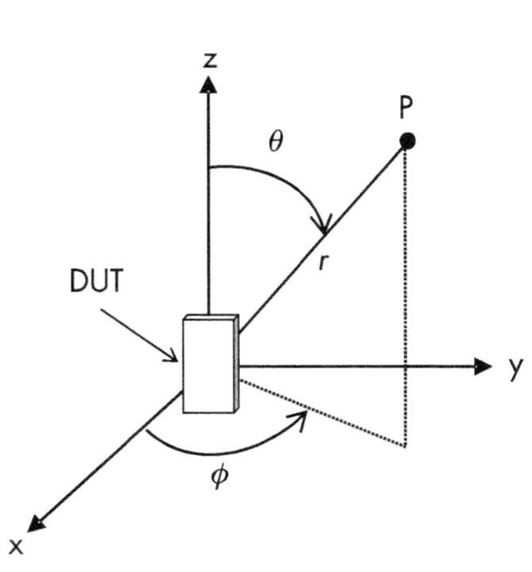

Figure 7.9 Coordinate system in measurement.

Introducing the isotropic antenna gain of the DUT (G_i) defined in (1.36), (7.16) is rewritten using TRP as:

$$EIRP = TRP \times G_i \tag{7.17}$$

TRP and EIRP are parameters to characterize the power level of DUTs in the transmit mode. A measure of the average sensitivity of a DUT is given by averaging over the entire space as total isotropic sensitivity (TIS) as:

$$TIS = \frac{4\pi}{\int_0^{2\pi}\int_0^{\pi}\left(\frac{1}{EIS_\theta(\theta,\phi)} + \frac{1}{EIS_\phi(\theta,\phi)}\right)\sin\theta \, d\theta \, d\phi} \tag{7.18}$$

where $EIS_i(\theta, \phi)$ ($i = \theta, \phi$) is the minimum power in reception mode for a fixed angle (θ, ϕ) and for a fixed polarization component denoted as i. EIS often becomes very small in the integral of (7.18), and the harmonic mean is used to eliminate the singularity in the calculation.

The minimum power is given by lowering the illuminated power to the DUT until the bit error ratio (BER) reaches the threshold. The BER determines the quality of a digital communication system. The BER is calculated using the error ratio of the received sequence of bits to the transmitted ones. Another quality factor is defined by error vector magnitude. An ideal signal sent by a transceiver has all constellation points precisely at the ideal location, as shown in Figure 7.10.

Actual constellation points deviated by noise factors are evaluated by how far the points are from the ideal locations as:

$$EVM = \frac{1}{N}\sum_{k=1}^{N}\sqrt{\frac{\left|\boldsymbol{P}_{ek}\right|}{\left|\boldsymbol{P}_{ik}\right|}} \tag{7.19}$$

where N represents the number of symbols and $\boldsymbol{P}_{ik}$ and $\boldsymbol{P}_{ek}$ represent a symbol and error vector, respectively.

7.2.2 OTA Measurement

The OTA measurement requires a two-axis positioner to scan a probe antenna in an entire space. Figure 7.11 shows a conical cut method,

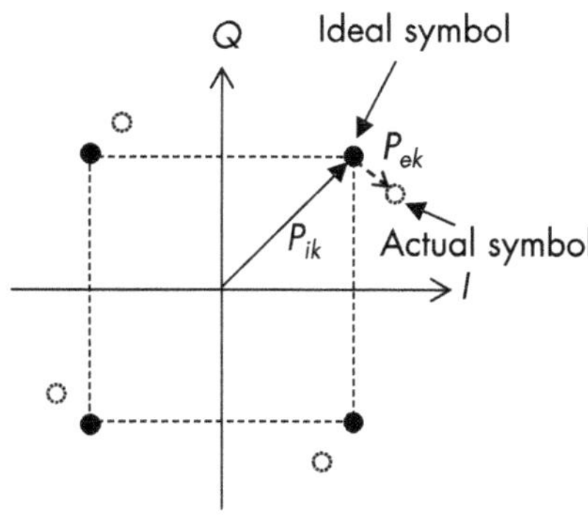

Figure 7.10 Constellation diagram and EVM.

where the receiving probe is moved in a hemispherical pattern. A test antenna is stationary, whereas the probe moves on a semicircle to measure the fields (i.e., the elevation angle θ is changed from 0° to 180°). This measurement gives a single cut in the elevation plane, and then the test antenna is rotated by $\Delta\phi$ in the azimuth plane to obtain the second step data. This measurement procedure is repeated to rotate the azimuth angle from 0° to 360°.

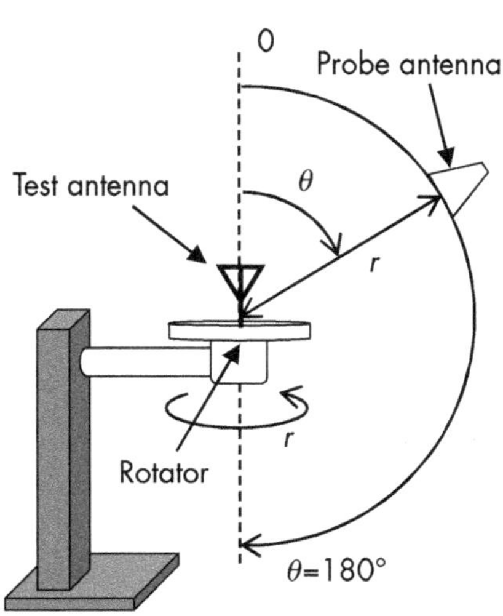

Figure 7.11 Conical cut method.

Another probe scanning is given by a great circle method, as shown in Figure 7.12. The probe antenna is fixed on the pole top, and the test antenna is mounted on the pole at the rotator disk edge. The test antenna is rotating for azimuth scan ($0° \le \phi \le 360°$), and the rotator provides the elevation scan ($0° \le \theta \le 180°$).

The Cellular Telecommunications and Internet Association recommends a sampling interval of 15° in the entire space measurement [8]. The number of divisions in the θ and ϕ axes is $N_\theta = 12$, $N_\phi = 24$, and the total number of sampling points is $N = (N_\theta - 1), N_\phi = 264$ by the equiangle method. The sampling interval has a weight factor of $\sin\theta$, which provides nonuniform sampling, as shown in Figure 7.13(a), and the sampling interval is very small near the top and bottom poles. An equal sampling interval along the vertical axis is obtained by the modified equiangle method in the elevation angle as:

$$\theta_i = \cos^{-1}\left\{1 - (i-1)\frac{2}{N_z}\right\} \tag{7.20}$$

where N_z represents the number of divisions in the vertical axis z. Figure 7.13(b) depicts the sampling points of the modified equiangle method. The errors of the modified method in TRP given by (7.15) are almost the same as the equispace angle method, and the total number of sampling data is not reduced.

Uniform weight factors for both the azimuth and elevation angles are given by the generalized helix method, as shown in Figure 7.13(c) [9]. The sampling points are plotted along a helix line on the sphere with an equal interval in this method as follows:

$$h_k = 1 - 2\frac{k-1}{N-1},\ \theta_k = \cos^{-1} h_k,\ \phi_k = \phi_{k-1} + \frac{3.6}{\sqrt{N}}\frac{1}{\sqrt{N - h_k^2}} \tag{7.21}$$

where N represents the total number of sampling points and k represents an integer greater than 1. The total sampling data can be reduced by 25% to maintain the same accuracy as the equispace method [7].

Mechanical probe scanning takes much time in measurement. The probe scanning in the elevation angle is replaced by a probe array, as shown in Figure 7.14 [10]. Received signals of the probe array

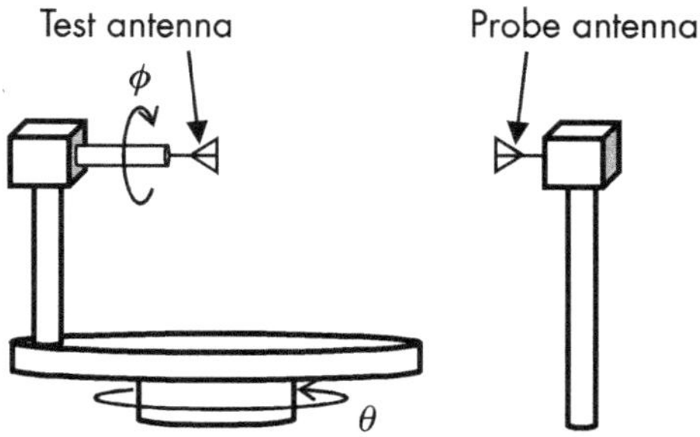

Figure 7.12 Great circle method.

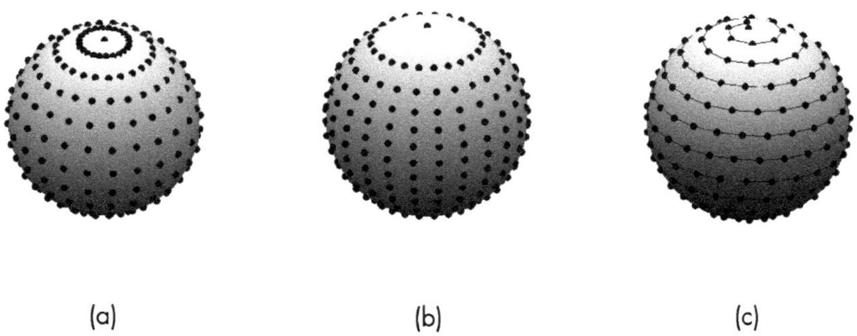

Figure 7.13 Sampling points on spherical surface: (a) equispace angle, (b) modified equiangle, (c) generalized spiral.

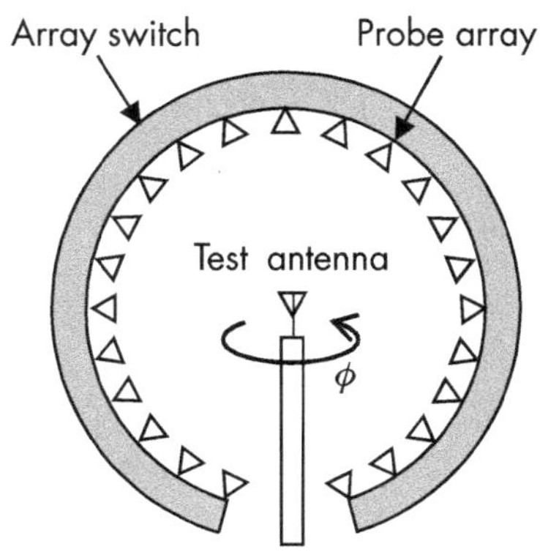

Figure 7.14 Conical cut method with probe array.

are detected by switching probe elements, and scanning in the elevation angle is performed quickly. The azimuth scanning is performed by rotating the test antenna; however, it reduces the measurement time by approximately $1/N_\theta$.

OTA measurements of TRP and TIS are affected by the radiation pattern of DUT, and pattern measurements in an entire space are required. To remove these effects, TRP and TIS measurements are carried out under an artificial fading environment described in Section 8.3.3. The generation of a random field can eliminate the effect of different radiation patterns in the antenna of a DUT, which reduces the measurement time.

Beam steering antennas are used to overcome large propagation loss in both BS and mobile terminals at millimeter frequency bands. The narrow beamwidth of a beam steering antenna necessitates a large scanning angle to steer the beam to create an optimal link between transmission and reception antennas. The range of solid angles of the mobile terminal should be spherical coverage.

EIRP toward a specific direction is unsuitable for characterizing the beam steering capability of a DUT, so the 3GPP 5G specification provides a measure of the uplink spherical coverage of DUT by the cumulative distribution function (CDF) of EIRP [11, 12]. The definition of the CDF is described in Section 2.7. The CDF of EIRP is given by the following procedure. The EIRP at a specific direction (θ_i, ϕ_i) is measured by steering the beam of DUT in all solid angles. Then, the EIRP of the measurement of other directions is repeated by changing the direction to cover the entire space. In each direction, the beam of the DUT is steered. The measured EIRP data are plotted as a CDF curve, and the EIRP is determined by a value less than or equal to a threshold probability value (Figure 7.15).

7.2.3 MIMO Performance Measurement

OTA measurements provide the static performance of mobile terminals. Diversity and MIMO performance must be tested in multipath environments. This section presents a spatial fading emulator based on Clarke's model [13] dedicated for performance evaluation of mobile terminals in multipath propagation environments reproduced by artificial fading in an anechoic chamber.

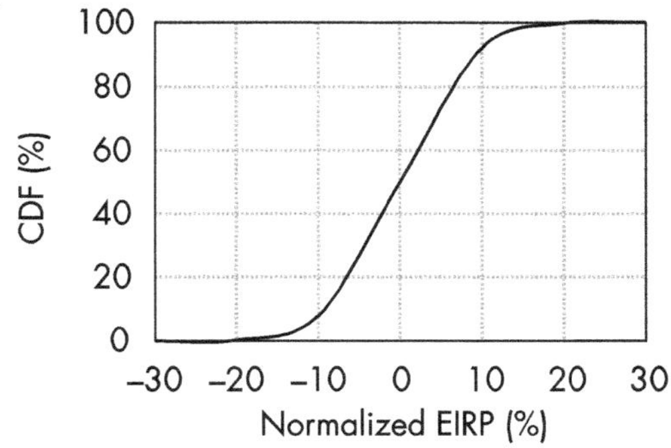

Figure 7.15 CDF of normalized EIRP.

The emulator consists of N-scatters located on the circumference of the ring at an equal interval, and the amplitude and phase of each wave radiated from scatters are controlled by phase shifters and attenuators to emulate a multipath propagation environment such as Rayleigh or Rician fading channel in the vicinity of its center [14].

Figure 7.16 shows an example of the emulator. N half-wavelength dipole antennas used as scatters are fixed on the top of resin pillars, which are located on the circumference with a radius of r in the

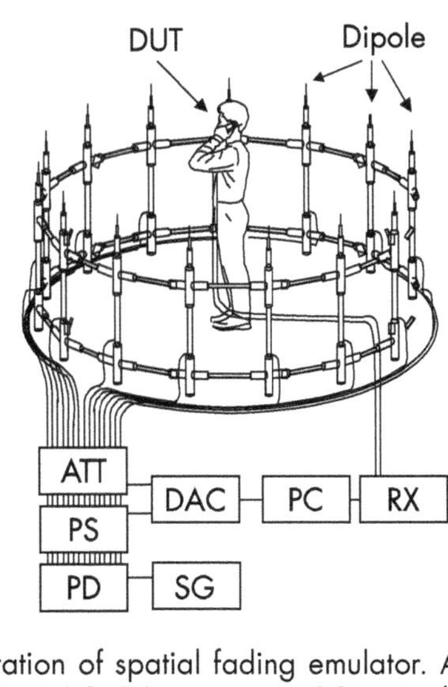

Figure 7.16 Configuration of spatial fading emulator. ATT: attenuator, PS: phase shifter, PD: power divider, DAC: DA convector, SG: signal generator, PC: computer, RX: receiver.

same interval. An operator holding a DUT in a talk or viewer position is located at the center of *N*-scatters. A transmitted wave from a signal generator is separated into *N* subchannels by a power divider and then radiated from scatters through phase shifters and attenuators. The amplitude and phase of signals radiated from scatters are controlled using attenuators and phase shifters operated by a digital-to-analog converter to emulate fading channels.

The transmitted signal forms the ith scatter, which is given as:

$$p_i(t) = 2\pi t f_D \cos(\phi_D - \phi_i) + \alpha_i \tag{7.22}$$

where ϕ_D is the angle between the 1st and i^{th} scatter, ϕ_i is the virtual moving direction of the operator, α_i is given by random numbers in the initial phase of each scatter, and the maximum Doppler frequency $f_D = v/\lambda$ is given by the mobile speed of the operator v and the wavelength of the carrier frequency λ.

Figure 7.17 shows the coordinate system. The received signal at the ring center is given by (7.23), assuming an omnidirectional radiation pattern in the horizontal plane for the mobile terminal as [15]:

$$P(t) = \sum_{i=1}^{L} \frac{\lambda}{4\pi r} \Omega(\phi_i) \exp\left[-j\{kr + p_i(t)\}\right] \tag{7.23}$$

where k represents the wave number in free space. To control the angular distribution of incoming waves at the handset, the amplitude coefficient $\Omega(\phi_i)$ is adjusted to satisfy the following equations:

$$\Omega(\phi_i) = 1 \tag{7.24}$$

$$\Omega(\phi_i) = \frac{1}{2\sigma} \exp\left\{-\frac{|\phi_i - \mu|}{\sigma}\right\} \tag{7.25}$$

where (7.24) represents a uniform angular distribution and (7.25) is the Laplace distribution with a central angle μ of the incoming wave and an angular spread σ.

This emulator provides Rayleigh fading based on Clarke's model and spatial correlation characteristics within a branch separation of

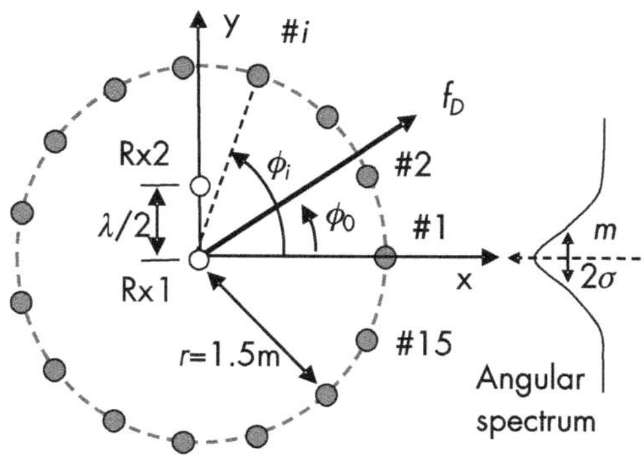

Figure 7.17 Arrangement of scatters and receiving antenna.

1.7λ for 15 scatters with a ring radius of $r = 10.7\lambda$ for vertical polarization [15].

A single-input single-output received signal in (7.23) is extended to MIMO transmission signals by using several signals in this emulator. The spatial correlation between transmitted signals is controlled by the initial phase α_i, and a channel response h_{nm} is given for N terminal antennas and M BS antennas.

$$h_{nm} = \sum_{i=1}^{L} E_n \sqrt{\Omega(\phi_i)} \frac{\lambda}{4\pi r} \exp\left[-j\left\{kr + 2\pi t f_D \cos(\phi_o - \phi_i) + \alpha_{mi}\right\}\right] \tag{7.26}$$

where L represents the number of scatters, E_n represents the complex radiation pattern of the nth terminal antenna, and α_{mi} is the initial phase of the ith dipole antenna, given as m^{th} BS antenna [16]. This MIMO emulator can control the angular distribution of incoming waves and provide almost equivalent channel capacity with outdoor propagation measurement.

7.3 EMC AND SAR/PD MEASUREMENT

EMC measurements are used to ensure that electric devices do not interfere with other devices by emitting EM waves or that they do not interfere with the illumination from other devices. EM interference

(EMI) assessment is used to reduce the emission of EM waves by devices to a specified level, and EM susceptibility assessment is used to ensure the operation of devices in the presence of EM wave illumination from other devices. This section focuses on EMI assessment to measure EM wave emission in space over 30 MHz; however, measurements of conducted interference on power lines of devices below 30 MHz are not presented here. The first part describes EMI measurements using an open site direct wave method. In this measurement, both a large measurement area and a long measurement time are required. To reduce both the space and time required in measurements, several measurement systems suitable for use in confined spaces are proposed. The second part of the section describes EMC evaluation procedures using compact anechoic boxes or chambers. Finally, human body exposure to EM wave radiation from devices is presented as the SAR/PD measurement.

7.3.1 Open Site EMI Measurement

All electronic equipment radiates unwanted EM waves and potentially causes interference to other devices. Each country has a standard for the acceptable level of EMI from DUT emission. These national standards are based on the International Special Committee on Radio Interference published regulatory standards [17].

Figure 7.18(a) shows a standard open site measurement setup, and Figure 7.18(b) shows a large chamber. In this system, the direct wave from the DUT and the wave reflected from the conducting ground plane are measured by the receiving antenna.

Depending on the type of DUT under test, the distance D from the DUT to the receiving antenna is 3, 10, or 30m. There are no other spurious reflections from scattering objects inside the measurement area, which is $2D$ in length and $\sqrt{3}D$ in width, for the open site test facility. The site performance is evaluated by the site attenuation factor, which is similar to the propagation loss described in Section 2.2. When the transmit cable is directly connected to the DUT antenna port and the receiving voltage is denoted as V_d, the height of the transmitting antenna is fixed at H and the receiving antenna height, h, is changed until the maximum receiving voltage V_r is found. The site attenuation factor is then defined as:

$$SA = V_d + V_r - (\alpha_t + \alpha_r) \tag{7.27}$$

where α_t and α_r represent the feeding losses at the transmitting and receiving antennas. All measurements are in decibel units. An acceptable error for the site attenuation is ±4 dB compared with the expected theoretical value [18]. This error value indicates the difference between an ideal site and a real test site. To simplify the measurement procedure, the DUT is mounted on a rotating table, thereby

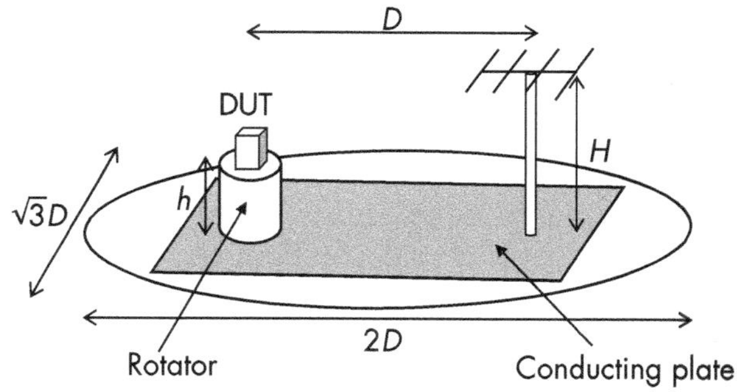

(a)

(b)

Figure 7.18 EMC measurement site: (a) open site, and (b) EMC chamber. (Courtesy of Microwave Factory.)

enabling the field radiated by the DUT to be measured over a 360° circle in the azimuth plane.

To obtain clear interference between the direct and reflected waves from the DUT, the receiving antenna height (H) is adjusted until the maximum receiving voltage V_m is obtained. For a distance within the range $D = 3$–6m, H is in the range of 1–4m, whereas, for a distance $D = 30$m, H is found to be between 2 and 6m.

The azimuth angle of the DUT is also adjusted until a maximum V_m is found. For this situation, the incoming electric field level E is calculated using the antenna factor (AF) as:

$$E = AF \times V_m \tag{7.28}$$

The AF of a half-wavelength dipole antenna is equivalent to the effective length λ/π of the antenna. However, cable and feed connector losses should be included in the measurement. In actual measurement systems, AF is evaluated by comparing the antenna under measurement with a calibrated standard antenna. AF is then given in terms of the calibrated antenna factor AF_o and the received voltage V_o for the standard antenna as:

$$AF = \frac{AF_o \times V_o}{V_m} \tag{7.29}$$

In this case, AF is defined only as a constant value for the horizontal polarization component since the radiation pattern of the vertical dipole changes with the elevation angle, and AF becomes a function of height. To exclude the AF from site attenuation, normalized site attenuation (NSA) is defined here:

$$NSA = V_d + V_r - \left(AF_t + AF_r\right) - \Delta NSA \tag{7.30}$$

where AF_t and AF_r are the antenna factors of the transmitting and receiving antennas, respectively, and ΔNSA is a correction factor for mutual coupling between the antennas and the effect of the earth's surface [19].

Depending on the frequency range of the EMC measurement, different types of antennas are used [20]. For example, in the 30–1000

MHz range, a half-wavelength dipole antenna or a log-periodic dipole array may be used, and a biconical antenna for the 30–300 MHz range, as well as 1 GHz. For frequency ranges above 1 GHz, a standard antenna is selected by the request of radiation pattern or directivity, and a typical one is a horn antenna.

7.3.2 Reverberating Enclosure Measurement

The open site measurement method is time-consuming because the DUT should be rotated to scan the full 2-D radiation pattern. Additionally, the height of the receiving antenna is adjusted until the maximum signal level is obtained. An alternative measurement system for EMI measurements is provided by a reverberating chamber, as presented in Section 8.3.3, to generate a Rayleigh fading environment. The same facility eliminates the need to search for the maximum radiation direction of the DUT.

The measurement system consists of two flat vanes installed on adjacent walls of the chamber with the transmitting and receiving antennas mounted on the remaining two walls of the chamber [21]. The vanes are rotated at speeds of 90 revolutions per minute (rpm) and 120 rpm, respectively. The differing speeds of rotation generate a random field inside the chamber.

The chamber size for this particular experimental setup is 2m × 2m × 2m with the DUT mounted on a fixed pedestal in the center of the chamber. The frequency range for this size of the chamber is above 1 GHz. The transmitting and receiving antennas are broadband log-periodic antennas with frequency ranges of 1–10 GHz.

Figure 7.19 shows an example of site attenuation as a function of frequency, with the mean value calculated in units of decibels. The rate of collection of data points is 10 per minute. The measurement method generally has an acceptable error of approximately a few decibels. The advantage of the reverberating system is that a short time is required to perform the measurement. However, the disadvantage is that the measurement error increases to a few decibels.

In situations where test space is even more severely restricted, a measurement system with a compact shield box may be used. Figure 7.20 shows the measurement setup. This measurement system can generate a random EM field distribution inside the box through reflections from the inner box walls. The DUT is mounted on a

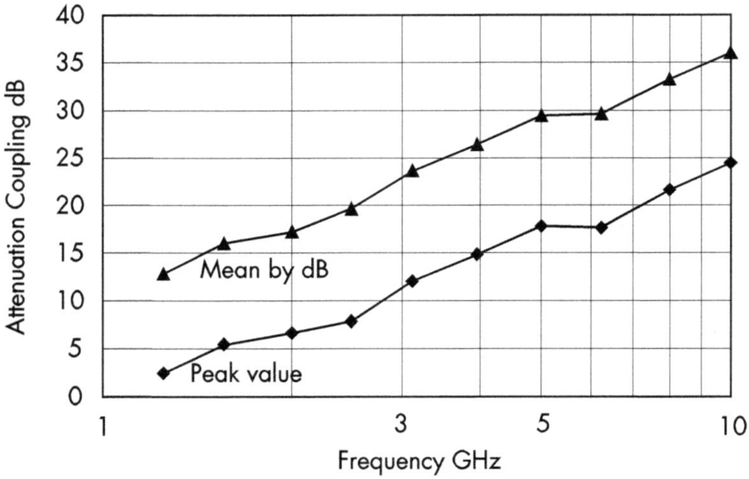

Figure 7.19 Site attenuation.

rotating pedestal offset from the rotation axis to reduce the effects of the terminal antenna directivity on the measurement.

The size of the shield box is only 50 cm × 50 cm × 100 cm, approximately one-thirtieth the size of chambers used in other methods. Its operating frequency is above 1 GHz. The upper limit of the operating frequency depends on the frequency band of the receiving antenna. The receiving antenna is a standard half-wavelength dipole antenna mounted with the element tilted at 45° to the vertical. The DUT is also mounted at a 45° angle on the rotating pedestal. The tilt

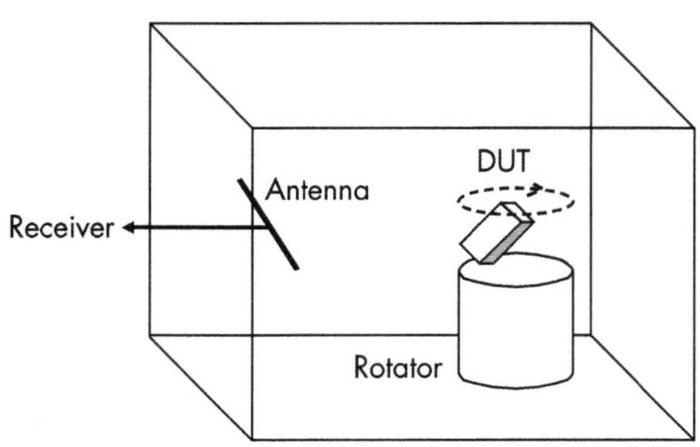

Figure 7.20 Compact shield box.

angle is used to suppress the direct path signal from the transmitter to the receiver. The measurement error for this method can be less than 1 dB, and a measurement time of only a few minutes, allowing data acquisition of thousands of data points at each frequency.

7.3.3 SAR/PD Measurement

As EM waves can warm up human body tissues, a safety standard is introduced to indicate an acceptable level of human body exposure. The safety standard level of the human body surface is defined by the SAR (in W/kg) for the irradiation level of EM waves on living bodies in the frequency range of 100 kHz to 6 GHz. The PD (in mW/cm^2) is used for the frequency above 6 GHz because the EM waves are mainly absorbed on the body surface in higher frequency [22]. The adopted PD level for a MT is defined in that a separation distance of at least 20 cm is maintained between the radiating structures and the user body or nearby persons. The maximum PD level is measured by scanning at the plane of the specified distance from the MT and is determined assuming square-shaped averaging area. A basic restriction on PD of 10 W/m^2 is defined by Federal Communications Commission (FCC), where the value is given by averaging an area of 1 cm^2 [22].

The SAR is a measure of heat energy absorption in unit time by a living body. There are three definitions of SAR: (1) the absorption ratio by short pulse waves, (2) the whole-body average SAR, and (3) the localized SAR. The latter two are defined as a mean value for a given time. When the temperature rises by more than 1°C due to an external heat source, it is assumed that there is some influence on the living body. Then, the corresponding whole-body average SAR is 2 W/kg.

To evaluate the influence of EM wave irradiation on the living body, the SAR is defined as an amount of EM energy absorption in a unit mass as follows:

$$SAR = \frac{\sigma E^2}{\rho} \tag{7.31}$$

where conductivity, the effective electric field amplitude, and the material density are denoted by σ (S/m), E (V/m), and ρ (kg/m^3), respectively. The localized SAR is mainly used as the standard of irradiation level for portable terminals. Conversely, an incident power level at the body surface is often used for the safety standard besides the SAR. A standard example of 1 mW/cm^2 is determined by one-tenth of sun exposure in summer to avoid the effect of heat on the body. The safety standards for the SAR used in Japan, the United States, and Europe are shown in Table 7.2.

The SAR is based on a temperature rise in unit time, so the heat diffusion from the sample during the measurement time can be neglected. The SAR at any time is defined by the specific heat capacity c (J/kg K) and the temperature rise ΔT (°C) at the observation point of the sample during the measurement time Δt.

$$SAR = c\frac{\Delta T}{\Delta t} \tag{7.32}$$

For the SAR measurement using this definition, EM waves are generated with an antenna installed either inside or outside the liquid phantom, and the temperature rise inside the liquid can be measured directly with a thermometer. There are internationally standardized methods for measuring spatial-average SAR. In these standard SAR methods, an electric field probe (E-probe) scans in a liquid phantom.

The internal SAR can be measured accurately because the interior of the phantom can be scanned in three dimensions using an E-probe, as shown in Figure 7.21 [23]. A tested mobile terminal is controlled by a signal from the BS simulator during the SAR measurement. The measurement of the peak spatial-average SAR is time consuming because the E-probe scans the phantom in three dimensions. To reduce the measurement time, a probe array embedded in a gel phantom, as shown in Figure 7.22, is used [24]. The gel phantom with electrical parameters identical to those of the human body is considered an alternative to a liquid phantom.

Table 7.2
SAR Standards

Region	SAR (mW/kg)	Mass to Average (g)	Average Time (min)
Japan	8.0	1	6
USA	1.6	10	30
EU	2.0	1	6

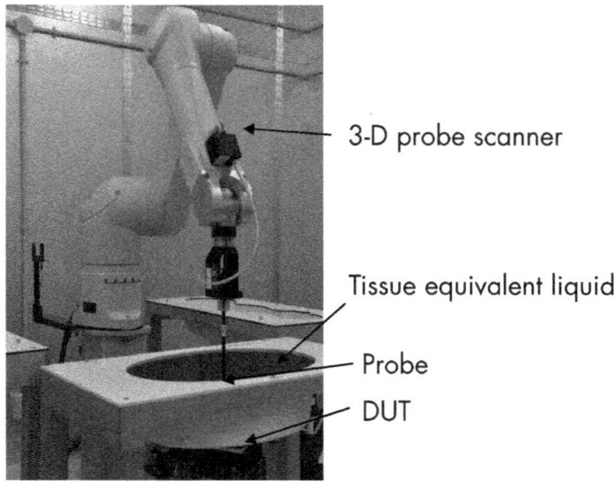

Figure 7.21 SAR measurement by probe scanner and liquid phantom. (Courtesy of Microwave Factory.)

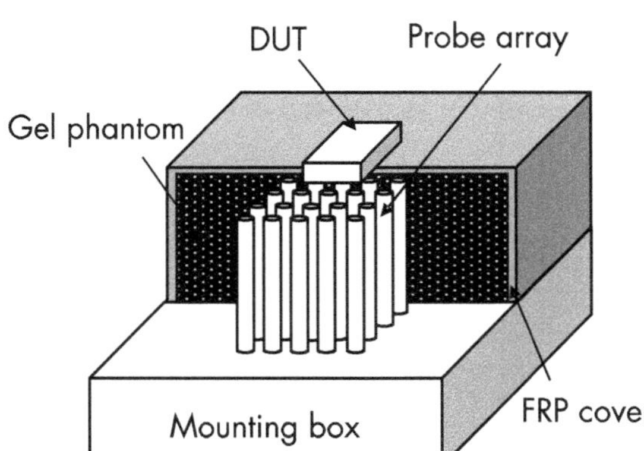

Figure 7.22 SAR measurement by probe array.

References

[1] Forged, L. J., et al., "Antenna Measurement Systems Using Multi-Probe Technology," *Proc., 2015 IEEE International Conference on Antenna Measurements & Applications*, December 2015.

[2] Salmeron-Ruiz, T., et al., "A Fast Single Cut Spherical Near-Field-to-Far-Field Transformation Using Cylindrical Modes," Proc., *European Conference on Antennas and Propagation*, April 2014, pp. 2476–2480.

[3] Wong, K. T., et al., "Electrically Long Dipole in a Collocated Orthogonal Triad—for Direction Finding and Polarization Estimation," *IEEE Trans., Antennas and Propag.*, Vol. 65, No. 11, November 2017, pp. 6057–6067.

[4] Sugimoto Y., et al., "Fast Far-Field Estimation Method by Compact Single Cut Near-Field Measurements for Electrically Long Antenna Array," *IEEE Transactions on Antennas and Propagation*, Vol. 66, Issue 11, November 2018, pp. 5859–5868.

[5] Hansen J. A., "On Cylindrical Near-Field Scanning Techniques," *IEEE Trans., Antennas and Propag.*, Vol. AP-28, No. 2, March 1980, pp. 231–234.

[6] Lang, M. "The Intermodulation Problem in Mobile Communications," *Microwave Journal*, Vol. 38, No. 5, May 1995, pp. 20–22.

[7] Ishii, N., and H. Arai, "Research Trend of Measurement Methods for Total Radiated Power from an Antenna in Japan," *IEICE Trans. Commum.*, Vol. J95-B, No. 5, May 2012, pp. 607–617.

[8] CTIA Certification: Test Plan for Mobile Station Over the Air Performance—Method of Measurement for Radiated RF Power and Receiver Performance, Rev. 2.2.2., December 2008.

[9] Rakhmanov, E. A., "Minimal Distance Energy on Sphere," *Mathematical Research Letters*, Vol. 1, 1994, pp. 647–662.

[10] Microwave Vision Group, https://www.mvg-world.com/en/products/field_product_family/antenna-measurement-2/sg-128 (accessed February 13, 2022).

[11] TS38.101-2 v15.0.0 User Equipment (UE) Radio Transmission and Reception, Part 2: Range 2 Standalone (Release 15), June 2018.

[12] Zhao, K., et al., "Spherical Coverage Characterization of 5G Millimeter Wave User Equipment with 3GPP Specifications," *IEEE Access* PP(99):1-1, December 2018.

[13] Clark, R. H., "A Statistical Theory of Mobile-Radio Reception," *Bell Syst. Tech., J.*, Vol. 47, No. 6, July–August, 1968, pp. 957–1000.

[14] Iwai, H., et al., "Handset Antennas Using Spatial Fading Emulator Based on Clarke's Model," *IEICE Trans. on Comm.*, Vol. E95-B, No. 1, January 2012, pp. 10–17.

[15] Ogawa, K., et al., "Channel Capacity of a Handset MIMO Antenna Influenced by the Effects of 3D Angular Spectrum, Polarization, and Operator," *IEEE AP-S Intl. Symp. Digest*, July 2006, pp. 153–156.

[16] Sakata, T., et al., "MIMO Channel Capacity Measurement in the Presence of Spatial Clusters Using a Fading Emulator," *2009 IEEE 20th International Symposium on Personal, Indoor and Mobile Radio Communications*, September 2009, pp. 97–101.

[17] CISPR 16-2-3:2016, https://webstore.iec.ch/publication/25877 (accessed February 13, 2022).

[18] Smith, A. A., Jr., "Calculation of Site Attenuation from Antenna Factors," *IEEE Trans. Electromagn. Compat.*, Vol. EMC-24, No. 3, 1982, pp. 301–316.

[19] Sugiura, A., et al., "An Improvement in the Standard Site Attenuation Method for Accurate EMI Antenna Calibration," *Trans. IEICE Japan*, Vol. E78, No. 8, 1995, pp. 1229–1237.

[20] MacNamara, T., *Handbook of Antennas for EMC*, Norwood, MA: Artech House, 1995, Chapter 5.

[21] Okamura, M., and A. Sugiyama, "Evaluation of the Performance of a Reverberating Enclosure Used for Making the Measurement of Total Radiated Power from Microwave Apparatus Operating in the Microwave Frequency Range," *International Symposium on EMC*, AA3, No. 18,1984, pp. 594–598.

[22] Cleveland, R. F., Jr., et al., "Evaluating Compliance with FCC Guidelines for Human Exposure to Radiofrequency Electromagnetic Fields," *OET Bulletin 65*, Washington, DC: FCC, August 1997.

[23] Speag, https://speag.swiss/products/dasy8/overview (accessed February 13, 2022).

[24] Kuehn, S., and N. Kuster, "Experimental EMF Exposure Assessment" in *Handbook of Biological Effects of Electromagnetics*, F. S. Barnes and B. Greenebaum (Eds.), CRC Press, 2007, pp. 381–409.

8

PROPAGATION ESTIMATION AND EMULATION OF FADING

In this chapter, the empirical estimation of propagation losses based on measurement data is described using Okumura-Hata curves. The prediction of propagation losses using this method together with a discussion of the requirements of the cellular system is the first topic. In an urban area, the diffraction and reflection of waves by buildings determines the propagation characteristics. Ikegami and multiscreen models for describing these phenomena are presented together with other estimation methods that can be applied to various indoor and outdoor scenarios. The selection of the base station (BS) antenna pattern that is needed to provide the required cell shape is discussed. Ray tracing, a powerful simulation tool for propagation estimation, is also explained in detail. Finally, methods of producing artificial propagation in indoor environments for the evaluation of diversity effects are discussed.

8.1 PREDICTION OF PROPAGATION LOSSES IN CELLULAR SYSTEM

Link budgets for cellular systems are based on predictions of the propagation loss. The propagation loss is determined by the distance between a BS and a mobile terminal (MT) and is affected by reflection, diffraction, and scattering by buildings, trees, and other obstacles. As described in this section, predictions of the propagation loss can be made by extracting the dominant parameters from the measurement data. Another approach to loss prediction is to use a physical model of a typical propagation environment. The dominant loss factor related to the buildings is determined for knife-edge diffraction and the reflection of a plane wave. The second topic in this section is the description of physical models. The loss factors from a microscopic viewpoint are determined by considering the penetration of electromagnetic waves into buildings and the shadowing effects produced by obstacles. The third part of this section concerns loss prediction based on these factors.

8.1.1 The Okumura-Hata Model

In mobile communication systems, the propagation characteristics can be predicted using estimation methods based on large quantities of data obtained from propagation loss experiments. The geographical features within a range of several tens of kilometers are roughly divided into the three different types, and the propagation loss is estimated for each type. This method involves the use of the curve known as Okumura's curve, which was determined from measurements of the propagation losses in the 150-MHz to 1500-MHz range in the Tokyo area [1].

Geographical features can be roughly divided into irregular and semiflat features. Irregular features include hills, isolated mountains, and inclined geographical features. Terrestrial mobile communication services are mainly used in areas containing semiflat geographical features. Within areas of semiflat geographical features, an undulating plain is defined as an area where the height does not exceed 20m. Local geographical features and buildings are described by a compensation value based on a place where the undulations are small. These features and buildings are classified as being in urban, suburban, or open areas. An urban area is defined as an area where the buildings have more than two stories and cover at least 23% of the area [2].

Figure 8.1 shows an example of Okumura's curve for the 900-MHz band in an urban area. At distances less than 10 km from the BS, there is a logarithmic relationship between the distance and the propagation loss whereas, at distances greater than this, the loss is

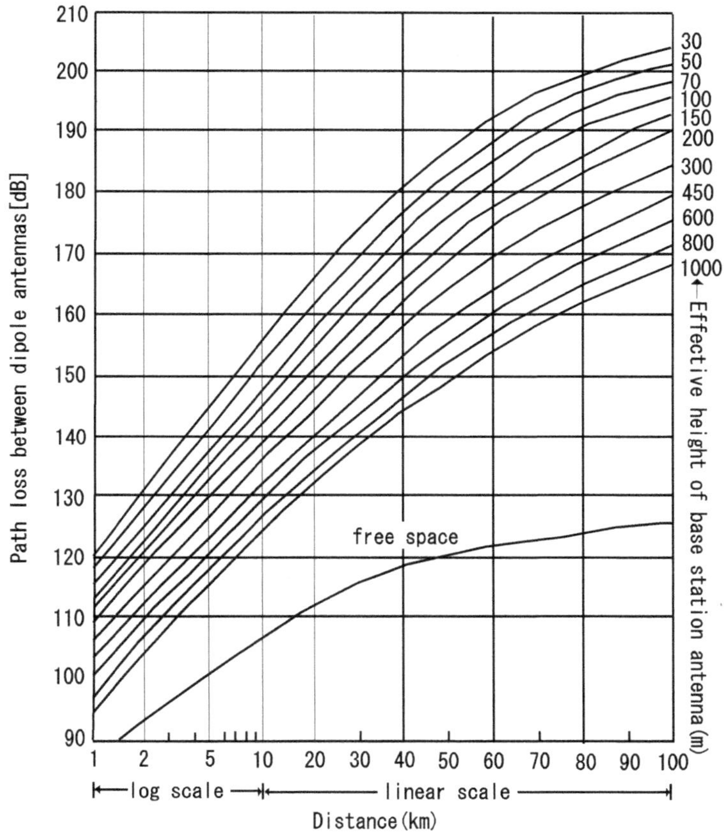

Figure 8.1 Okumura curve.

linear; this is a unique feature of Okumura's curve. It is convenient to calculate the propagation loss using the closed-form equation derived by Hata that is described next [3].

The propagation loss (L_P dB) in an urban area can be expressed for a frequency f MHz (in the range 150–1500 MHz) for a BS antenna of height h_b m (in the range 30–200m) and a MT antenna of height h_m m (in the range 1–10m), where the distance between the BS and MT is d km (in the range 1–20 km). The equation is:

$$L_p = 69.55 + 26.16\log f - 13.82\log h_b - a(h_m) + (44.9 - 6.55\log h_b)\log d \tag{8.1}$$

where the effective height of the antenna, $a(h_m)$, is defined as the compensated antenna height of the mobile terminal:

$$a(h_m) = (1.1\log f - 0.7)h_m - (1.56\log f - 0.8) \tag{8.2}$$

For large cities:

$$a(h_m) = 8.29(\log 1.54h_m)^2 - 1.1 \qquad f \le 400 \text{ MHz} \tag{8.3}$$

$$a(h_m) = 3.2(\log 11.75h_m)^2 - 4.97 \qquad f \ge 400 \text{ MHz} \tag{8.4}$$

As shown in Figure 8.2, the effective height of the antenna is approximated as the height of the average surface elevation of the area within the range of 3 km and 15 km from the BS antenna. The equation for calculating the propagation loss can be extended to the frequency range 1500–2000 MHz [2]:

$$L_p = 46.3 + 33.9\log f - 13.82\log h_b - a(h_m) + (44.9 - 6.55\log h_b)\log d + C_m \tag{8.5}$$

where C_m is a correction factor, which is taken as 0 for cities and 3 for large cities. The detail description of the cites are defined by adding street and building parameters [4].

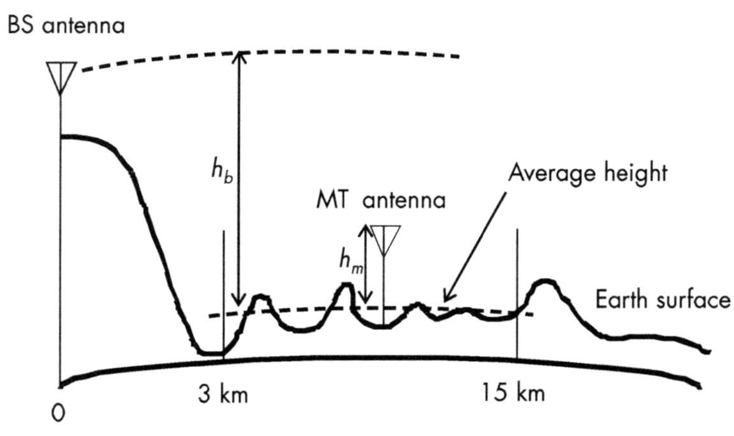

Figure 8.2 Antenna effective height.

Next, the propagation loss is considered for areas that can be classified as suburban or farmland; these include areas such as those alongside a road where trees and houses disturb the propagation of electromagnetic waves. Then the equation is given as:

$$L_p = L_p\{\text{city}\} - 2\left\{\log\left(\frac{f}{28}\right)\right\}^2 - 5.4 \quad (8.6)$$

In open areas where high trees, buildings, and other obstructions do not exist within 300–400m of the MT in the direction of an arrival wave, the propagation loss is given as:

$$L_{ps} = L_p\{\text{city}\} - 4.78\{\log f\}^2 + 18.33\log_{10} f - 40.94 \quad (8.7)$$

The propagation loss as a function of frequency estimated by the Okumura-Hata method is shown in Figure 8.3. Estimates for areas

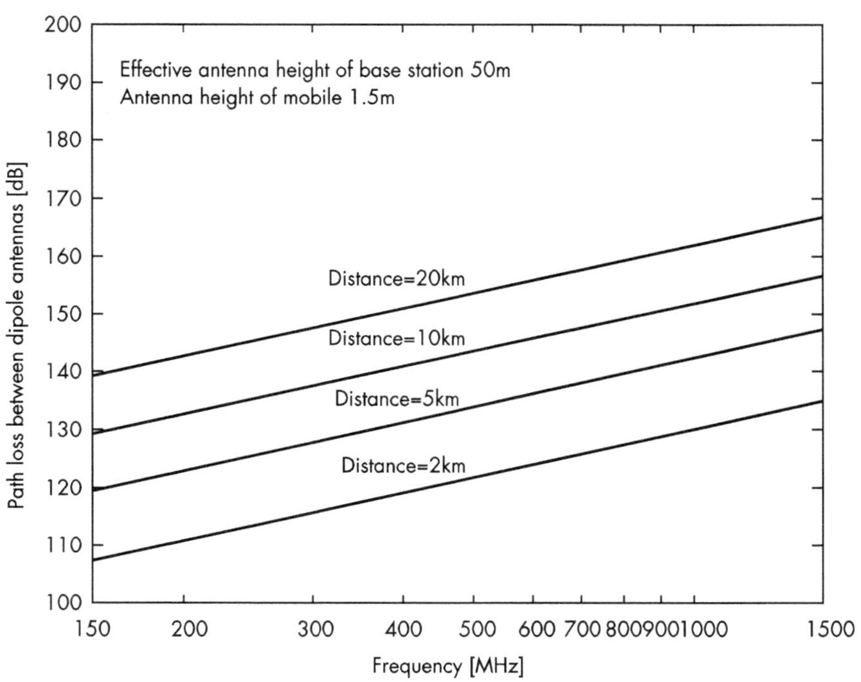

Figure 8.3 Propagation loss by Okumura-Hata model.

more than 1 km from a BS can be obtained using this method. The standard deviation of the error in the predictions produced by these equations is about 6 dB.

Okumura's curve provides the estimated median value of the propagation loss for a square area of side 500m. Sakagami made predictions for smaller areas of 80m square; Fujii [5] extended these predictions to an upper frequency of 2600 MHz. For urban areas, the predictions were based on parameters including building heights and street widths. In addition to the parameters in (8.1), the propagation loss, L_p, can be expressed for a street width W (in the range 5–50m), an average building height H (in the range 5–50m), a BS antenna height above the ground h_b (in the range 20–100m), and an average building height near the BS, H (where $H \leq h_b$ m):

$$L_p = 101 - 7.1\log W + 7.5\log\langle H\rangle - \left\{24.37 - 3.7\left(\frac{H}{h_b}\right)^2\right\}\log h_b$$
$$+ \left(43.42 - 3.11\log h_b\right)\log d - \left[3.2\left\{\log\left(11.75h_m\right)\right\}^2 - 4.97\right] + 20\log f \quad (8.8)$$

where the frequency range is 450 to 2600 MHz and the street angle is defined in Figure 8.4. Equation (8.8) produces prediction errors with a standard deviation of about 5 dB for a MT height of 1.5m.

8.1.2 Physical Model

Propagation loss prediction methods derived from Okumura's curve are based on measured data in actual propagation environments. The physical propagation models described in this section focus on local conditions around mobile terminals. Figure 8.5 shows the parameters and paths used in the Ikegami model [6]. A mobile terminal is moving on a street surrounded by tall buildings. A direct path from a BS, E_1, exists; for free-space propagation, the loss, L_o, is then given by (2.2). Waves that follow this path are diffracted at the top of the building, then follow two main paths at the MT, L_{ex}. One of these paths is a direct path, E_2; waves that follow the other path, E_3, are reflected at the building wall. The geometry of these two paths is shown in Figure 8.5(b) as a top view; the angle between the incident wave and the street is denoted by ϕ. The diffraction coefficient is

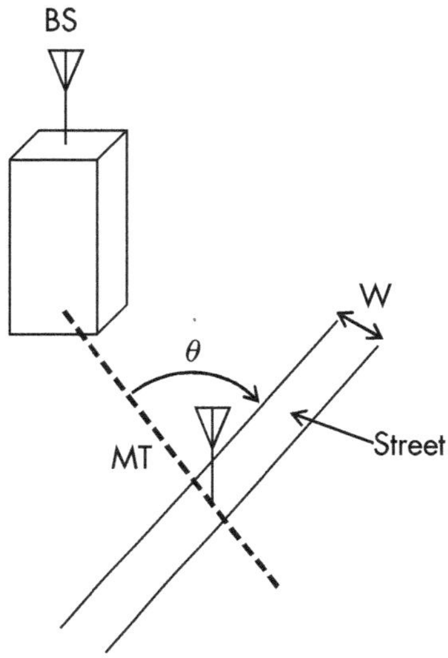

Figure 8.4 Parameters of street.

calculated by assuming that the building top forms a knife-edge (see Section 2.6). The propagation loss, L, between the BS and the MT is then given by

$$L = L_o + L_{ex} \tag{8.9}$$

where

$$L_o = 20\log\frac{4\pi d}{\lambda} = 32.4 + 20\log d + 20\log f \tag{8.10}$$

and

$$L_{ex} = -5.8 - 10\log\left(1 + \frac{3}{L_r^2}\right) - 10\log W + 20\log\left(h_r - h_m\right) + 10\log(\sin\phi) + 10\log f \tag{8.11}$$

Here, the units of the distance, d, and frequency, f, are kilometers and megahertz, respectively; the heights of the building and MT are

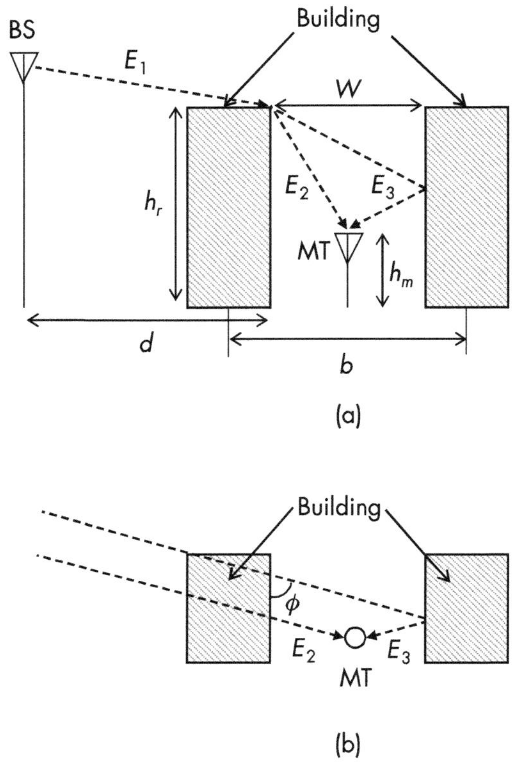

Figure 8.5 Ikegami model: (a) side view, and (b) top view.

measured in meters. The reflection loss at the building wall is given by L_r and is generally of the order 4 to 10 dB.

In European countries, the heights of buildings in most cities are generally quite uniform and the situation can be modeled using a multiscreen model, as shown in Figure 8.6. The propagation path above the buildings, E_1, can be modeled by assuming that the waves propagate as predicted by Huygens' principle and the MT receives waves that propagate along a diffracted path at the screen top E_2. This method of predicting the propagation loss is known as the Walfisch model and describes multiple diffractions at the top of the buildings; the diffraction results in an overall loss factor of $e^{-3.8d}$ [7]. The extra loss given by (8.11) can be rewritten as:

$$L_{ex} = 57.1 + A + \log f + 18\log d - 18\log\left(h_b - h_r\right) - 18\log\left(1 - \frac{d^2}{17\left(h_b - h_r\right)}\right) \tag{8.12}$$

where

$$A = 5\log\left\{\left(\frac{s}{2}\right)^2 + \left(h_r - h_m\right)^2\right\} - 9\log s + 20\log\left\{\tan^{-1}\frac{2}{s}\left(h_r - h_m\right)\right\} \tag{8.13}$$

Here, A represents a factor that depends on the building geometries, and s is the multiscreen spacing.

The European Cooperation in Science and Technology (COST) organization has extended and merged these two models to produce the COST-231 Walfisch-Ikegami model. In this model, the propagation loss is expressed as the sum of the free-space loss, L_o, the extra loss due to the multiscreen diffraction, L_{ms}, described by the Walfisch model, and the loss near the MT, L_{rt}, described by the Ikegami model:

$$L = L_o + L_{ms} + L_{rt} \tag{8.14}$$

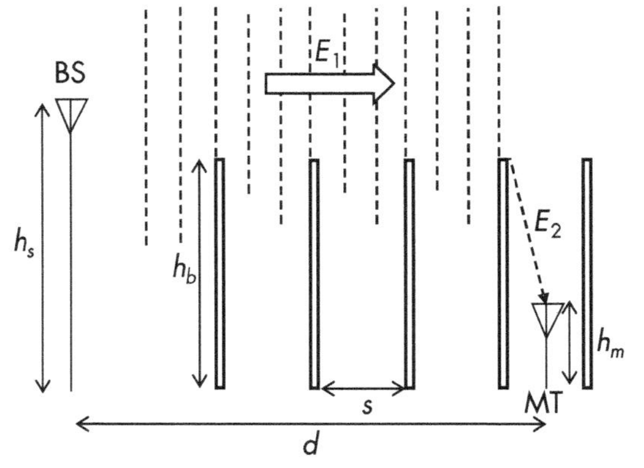

Figure 8.6 Walfisch model.

where

$$L_{ms} = -18\log\{1 + (h_b - h_r)\} + 54 + 18\log d - 9\log b$$
$$+\begin{cases} \left\{-4 + 0.7\left(\dfrac{f}{925} - 1\right)\right\}\log f & \text{for cities and suburban areas} \\ \left\{-4 + 1.5\left(\dfrac{f}{925} - 1\right)\right\}\log f & \text{for large cities} \end{cases} \tag{8.15}$$

and

$$L_{rt} = -8.2 - 10\log W + 10\log f + 20\log(h_r - h_m)$$
$$+\begin{cases} -10 + 0.354\phi & (0 \le \phi \le 35^\circ) \\ 2.5 + 0.075(\phi - 35) & (35^\circ \le \phi \le 55^\circ) \\ 4.0 + 0.114(\phi - 55) & (55^\circ \le \phi \le 90^\circ) \end{cases} \tag{8.16}$$

where b (m) is the building spacing as shown in Figure 8.5(a). The parameter values are restricted to the ranges 4m $\le h_b \le$ 50m, 1m $\le h_m \le$ 3m, 800 MHz $\le f \le$ 2000 MHz, and 0.02 km $\le d \le$ 5 km.

Figure 8.7 shows the results of propagation loss predictions based on the five methods described earlier. The propagation loss

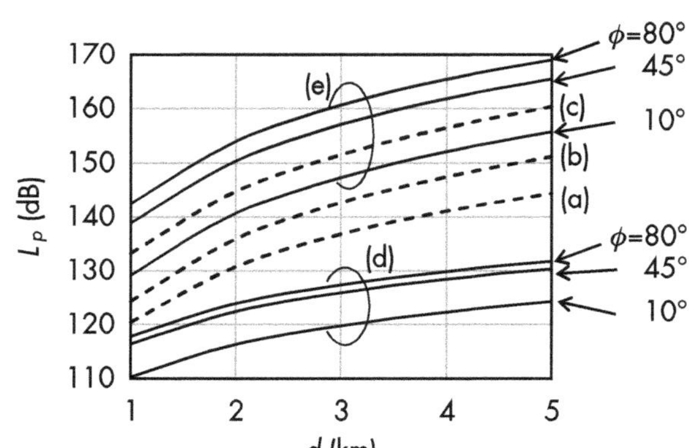

Figure 8.7 Propagation loss prediction of big cities by: (a) Okumura, (b) Fujii, (c) Walfisch, (d) Ikegami, (e) Walfisch-Ikegami. f = 800 MHz, h_b = 40m, h_m = 1.5m, h_r = 20m, W = 10m, $s = b$ = 10m.

strongly depends on the local environment, and the propagation loss prediction method that is selected for application to cellular systems should be appropriate for the local conditions.

8.1.3 Standardization of Loss Predictions

Mobile communication systems are built using integrated technologies and are standardized for worldwide use. The International Mobile Telecommunications-2000 (IMT-2000) specifications produced by the International Telecommunication Union (ITU) provide the first set of global standards for mobile communication service. These standards are designed to meet the requirements of third-generation (3G) services. This section summarizes the propagation loss prediction methods that can be applied to 3G and IMT-Advanced or 4G systems.

In 1997, the ITU released its "Guidelines for Evaluation of Radio Transmission Technologies for IMT-2000 (M.1225)," in which the propagation models described in the previous section were extended. The three settings considered in the guidelines are (a) an indoor office, (b) indoor and outdoor pedestrian environments, and (c) inside vehicles environments. Setting (a) consists of a base station and a terminal carried by a user on foot inside a small indoor office. The devices used are low-power devices, and the propagation model is based on the results of applying COST-231. Outdoor microcellular sites are used for setting (b); in this setting, the antenna height of the BS is lower than the height of the surrounding buildings. Terminals are carried by users on foot inside and outside the office, and the degree of penetration from the outdoors to the indoor office is included in the modeling. The propagation model used for setting (b) is equivalent to the Walfisch model or the Walfisch-Ikegami model. In setting (c), it is assumed that propagation occurs within macrocells, the user terminals are carried inside vehicles, and that BS antennas are higher than the surrounding buildings. Details of the propagation loss predictions are given in [8].

The systems that succeeded 3G are known as IMT-Advanced or 4G systems, and the appropriate propagation models are presented as ITU guideline M.2135 [9]. The five models presented for the frequency range 2 to 6 GHz are as follows.

The indoor hotspot (InH) model describes propagation inside a building with a height of 6m, a width of 50m, and a length of 120m (see Figure 8.8). A central corridor with a width of 20m is surrounded by eight rooms on each side. Two BS antennas are positioned 30m away from the ends of the corridors; it is assumed that the terminals are either static or carried by users on foot.

The urban micro (UMi) model describes the mobile communication between an outdoor BS and indoor or outdoor user terminals where the heights of both antennas are lower than those of the surrounding buildings. Hexagonal cells and grid-like street networks are assumed in this model.

The urban macro (UMa) and suburban macro (SMa) models cover larger areas than the UMi model and describe situations in which the BS antenna is higher than the surrounding buildings and the user terminals are in motion on the street. Buildings more than four stories high are assumed in the UMa model; in the SMa model, the buildings are assumed to be detached houses.

The final model is the rural macro (RMa) model, which can apply in areas where the density of buildings is lower than in the other models. The area covered by this model has a radius of up to 10 km, and the height of the BS antenna is assumed to be in the range of 20m to 70m. The user terminals are assumed to be moving at very high speeds of up to 350 km/h. The RMa model can be applied to frequencies in the range 450 MHz to 6 GHz. Propagation loss predictions for the five models described here are given in [10].

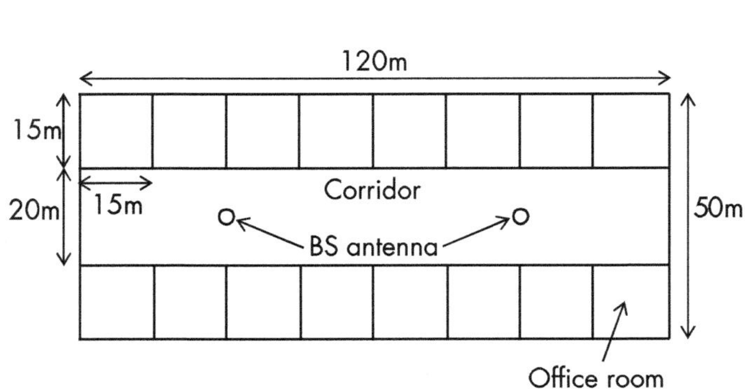

Figure 8.8 Layout of indoor hotspot.

8.2 RAY-TRACING METHODS

The equations for propagation loss that were given in the previous section are based on indoor and outdoor propagation measurements. It requires a lot of time and equipment to obtain the huge amount of data in propagation measurements, and licensing requirements may make this difficult. However, a virtual propagation environment can easily be produced on a computer using electromagnetic analysis methods. This section describes some numerical simulation methods that can be used to predict the propagation loss.

8.2.1 The FDTD Method and the Ray-Tracing Method

The finite domain and time difference (FDTD) method is a well-known numerical simulation technique that can be used to characterize antenna performance and propagation phenomena [11, 12]. The FDTD method is a differential numerical modeling method that discretizes the time-dependent Maxwell equations using central-difference approximations of the spatial and temporal partial derivatives. The electric and magnetic field components are allocated to a Yee cell to discretize the equations; a cell size of less than $\lambda/10$ is needed for the results of the calculations to be stable. A very large amount of memory and a long computation time are needed to calculate a propagation environment because millions of cells are needed to simulate just a small empty room. To estimate the propagation loss precisely, the simulation model would also need to include the desks, chairs, other equipment, and people in the room. As modeling all of these is very complicated and time consuming, the ray-tracing method is also widely used to predict the propagation characteristics in such settings.

Ray tracing is based on an approximate solution of Maxwell's equations for the light waves propagating through and around objects that have dimensions much greater than the wavelength of light. The solutions do not include the effects of reflection and diffraction, which can be obtained using plane-wave theory and the geometrical theory of diffraction (GTD) [13]. Figure 8.9 shows three paths—(a) a direct path, (b) a reflected path, and (c) a diffracted path—from a BS to a MT. These paths are determined by Fermat's principle so that waves propagate from the BS to the MT by the shortest route. Ray-tracing methods can be classified as either image or ray-launching methods to determine the paths used in the propagation loss predictions.

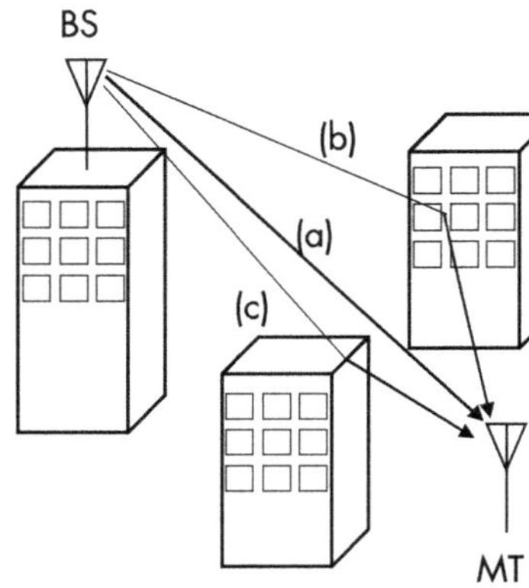

Figure 8.9 Paths from base station to mobile terminal: (a) direct path, (b) reflected path, and (c) diffracted path.

8.2.2 Basis of the Ray-Tracing Method

An electromagnetic wave radiating from a BS spreads out across space, and its intensity decreases with distance. Figure 8.10 illustrates the geometry of the situation. The electromagnetic energy contained within the surface S_1 with radius r_o spreads out so that it is contained within the surface S_2 with radius r. Curvatures in the u and v axes are different from each other and are defined as r_1, r_2 on S_1, and $r_1 + s$, $r_2 + s$ on S_2, respectively. As r_1 and r_2 are both very large, the amplitude of the electric field on both surfaces is assumed to be uniform. The law of conservation of energy then gives:

$$\frac{1}{2Z_o}\left|E_1\right|^2 r_1\phi_1 r_2\phi_2 = \frac{1}{2Z_o}\left|E_2\right|^2 (r_1 + s)\phi_1(r_2 + s)\phi_2 \tag{8.17}$$

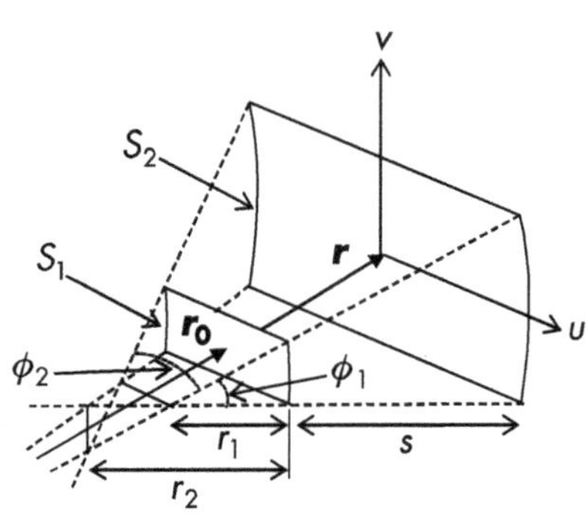

Figure 8.10 Spreading factor.

where Z_o is the characteristic impedance of free space and the areas of S_1 and S_1 are approximated as $r_1\phi_1 r_2\phi_2$ and $(r_1 + s)\phi_1(r_2 + s)\phi_2$, respectively. The field strength, E_2, is obtained by solving (8.17):

$$E_2 = \sqrt{\frac{r_1 r_2}{(r_1 + s)(r_2 + s)}}E_1 \tag{8.18}$$

The coefficient of E_1 is defined as the spreading factor, which can be used to express the details of a local diffracted field. In general, when using the ray-tracing method, a BS antenna can be regarded as a point source of radiation. The far-field equation that was given as (1.32) can be rewritten to include the antenna directivity gain function, $\mathbf{G}(\theta, \phi)$:

$$\mathbf{E}(r,\theta,\phi) = \sqrt{\frac{Z_o P_t}{2\pi}}\frac{e^{-j_o r}}{r}\mathbf{G}(\theta,\phi) = \sqrt{\frac{Z_o P_t}{2\pi}}\frac{e^{-j_o r}}{r}\left\{G_\theta(\theta,\phi)\mathbf{e}_\theta + G_\phi(\theta,\phi)\mathbf{e}_\phi\right\} \tag{8.19}$$

where Z_o and P_t are the characteristic impedance of free space and the transmission power of the antenna, respectively. G_i $(i = \theta, \phi)$ denotes the directivity gain of component i of the BS antenna. The path of a ray propagating from the antenna in the direction (θ, ϕ) is given by (8.19); if this ray is reflected, transmitted, or diffracted by an object, the direction of propagation will change.

The reflection and transmission are determined by the law of reflection and Snell's law, as described in Section 2.3; the reflection and transmission coefficients were defined in Section 2.3. The diffracted field at the edge of the object can be estimated by applying the GTD for a wedge shape (see Figure 8.11). The incident and diffracted wave vectors, $\mathbf{r}_i$ and $\mathbf{r}_d$, should satisfy the following extension of Fermat's principle:

$$\hat{\mathbf{r}}_i \cdot \hat{\mathbf{l}} = \hat{\mathbf{r}}_d \cdot \hat{\mathbf{l}} \text{ and } \cos\beta_i = \cos\beta_d \tag{8.20}$$

where $\mathbf{l}$ is a vector directed along the thin end of the wedge and $\hat{\boldsymbol{a}}$ is the unit vector with the same direction as $\boldsymbol{a}$. Diffracted rays are on the cone edge of vertex angle β_d in Figure 8.11 to satisfy (8.20). An

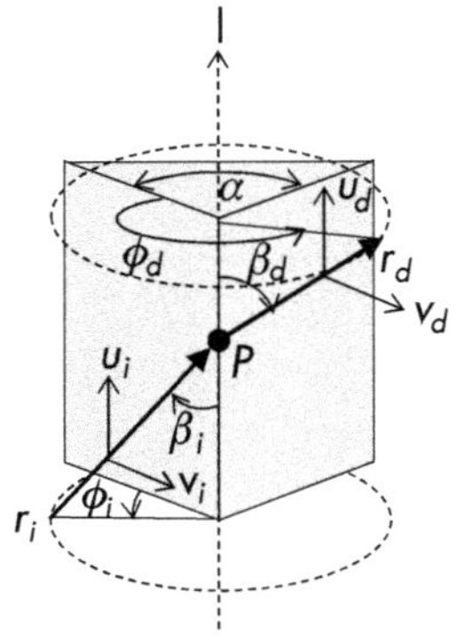

Figure 8.11 Local coordinate system at diffraction point *P*.

incident and an outgoing surface are determined by the vectors $\boldsymbol{l}$ and $\boldsymbol{r}_j$ ($j = i, d$); the normal vectors to the two surfaces, $\boldsymbol{v}_j$, and the orthogonal vectors, $\boldsymbol{u}_j$, are defined as:

$$\boldsymbol{v}_j = \frac{\boldsymbol{r}_j \times \boldsymbol{l}}{\left|\boldsymbol{r}_j \times \boldsymbol{l}\right|} \text{ and } \boldsymbol{u}_j = \boldsymbol{v}_j \times \boldsymbol{l}, \text{ respectively.} \tag{8.21}$$

The incident and outgoing electric fields are described by vectors $\boldsymbol{u}_j$ and $\boldsymbol{v}_j$ ($j = i, d$) as follows:

$$\boldsymbol{E}_j(\boldsymbol{r}) = E_u^j(\boldsymbol{r})\boldsymbol{u}_j + E_v^j(\boldsymbol{r})\boldsymbol{v}_j \tag{8.22}$$

where $\boldsymbol{r}$ denotes the position vector of an observation point. The electric field diffracted at point P is given using the spreading factor and the position vector, $\boldsymbol{r}_P$, of point P as:

$$\boldsymbol{E}_d\left(\boldsymbol{r}_P\right) = \boldsymbol{E}_i\left(\boldsymbol{r}_P\right) \cdot \overline{\overline{\boldsymbol{D}}} \sqrt{\frac{r_d}{r_d\left(r_i + r_d\right)}} e^{-jk_o r_d} \tag{8.23}$$

where $\overline{\overline{\boldsymbol{D}}}$ is the dyadic diffraction coefficient:

$$\overline{\overline{D}} = -\begin{bmatrix} \hat{v}_l & \hat{u}_l \end{bmatrix} \begin{bmatrix} D_a & D_b \\ D_c & D_d \end{bmatrix} \begin{bmatrix} \widehat{v_d} \\ \widehat{u_d} \end{bmatrix} \tag{8.24}$$

The dyadic diffraction coefficient can be obtained using the uniform theory of diffraction by treating the wedge as a perfectly conducting surface [14]. Coefficients for materials with a finite conductivity are given in [15] and can be used to predict the propagation loss.

8.2.3 Image and Ray-Launching Methods

The image method and the ray-launching method are used to find the paths of a ray between the transmitting (TX) and receiving (RX) points. Figure 8.12 show paths estimated using the image method. The images of TX and RX produced by reflection are denoted as TX_i and RX_i, respectively, and the paths from TX to RX are denoted as a direct path, P_1, and reflected paths, P_2 and P_3. A reflection point of the path P_2 is denoted as R_1 and marks where the line connecting TX and TX_i intersects the reflecting surface. Reflection points R_2 and R_3 lie on the line connecting TX_i and RX_i. As shown in Figure 8.12, a ray following path P_3 from TX to RX is reflected at both these points.

Using the image method, exact paths that can be used to predict the propagation loss can be derived; the total number of paths, N_T, is obtained using the equation

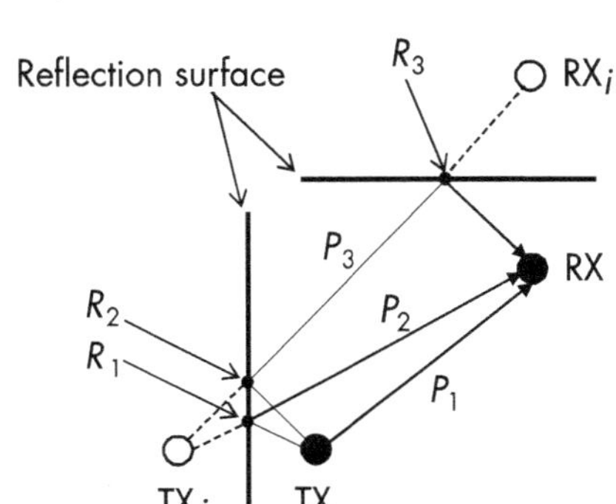

Figure 8.12 Image method.

$$N_T = 1 + \begin{cases} 2 & M = 1 \\ 2N + 1 & M = 2 \\ \dfrac{M(M-1)^N - 2}{M - 2} & M > 2 \end{cases} \tag{8.25}$$

where M is the number of reflecting surfaces and N is the total number of reflections. N_T increases exponentially for $N > 2$, and the computational procedure is cumbersome. Because of this, the application of the image method is limited to cases involving a small number of orthogonal reflecting surfaces. Figure 8.13 shows an example of a two-dimensional rectangular reflecting wall where overlapping image sources reduce the total number of paths in the calculation.

The ray-launching method is another type of path-finding method. Figure 8.14 shows how the radiation from TX can be replaced by the sum of individual rays and how the rays reaching area S surrounding RX can be used to estimate the received field. The paths used for the estimation are marked as solid lines and those that miss the area S by dotted lines. The radius of the receiving area is given by $r\Delta\Omega/\sqrt{3}$ in the three-dimensional model [16] and by $r\Delta\Omega/\sqrt{2}$ in the two-dimensional model [17]. Here, the angle between the rays is

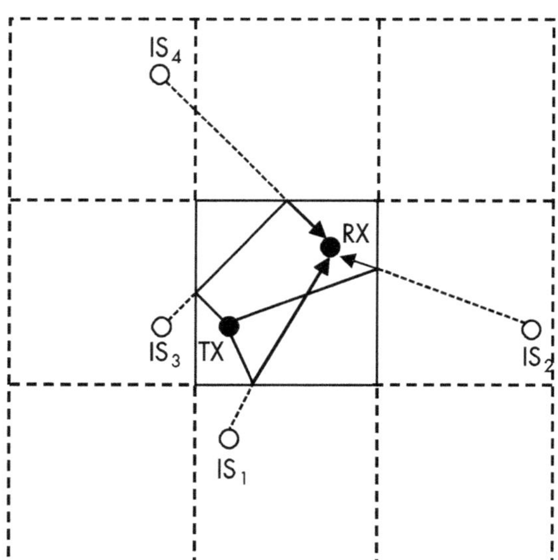

Figure 8.13 Image sources by rectangular reflection surface.

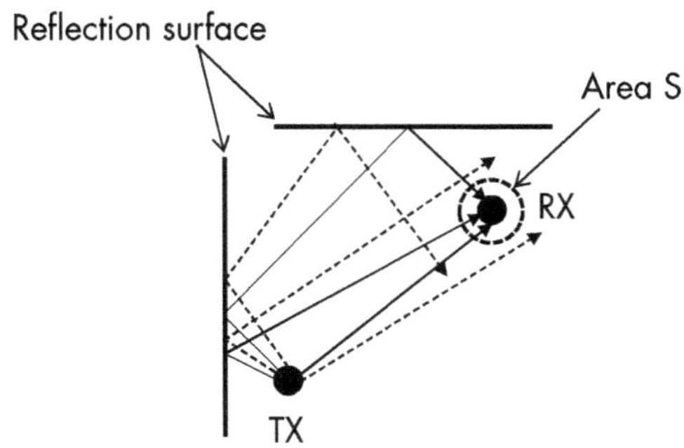

Figure 8.14 Ray-launching method.

$\Delta\Omega$; the distance r is defined as the length between RX and the last reflection or diffraction point on the path before RX.

8.2.4 Examples of Ray-Tracing Simulations

In this section, electric field patterns produced by a ray-tracing simulation method are presented [18]. Figure 8.15 shows the simulation model: it consists of two buildings approximated by homogeneous dielectric blocks placed on level ground. A vertical half-wavelength dipole antenna—the transmitting antenna—is installed at the edge of the roof of block #1. Electrical parameters for the blocks and the

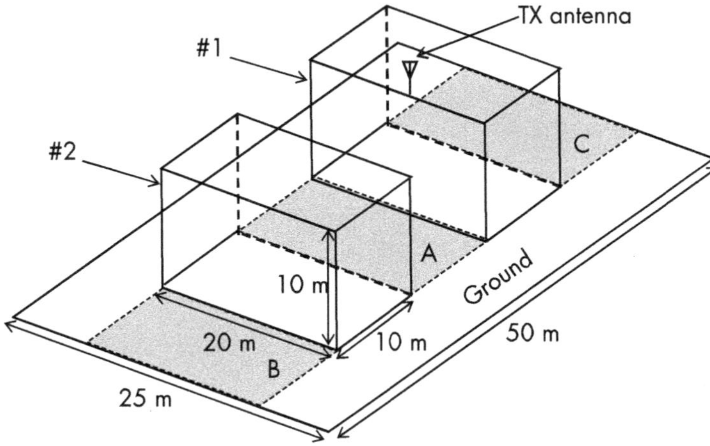

Figure 8.15 Simulation model by ray-tracing method; dielectric blocks $\varepsilon_r = 5$, $\tan\delta = 0.1$, ground $\varepsilon_r = 10$, $\tan\delta = 0.2$.

ground are given in the figure caption. The electric field pattern on the plane 1.5m above the ground was calculated using the ray-tracing method. The total number of rays was ten; four different combinations of rays were considered in the simulation.

Figure 8.16(a) shows the situation for a single direct ray emitted at TX without any reflection or diffraction. The electric field is mainly limited to the LOS area dented by A in Figure 8.15 and has a minimum value directly under TX due to the radiation pattern produced by the dipole. A single reflection at the ground or the building

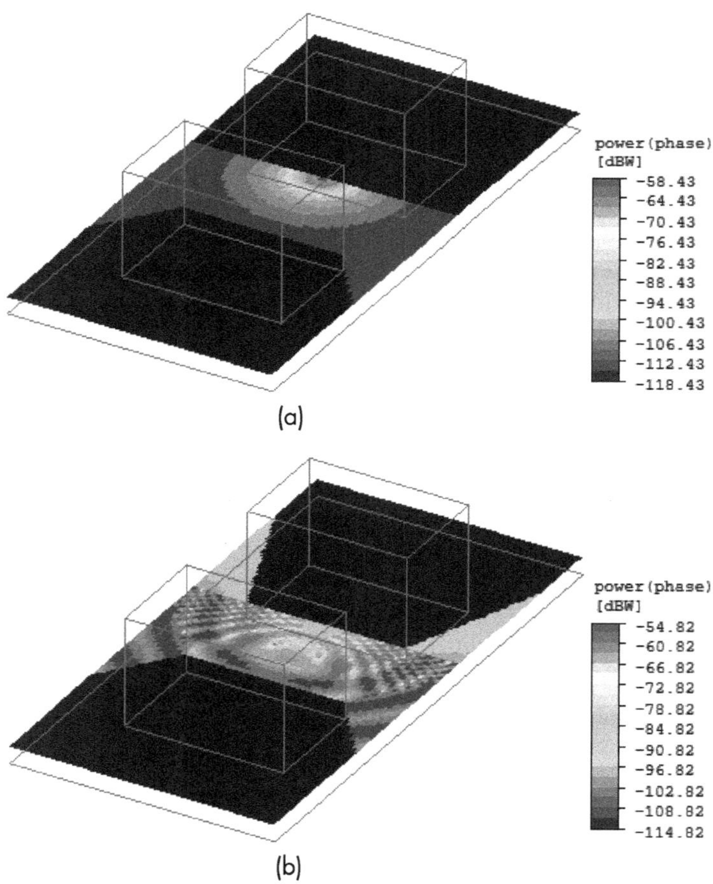

Figure 8.16 Electric field distributions by ray-tracing method: (a) direct ray, (b) direct and one-reflection rays, (c) direct and two-reflection rays, and (d) direct and two-reflection, or one-reflection and one-diffraction, or two-diffraction rays.

wall is considered in Figure 8.16(b). The LOS area is extended as result of the reflection at the wall, and standing waves are observed in area A. The phase difference between the direct and reflected waves produces the observed electric field patterns. The results for two reflected rays are shown in Figure 8.16(c). The standing wave patterns are similar to the results shown in Figure 8.16(b), and the LOS area is extended a little toward block #2 side. Diffraction at the building edges is considered in Figure 8.16(d); here, either one reflected and one diffracted ray or two diffracted rays are added to the rays

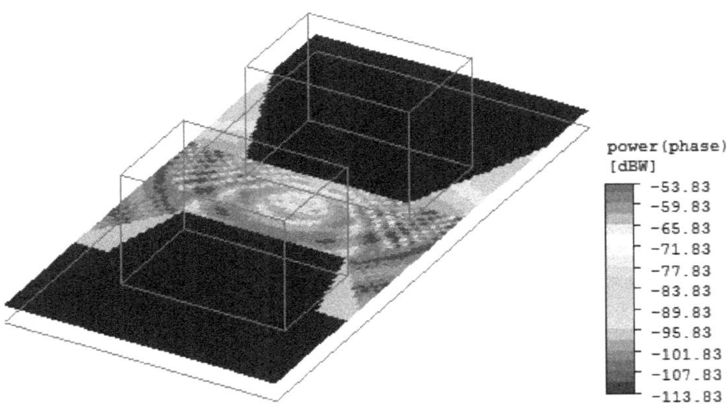

(c)

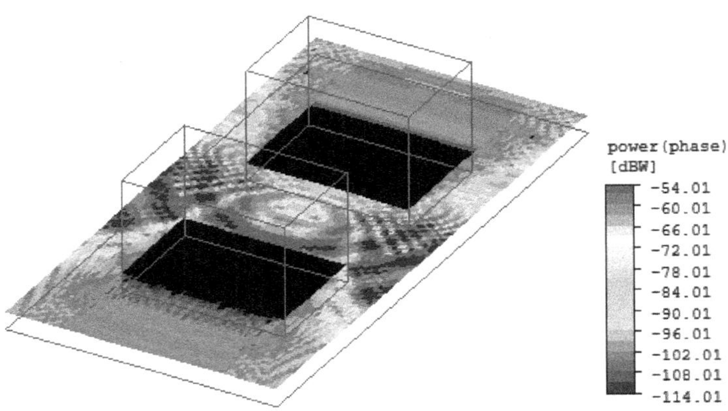

(d)

Figure 8.16 *(continued)*

considered in Figure 8.16(c). Diffracted rays arrive in the shadow of the blocks denoted as B and C in Figure 8.15, which extends the area covered by the TX. Thus, as illustrated in Figure 8.16, ray-tracing simulations can be used for different types of rays and help with the understanding of propagation phenomena and the determination of the area covered by a BS.

8.3 MULTIPATH FADING EMULATION

Mobile terminals are tested at the development stage to ascertain whether they can satisfy the necessary specifications when fading is present. These tests involve propagation measurements in a real outdoor multipath propagation environment. However, outdoor propagation measurements are not always stable over time, and it is also necessary to obtain a license to carry out the required tests. These considerations indicate the need for indoor facilities where real multipath propagation environments can be reproduced. The aim of the fading reproduction method is to generate fading waves as the RF output signal from a radio unit; this produces what is known as a fading simulator. Another method is to reproduce fading waves that are the sum of spatially distributed standing waves within an artificial multipath propagation environment in an indoor facility. This section describes the principles and operation of fading-field simulators, which are indispensable to the development of mobile terminals.

8.3.1 Fading Simulator

The received voltage at the input port of a radio unit can be regarded as a function of time for moving terminals because the RF signal transmitted from the BS is received at the terminal as a sum of waves that have followed multiple paths. A fading simulator should include the reproduction of the Doppler fading that results from the motion of the MTs, and also include more than one correlated received signal with fading, and have wideband characteristics so that all the allocated frequency bands are covered.

Assuming the Rayleigh fading distribution at a MT that was described in Section 2.8, the fading signal, $E_r(t)$, can be obtained as a function of the time, t, as the sum of the in-phase and orthogonal independent signals with carrier angular frequency ω_c:

$$E_r(t) = x(t)\cos\omega_c t - y(t)\sin\omega_c t \tag{8.26}$$

The amplitude coefficient of each signal is given by a Gaussian distribution and is identical to the Gaussian noise signal. Based on this relation, the fading signal is synthesized by modulating the in-phase and orthogonal carrier signals with Gaussian noise as shown in Figure 8.17. This type of modulation is known as double sideband suppressed carrier modulation. The power frequency spectra of the base-band noise signals used for the modulation can be written as functions of the velocity of the MT, v (m/s), and the maximum Doppler frequency, $f_D = v/\lambda$ (Hz)

$$x(f) = y(f) = \frac{\sigma^2}{\pi f_D}\frac{1}{\sqrt{1-\left(\frac{f}{f_D}\right)^2}},\quad 0 \le f \le f_D;\; x(f) = y(f) = 0,\; f_D \le f \tag{8.27}$$

where σ is the mean value of $x(f)$ and $y(f)$ that were defined in (2.33).

The Gaussian noise is obtained by restricting the frequency bandwidth of the M-sequence (maximum-length) code with a frequency of f_b to

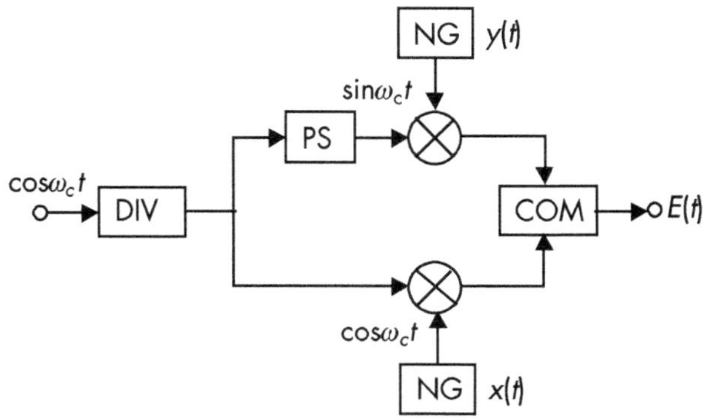

Figure 8.17 Block diagram of fading simulator. (DIV, power divider; PS, phase shifter; NG, Gaussian noise generator; COM, power combiner.)

$$f_D = \frac{f_b}{20} \tag{8.28}$$

Although the length of one cycle of the output signal of the fading simulator is the same as that of the M-sequence code, an effective length of more than 1600 wavelengths is sufficient for the measurement facility. The M-sequence is used to obtain random sequences. The M-sequence generator can easily be constructed from a shift register and an exclusive OR operator—this is an arrangement that is often employed in digital communication systems to synchronize input and output signals.

Fading effects can be reduced by diversity reception, as described in Section 2.8, and by using multiple input branches with small correlation coefficients. In order to test the diversity reception, it is necessary for the fading simulator to generate correlated output signals. Two correlated fading signals can be described by replacing the amplitude coefficient in (8.26) with the following equations:

$$x_i(t) = u_i(t)\cos\phi_i, \quad y_i(t) = v_i(t)\sin\phi_i, \qquad i = 1,2 \tag{8.29}$$

where ϕ_i denotes the phase of the signal. The correlation coefficient, ρ, for the two signal envelopes can be obtained from (3.33) as:

$$\rho = \cos(\phi_1 - \phi_2) \tag{8.30}$$

The correlation coefficient is arbitrarily defined by ϕ_1 and ϕ_2, which are adjusted at the fading simulator by adding an appropriate phase difference to the noise generator shown in Figure 8.17.

8.3.2 Delay Spread Simulation

In mobile communication systems, the propagation path from a transmitting source to a reception point has a different length for each frequency. This causes delay spread distribution, especially for high-speed digital data transmission. To be able to investigate the performance of MTs under this delay spread condition, a fading simulator needs to have frequency-selective fading characteristics. The amplitude of each delayed signal is adjusted by an arbitrary scaling coefficient, and all of the delayed signals are combined by a power combiner at the final stage [19].

In the case of frequency-selective fading in a real propagation environment, the delay time, the amplitude, and the phase of the delayed wave all behave as functions of time. Measured delay times for current digital cellular systems vary from a few hundreds of nanoseconds (ns) to tens of microseconds (μs). To produce real frequency-selective fading, the transverse filter implemented by LSI shown in Figure 8.18 must process the signal in the time domain. This signal-processing fills the base-band frequency with digital data, thus providing detailed parameter settings for use in delay spread simulation. The synthesized digital signal is input to a band-pass filter after D/A conversion and is modulated by Gaussian noise to generate fading waves [20]. This type of field simulator is used to evaluate the equalizer that suppresses the delayed waves in mobile terminals.

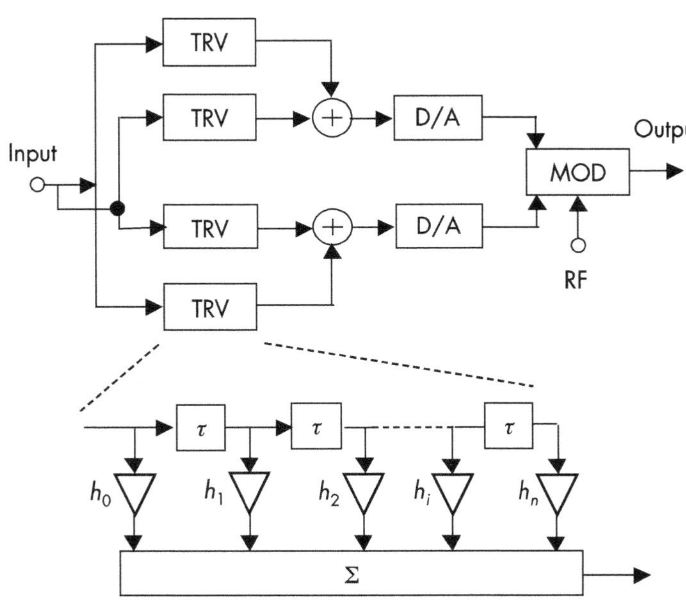

Figure 8.18 Block diagram for real time delay spread simulator: transversal filter (TRV), digital to analog converter (D/A), modulator (MOD), τ delay circuit, h_i filter coefficient.

8.3.3 Fading-Field Simulator

Using a fading simulator, the characteristics of mobile terminals can be evaluated by inputting a fading wave to the antenna input port with the antenna removed. This is an effective way of evaluating the radio unit of a mobile terminal that has had its antenna removed. However, it is also necessary to test the characteristics of the product with the antenna included. The product test performed inside an anechoic chamber that was described in Section 3.2.2 is not sufficient for mobile terminals intended for use in urban multipath propagation environments. In this section, the structure of a fading-field simulator that can be used to produce an artificial multipath propagation environment and the theory behind such a simulator will be presented. A fading simulator reproduces fading signals as RF signals fed through a coaxial cable, whereas a field simulator generates fading waves that are the sum of spatially distributed electromagnetic waves.

Simple fading may be obtained by moving the device under test (DUT) along a standing wave generated by a feed probe and reflectors as shown in Figure 8.19. The feed probe consists of a monopole antenna installed on a large area of flat ground and placed a distance $\lambda/4$ away from one reflector to excite the standing wave effectively. The distance between the two reflectors is set to $m\lambda/2$, where m is a natural number [21]. The number of incoming waves to a mobile

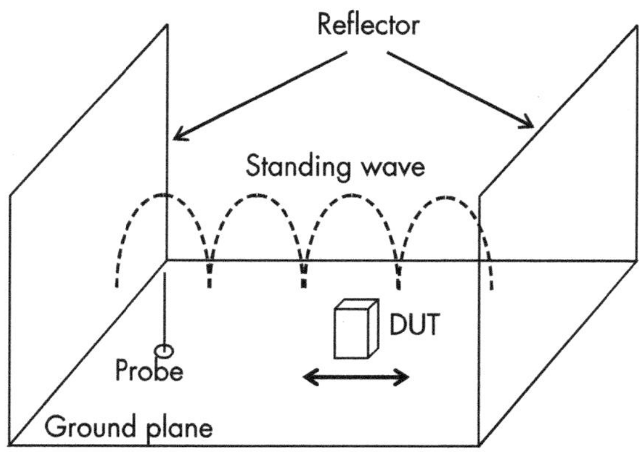

Figure 8.19 Fading measurement using standing waves.

terminal in a cellular system is seven or eight [22]; this is quite different from the case just described, where there is only one standing wave.

Multipath fading in an outdoor environment is caused by the arrival of randomly distributed waves from all directions. Under such conditions, the generation of a random field is necessary for testing devices. This technique is also used to eliminate the effect

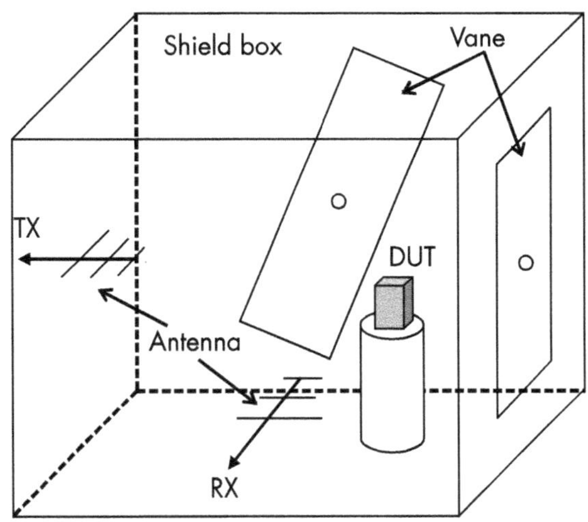

Figure 8.20 Reverberation chamber.

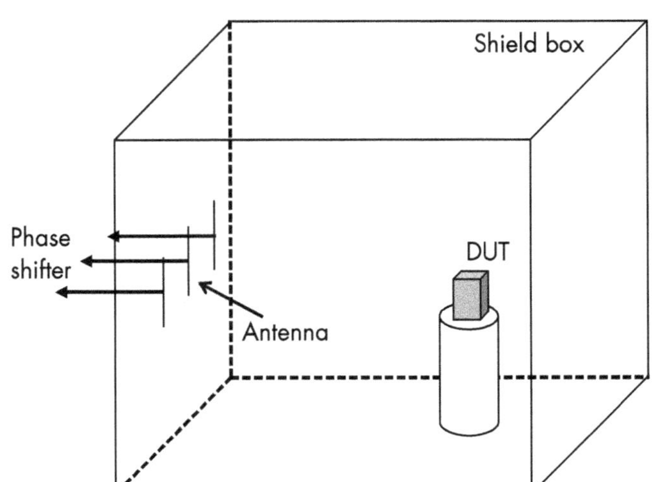

Figure 8.21 Field simulator.

of different radiation patterns in the antenna of a DUT. A reverberation chamber based on this principle is one of the facilities that can be used to generate a random field by using rotating vanes inside a shielded room. A setup of this type is shown in Figure 8.20 [23].

To generate interference fading using a limited number of sources, in the case of a field simulator, it is appropriate to use the waves reflected from the conducting reflector. One possible compact field simulator consists of the transmitting antennas installed on the inside walls of a shielded box as shown in Figure 8.21. The field distribution is regarded as a function of time, and the phase of each source is randomly changed electronically by a digital phase-shifter. An eight-bit digital phase-shifter is used to satisfy the requirement that the minimum number of bits should be more than five [24]. The random phase variation means that the field radiating from the sources is scanned electronically.

References

[1] "Propagation Data and Prediction Methods for the Terrestrial Land Mobile Service Using Frequency Range 30 MHz to 3 GHz," CCIR SG-5, Report 567-3 (MOD F).

[2] Kozono, S., and K. Watanabe, "Influence of Environmental Buildings on UHF Band Mobile Radio Propagation," *Commun., COM-25 IEEE Trans.*, Vol. 10, October 1977, pp. 1133–1143.

[3] Hata, M., "Empirical Formula for Propagation Loss in Land Mobile Radio Services," *Veh. Technol. IEEE Trans.*, Vol. 29, No. 3, 1980, pp. 317–325.

[4] Sakagami, S., and K. Kuboi, "Mobile Propagation Loss Prediction for Arbitrary Urban Environments," *Trans-B. on IEICE Japan*, Vol. J74-B-II, No. 1, January 1991, pp. 17–25 (in Japanese).

[5] Fujii T., "Path Loss Prediction Formula in Mobile Communication—An Expression of 'Sakagami' Patch Loss Prediction Formula," *Trans-B. on IEICE Japan*, Vol. J86-b, No. 10, October 2003, pp. 2264–2267 (in Japanese).

[6] Ikegami, F., et al., "Propagation Factors Controlling Mean Field Strength on Urban Streets," *Antennas Propagt IEEE Trans.*, Vol. 32, No. 8, August 1984, pp. 822–829.

[7] Walfisch, J., and H. L. Bertoni, "A Theoretical Model of UHF Propagation in Urban Environments," *Antennas Propagt IEEE Trans.*, Vol. 36, No. 12, December 1988, pp. 1788–1796.

[8] ITU R, Recommendation ITU-R M, "Guidelines for Evaluation of Radio Transmission Technologies for IMT-2000," Vol. 1997, 1225.

[9] ITU R, Recommendation ITU-R M. 2135-1, "Guidelines for Evaluation of Radio Transmission Technologies for IMT-Advanced," 2009.

[10] Ist-Winner I. I. Deliverable, 1.1.2 v 1.2 WINNER II Channel Models, IST-WINNER2. Tech. Rep., 2008.

[11] Yee, K. S., "Numerical Solution of Initial Boundary Value Problems Involving Maxwell's Equations in Isotropic Media," *IEEE Trans. Antennas Propag.*, Vol. 14, No. 4, 1966, pp. 302–307.

[12] Taflove, A., *Computational Electrodynamics: The Finite-Difference Time-Domain Method*, Norwood, MA: Artech House, 1995.

[13] Keller, J. B., "Geometrical Theory of Diffraction," *J. Opt. Soc. Am.*, Vol. 52, No. 2, 1962, pp. 116–130.

[14] Kouyoumjian, R. G., and P. H. Pathak, "A Uniform Geometrical Theory of Diffraction for an Edge in Perfectly Conducting Surface," *Proc. IEEE*, Vol. 62, No. 11, 1974, pp. 1448–1461.

[15] Luebbers, R., "Finite Conductivity Uniform GTD Versus Knife-Edge Diffraction in Prediction of Propagation Path Loss," *IEEE Trans. Antennas Propag.*, Vol. 32, No. 1, 1984, pp. 70–76.

[16] Seidel, S. Y., and T. S. Rappaport, "Site-Specific Propagation Prediction for Wireless In-Building Personal Communication System Design," *IEEE Trans. Veh. Technol.*, Vol. 43, No. 4, 1994, pp. 879–891.

[17] Sarkar T. K., et al., "A Survey of Various Propagation Models for Mobile Communications," *IEEE Antennas Propag. Mag.*, Vol. 45, No. 3, 2003, pp. 51–82.

[18] EEM R. T. M., http://www.e-em.co.jp (accessed February 13, 2022).

[19] Arnold, H. W., and W. F. Bodtmann, "A Hybrid Multichannel Hardware Simulator for Frequency-Selective Mobile Radio Paths," *IEEE Trans. Commun.*, Vol. 31, No. 3, 1983, pp. 370–377.

[20] Ueda, T., and H. Suzuki, "Selective Fading Simulator for Digital Mobile Radio," *IEICE Fall Conf.*, B-1-257-09, 1998 (in Japanese).

[21] Arai, H., et al., "A Flat Energy Density Antenna System for Mobile Telephone," *IEEE Trans. Veh. Technol.*, Vol. 40, No. 2, May 1991, pp. 483–486.

[22] Kuwabara, M., *Vehicular Telephone System*, edited by IEICE, Corona Co., Japan, 1985, p. 17 (in Japanese).

[23] Kildal, P. -S., and K. Rosengren, "Correlation and Capacity of MIMO Systems and Mutual Coupling Radiation Efficiency, and Diversity Gain of Their Antenna: Simulation and Measurements in a Reverberation Chamber," *IEEE Commun. Mag.*, Vol. 42, No. 12, December 2004, pp. 104–112.

[24] Arai, H., "Field Simulator for Rayleigh/Rician Fading Reproduction," *Conf. Dig, Symp.*, Baltimore, MD: IEEE, AP-S, Int, July 1996, pp. 1552–1555.

9

ANTENNA AND CHANNEL CAPACITY

Channel-capacity estimation is a key factor in high-speed data transmission in mobile communications. This chapter presents the definition of channel capacity for single-input and single-output (SISO) and multiple-input and multiple-output (MIMO) systems. First, the definition of channel capacity of MIMO antennas is presented. To maximize channel capacity, the optimization method is explained using the base station antenna location, pattern synthetization, and polarization selection in several propagation environments. Examples of channel-capacity estimation in typical outdoor and indoor environment are described. Antenna optimization of mobile terminals is explained using the radiation efficiency and mutual coupling among the antenna elements.

9.1 CHANNEL CAPACITY OF SISO AND MIMO

Mobile communications are seriously affected by multipath fading. Diversity schemes are used to overcome these problems using several transmitting and/or receiving antennas in SISO systems. Multiple transmitting and receiving antennas that are simultaneously used can provide several independent propagation paths, resulting in channel-capacity enhancement by allocating data streams to different paths. This technique is known as MIMO. This section presents the basics of MIMO and channel-capacity estimation.

9.1.1 System Model of MIMO

The channel capacity of mobile communication systems is determined based on the signal-to-noise ratio (SNR) and characteristics of the propagation channel. The SISO system is characterized using the antenna pattern and transmitting power of a transmitter (TX), and the propagation channel is determined using the propagation loss and TX antenna pattern. Figure 9.1 shows the system model of SISO. Received signal $y(t)$ is given by transmission power P_t and thermal noise $n(t)$ at the receiver as a function of time, that is,

$$y(t) = \sqrt{P_t}hs(t) + n(t) \tag{9.1}$$

where $s(t)$ is the input signal at TX, and h is the propagation channel. The time variation in channel h is sufficiently slow compared with that in the other parameters. Then, channel SNR = ρ is obtained using the ensemble average of noise $\sigma = E[n^2(t)]$ and normalized signal level $E[s^2(t)] = 1$ as:

$$\rho = \frac{P_t|h|^2}{\sigma^2} \tag{9.2}$$

where ensemble average x is expressed using $E[x]$.

Figure 9.2 shows the system model of MIMO. The numbers of transmitting and receiving antennas are N_t and N_r, respectively, and the noise at each receiving port is denoted by $n_j(t)$ ($j = 1,2\ldots N_r$). Equation (9.1) is extended to the following $N_t \times N_r$ matrix form as follows:

$$[\boldsymbol{Y}] = [\boldsymbol{H}][\boldsymbol{X}] + [\boldsymbol{N}] \tag{9.3}$$

$$[\boldsymbol{Y}] = \left[y_1(t), y_2(t), \ldots y_{N_r}(t)\right]^t \tag{9.4}$$

$$[\boldsymbol{H}] = \begin{bmatrix} h_{11} & h_{12} & \cdots & h_{1N_r} \\ h_{21} & h_{22} & \cdots & h_{2N_r} \\ \vdots & \cdots & \vdots\cdots & \cdots \\ h_{N_t1} & h_{N_t2} & \cdots & h_{N_tN_r} \end{bmatrix} \tag{9.5}$$

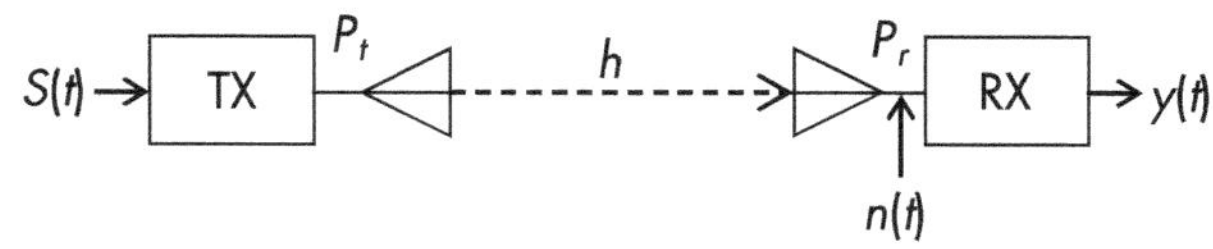

Figure 9.1 System model of SISO.

$$[\boldsymbol{X}] = \left[x_1(t), x_2(t), \ldots x_{N_r}(t)\right]^t = \left[\sqrt{P_1}s_1(t), \sqrt{P_2}s_2(t), \ldots \sqrt{P_{N_r}}s_{N_r}\right]^t \quad (9.6)$$

$$[\boldsymbol{N}] = \left[n_1(t), n_2(t), \ldots n_{N_r}(t)\right]^t \quad (9.7)$$

where t denotes the transpose matrix, and $\boldsymbol{H}$ is known as the propagation channel matrix.

The singular value decomposition of matrix $\boldsymbol{H}$ is expressed as $\boldsymbol{UDV}^{\mathrm{H}}$. $\boldsymbol{U}$ and $\boldsymbol{V}$ are complex unitary matrices, and $\{\cdot\}^{\mathrm{H}}$ represents a complex conjugate transpose. Diagonal matrix $\boldsymbol{D}$ is expressed as:

$$[\boldsymbol{D}] = \begin{bmatrix} \sqrt{\lambda_1} & 0 & \cdots & 0 \\ 0 & \sqrt{\lambda_1} & \cdots & 0 \\ \vdots & \cdots & \vdots\cdots & \cdots \\ 0 & 0 & \cdots & \sqrt{\lambda_{N_r}} \end{bmatrix} \quad (9.8)$$

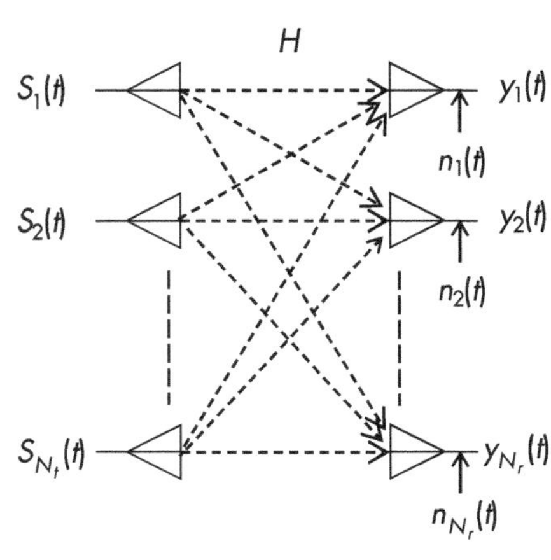

Figure 9.2 System model of MISO.

$\boldsymbol{U}$ and $\boldsymbol{V}$ are transformed into unitary matrix $\boldsymbol{I}$ using the relationship $\boldsymbol{U}\boldsymbol{U}^{\mathrm{H}} = \boldsymbol{V}\boldsymbol{V}^{\mathrm{H}} = \boldsymbol{I}$ and are used as weight matrices for transmission and reception to obtain the following formula:

$$\boldsymbol{D}\boldsymbol{s}(t) = \begin{bmatrix} \sqrt{\lambda_1} & 0 & \cdots & 0 \\ 0 & \sqrt{\lambda_2} & \cdots & 0 \\ \vdots & \vdots & \cdots & \cdots \\ 0 & 0 & \cdots & \sqrt{\lambda_{N_r}} \end{bmatrix} \begin{bmatrix} s_1(t) \\ s_2(t) \\ \vdots \\ s_{N_r} \end{bmatrix} = \begin{bmatrix} \sqrt{\lambda_1}s_1(t) \\ \sqrt{\lambda_2}s_2(t) \\ \vdots \\ \sqrt{\lambda_{N_r}}s_{N_r} \end{bmatrix} \tag{9.9}$$

This transformation indicates that a MIMO system independently transmits N_r orthogonal signals, which enhances the channel capacity of a system.

9.1.2 Channel Capacity Under Rayleigh Environment

The Shannon-Hartley theorem expressed the channel capacity of SISO in the units of bits per second (bps) as:

$$C = \log_2(1 + \rho) \tag{9.10}$$

The channel capacity of $N_t \times N_r$ MIMO is obtained using normalized channel matrix $\boldsymbol{H_n}$ as follows [1]:

$$C = \log_2\left(\det\left(\boldsymbol{I} + \frac{\rho}{N_t}\boldsymbol{H}_n\boldsymbol{H}_n^H\right)\right) \tag{9.11}$$

Equation (9.11) is transformed into the following [2]:

$$C = N_r \log_2\left(\frac{\rho}{N_t}\right) + \log_2\left(\det\left(\boldsymbol{I}\cdot\frac{N_t}{\rho} + \boldsymbol{H}_n\boldsymbol{H}_n^H\right)\right) \tag{9.12}$$

$$\approx N_r \log_2\left(\frac{\rho}{N_t}\right) + \log_2\left(\det\left(\boldsymbol{H}_n\boldsymbol{H}_n^H\right)\right) = C_{SNR} + C_{COR} \tag{9.13}$$

The aforementioned approximation is given by condition $N_r > N_t$, a well-conditioned channel, and large ρ. Equation (9.13) indicates that the channel capacity contains elements that depend on SNR (C_{SNR}) and the correlation (C_{COR}).

Figure 9.3 shows the channel capacity as a function of SNR for four configurations of antenna elements, where 1,000 patterns of channel matrices are investigated in each SNR and the average is used to evaluate the channel capacity. These patterns are random under the assumption of a Rayleigh propagation environment. The result indicates that the channel capacity is proportional to SNR and the number of antennas. We need to note that (9.13) is effective in high SNR values of more than 20 dB.

To individually evaluate the contribution of SNR and the correlation to the channel capacity using (9.13), we focus on 2 × 2 MIMO systems. C_{SNR} and C_{COR} are shown in Figure 9.4 where the channel capacity is evaluated not based on $\det(\boldsymbol{H}_n\boldsymbol{H}_n^H)$ but on spatial correlation coefficient γ_s, which is defined by:

$$\gamma_s = \frac{R_{r,12}}{\sqrt{R_{r,11}R_{r,22}}} \tag{9.14}$$

where $R_{r,ii}$ are the components of $\boldsymbol{R}_r = \boldsymbol{H}\boldsymbol{H}^{\mathrm{H}}$. According to (9.13), Figure 9.4(a) shows that a high SNR value increases the channel capacity. The C_{COR} values are scattered below the upper bound, as shown in Figure 9.4(b), because the power distribution to the first and

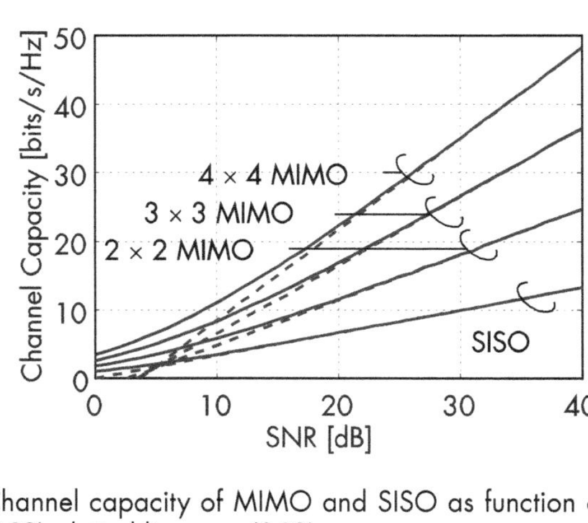

Figure 9.3 Channel capacity of MIMO and SISO as function of SNR; solid lines are (9.11) and (9.12), dotted lines are (9.13).

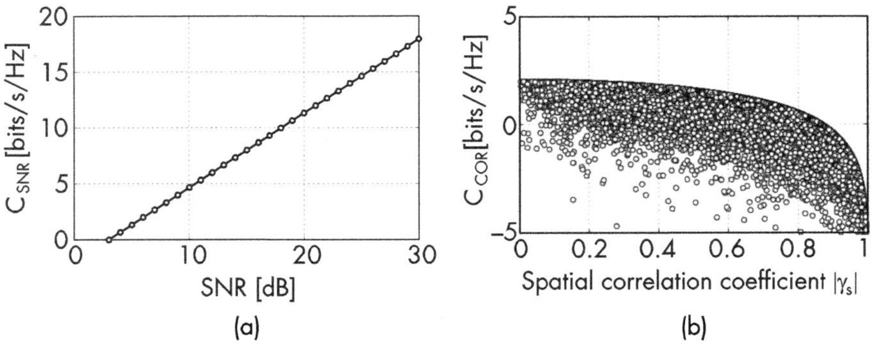

Figure 9.4 Performance of the components depending on (a) SNR, and (b) correlation in 2 × 2 MIMO channel capacity.

second eigenvalues are not optimized. The obtained upper bound curve indicates that a low correlation yields a high channel capacity in which the increasing ratio reaches saturation at $\gamma_s < 0.5$. The upper bound is given by an ideal propagation channel, particularly in the low-SNR regime, which means that the improvement rate in the channel capacity by the second term in (9.13) is high.

9.1.3 Channel Capacity Under Nakagami-Rice Environment

The condition discussed in the preceding section is based on Rayleigh distribution in the propagation environment. This section describes the influence of SNR and correlation on the channel capacity under a Nakagami-Rice distribution. Averaged spatial correlation coefficient $\overline{\gamma}_s$ has a one-to-one correspondence with Rician factor K, which is defined in Section 2.8, and it is expressed as follows [3]:

$$\overline{\gamma}_s = \frac{K}{K+1} \tag{9.15}$$

where $\overline{\gamma}_s$ is a monotonically increasing function of K. Figure 9.5 shows the channel capacity in (9.11) as a function of $\overline{\gamma}_s$. Here $\boldsymbol{H_D}$ is derived from channel matrix $\boldsymbol{H}$, which is expressed by the following equation [4]:

$$\boldsymbol{H} = \sqrt{\frac{K}{K+1}}\boldsymbol{H}_D + \sqrt{\frac{1}{K+1}}\boldsymbol{H}_s \tag{9.16}$$

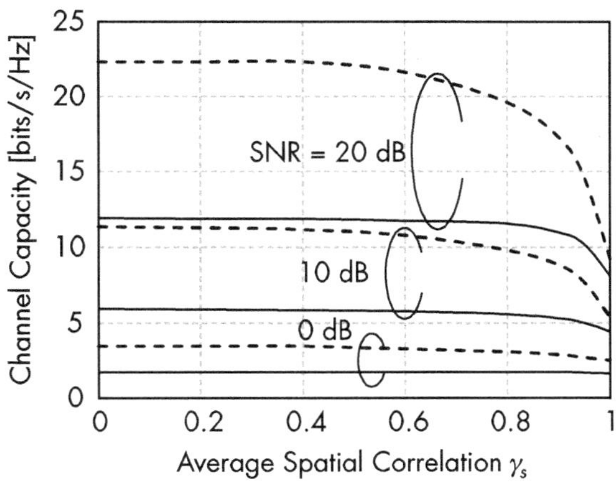

Figure 9.5 Channel capacity as a function of the averaged spatial correlation coefficient, solid lines are 2 × 2 MIMO, and dotted lines are 4 × 4 MIMO.

$$\boldsymbol{H}_D = \begin{bmatrix} 1 & \cdots & 1 \\ \vdots & \ddots & \vdots \\ 1 & \cdots & 1 \end{bmatrix} \tag{9.17}$$

where $\boldsymbol{H}_s$ is a matrix with a complex Gaussian distribution. The channel matrices, including K, are evaluated using a similar procedure presented in Section 9.1.2. Figure 9.3 shows that the channel capacity increases in the low-correlation regime (i.e., close to the Rayleigh distribution). The increase in elements (N_t, N_r) and a high SNR are effective for enhancing the channel capacity under a Rician environment. However, the capacity does not increase under large-K regimes ($\bar{\gamma}_s$ < 0.5) in the MIMO systems.

9.2 BS ANTENNA DESIGN AND CHANNEL CAPACITY

The channel capacity in MIMO systems is increased by high SNR and low spatial correlation in the propagation channel. To satisfy this requirement, high gains with low correlated radiation patterns are required in BS antenna designs. Actual propagation environments involve reflection, transmission, and diffraction waves from BSs that are often installed near the ceiling to realize a LOS indoor environment for MTs. This section describes the design requirements for obtaining a large channel capacity in a BS antenna for downlink transmissions and the measured channel-capacity performance of a fabricated antenna. In this section, *correlation* represents a spatial correlation, including the effect of angular spread.

9.2.1 BS Near Walls

To determine the design guidelines for MIMO antenna radiation patterns, parameter studies are conducted using the ray-tracing method based on the scenario shown in Figure 9.6. In this scenario, BS is installed near a wall, and the channel matrices at several MT positions are calculated. The environments between the BS and all MT positions are LOS environments. The transmitting and noise power are –5 dBm/ch and –85 dBm, respectively, and a 2 × 2 MIMO communication is assumed. The channel capacity of the directional antennas is larger than that of the antennas with omnidirectional patterns [5], and the BS radiation patterns are assumed to be a pencil beam with an infinite front-to-back ratio, as shown in Figure 9.6.

The half-power beam width (HPBW) is fixed at θ_h = 60°, and the space-averaged channel capacities are calculated at each angle between two beams (θ_s). Figure 9.7 shows the relationship between θ_s with a maximum space-averaged channel capacity and aspect ratio *t*. The channel-capacity distribution represented by the black circles follows the dotted line, which indicates that the main beam of BS

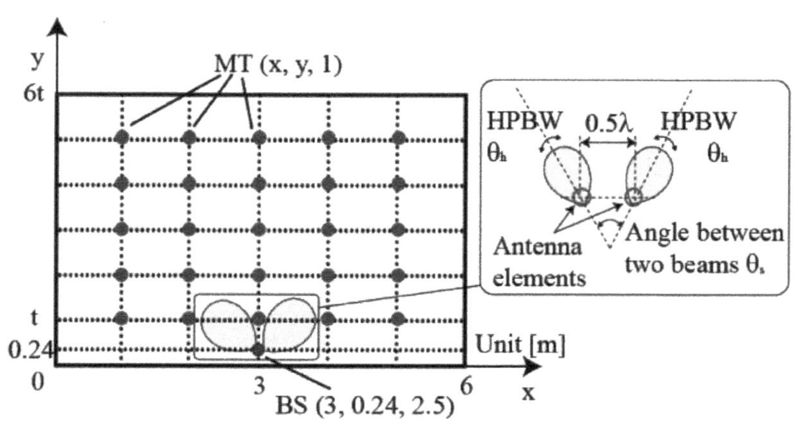

Figure 9.6 Room geometry and location of the BS and MT to determine suitable radiation patterns (overhead view). The room height is 2.7m, and *t* denotes the aspect ratio of the room.

should be directed toward the room corners to increase the channel capacity. This improvement effect is not limited to 2 × 2 MIMO systems because it is also observed in four streams [6].

The antenna that is built based on the abovementioned design guidelines consists of a two-element patch-antenna array with dual feeds, as shown in Figure 9.8(a). Two beams with orthogonal polarizations are radiated from each patch antenna, and these are tilted downward, as shown in Figure 9.8(b). Figure 9.9 shows the radiation patterns. The HPBWs in the *yz* and *zx* planes are approximately 80° and 90°, respectively, and the antenna gain is 7.0 dBi.

Figure 9.9 shows the positions of BSs and MTs in the environment shown in Figure 9.6. As a reference, a sleeve array is also used for a dual-polarization BS antenna where the spacing between the vertical- and horizontal-oriented sleeves is equal to 1λ. For the MT antennas, a single dipole antenna is moved to evaluate the impulse responses of 16 channels using four channels because the measured environment is quasistatic and the time variation is almost negligible. Under this measurement condition, the element spacing between the MT antennas is 0.5λ. The BS antenna is installed at the ceiling near a wall, and the channel matrices are measured at three typical positions (1–3) for the MT, as shown in Figure 9.9. At positions 1 and 3, signals from the four beams of BS are received at almost the same level, and SNR at position 1 is higher than that at position 3.

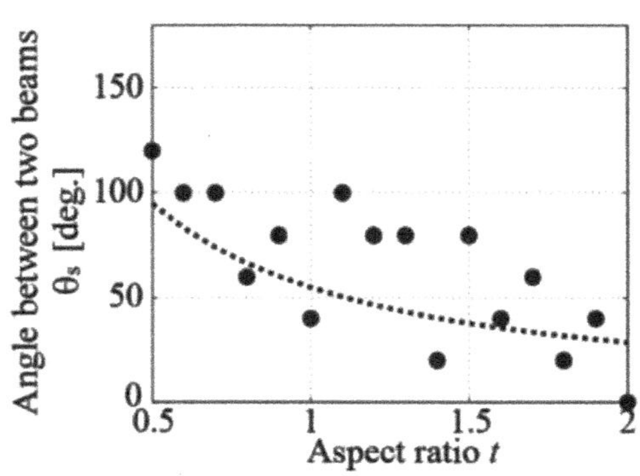

Figure 9.7 Angles between two beams leading to the maximum space-averaged channel capacity, and the dotted line indicates θs obtained when the main beams are pointed to the corner of the room.

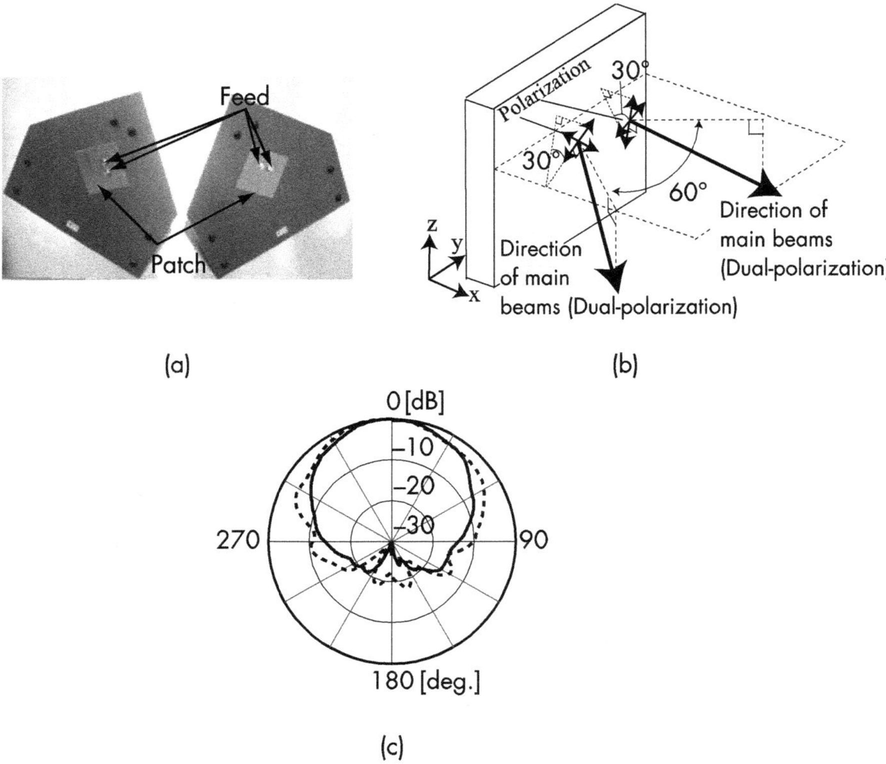

Figure 9.8 Geometry of patch array: (a) fabricated patch array, (b) directions of the main beams, and (c) radiation patterns in the vertical (solid line) and horizontal (dotted line) planes.

At position 2, the signals from the left patch are dominant. In this measurement, the frequency is 2.45 GHz.

Figure 9.10 shows the measured channel capacity and SNR at the three MT positions. In all positions, the SNRs of the patch array-based antenna are higher than those of the reference antenna that is based on the sleeve array. The channel capacities are enhanced by the patch array with improvements of 20% at positions 1 and 2. This 20% improvement is equivalent to an increase of 3 dB SNR, which indicates that the same channel capacity is obtained by transmitting at a half-power level. The channel capacity of the patch array is slightly smaller than that of the sleeve array at position 3 owing

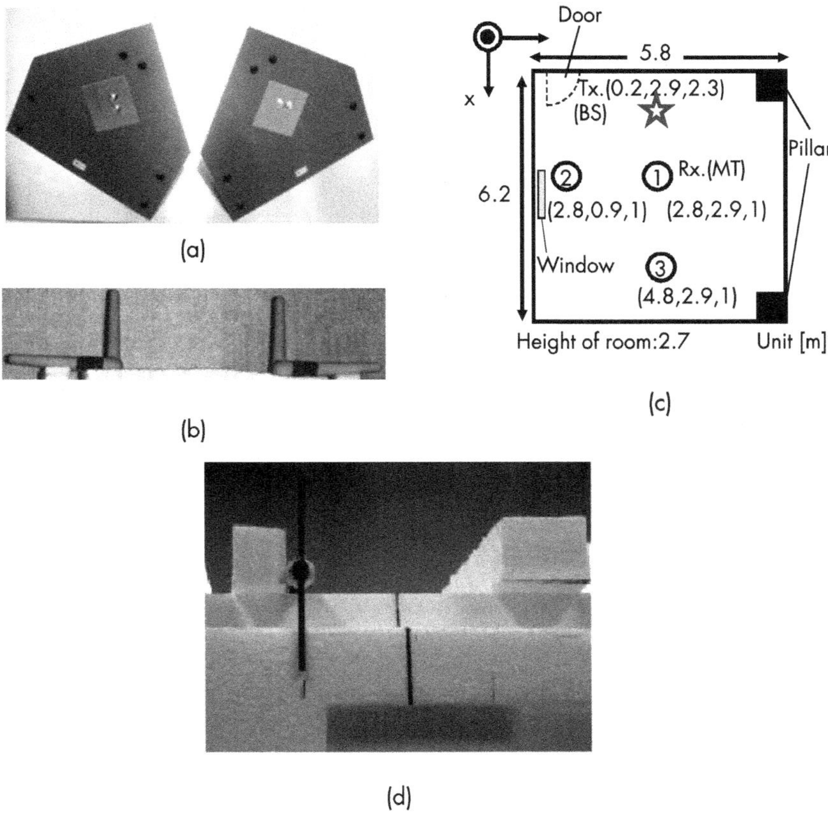

Figure 9.9 Measurement environment: (a) patch array, (b) sleeve array, (c) layout of site, and (d) dipole antenna.

to the degradation in the fourth eigenvalue, which is the minimum eigenvalue, in the spatial correlation matrix. However, the patch-antenna array enhances the channel capacity by having a high SNR at the MT positions.

9.2.2 BS at the Ceiling

This section presents the channel capacity of a BS at the ceiling. Under this condition, the suitable radiation pattern for 4 × 4 MIMO systems has an HPBW value of 80° and a downward tilt angle of 30° with vertical and horizontal polarizations to increase the channel capacity, as shown in Figure 9.11. These conditions are estimated

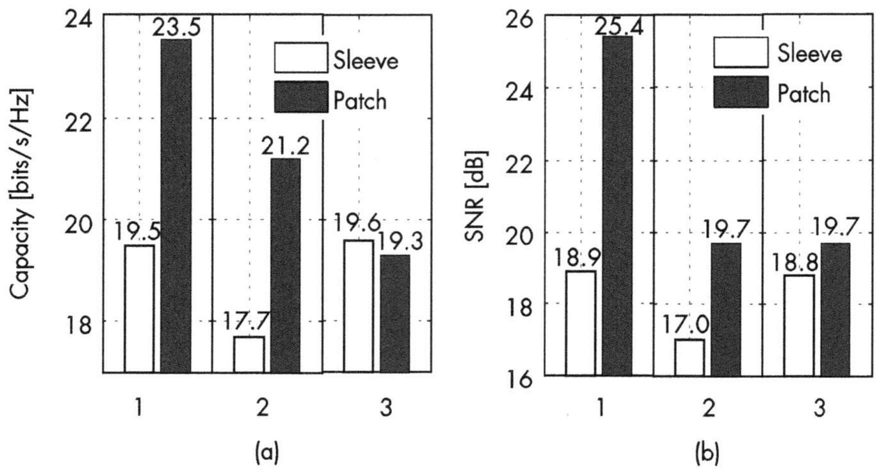

Figure 9.10 Channel capacities and SNRs at MT positions 1, 2, and 3: (a) channel capacity, and (b) SNR.

using the same ray-tracing analysis procedure presented in Section 9.2.1. This antenna design leads to an improvement of 10% in capacity compared with an isotropic antenna [6]. Figure 9.12 shows a MIMO antenna based on the aforementioned design guidelines using four dipole and four slot antennas [7].

In the radiation patterns shown in Figure 9.13(a), the xy (horizontal) plane is the ceiling, and the z axis represents the direction (vertical plane) toward the floor. In the vertical plane, the HPBWs values for the slot and dipole antennas are 84° and 85.5°, respectively.

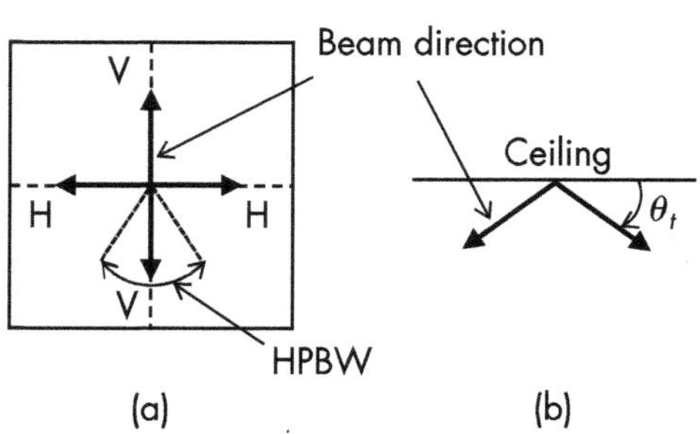

Figure 9.11 Simulation model and radiation patterns leading to large channel capacity; V is vertical polarization, H is horizontal polarization, tilt angle is θ_t: (a) top view, and (b) side view.

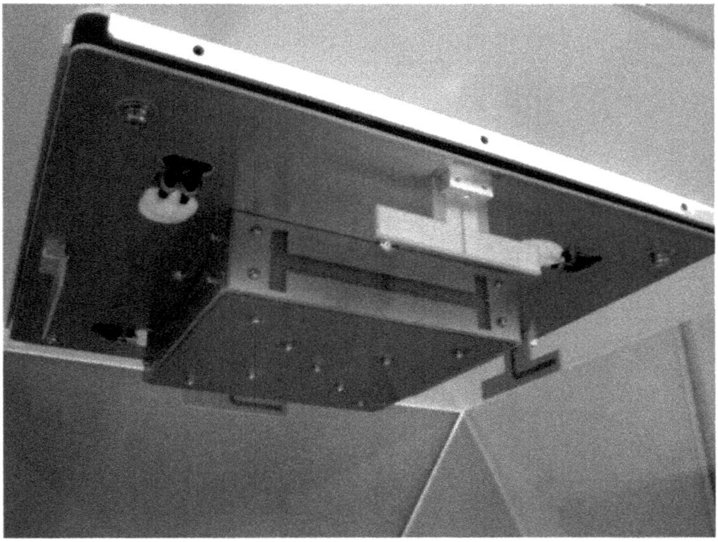

Figure 9.12 Photos of the MIMO antenna constructed using four slot and dipole antennas.

In the horizontal plane, the antenna beams are tilted, and the HPBWs values at 60° from the z-axis are 61° and 53°, respectively. Figure 9.14 shows the scenario of 2 × 2 MIMO channel measurements using this antenna.

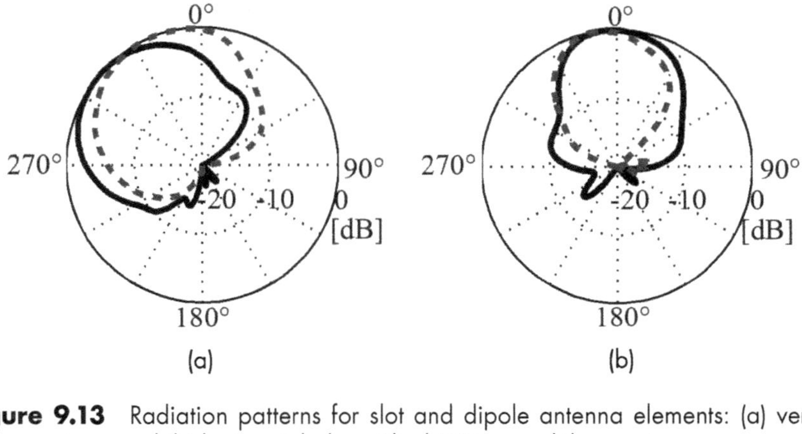

Figure 9.13 Radiation patterns for slot and dipole antenna elements: (a) vertical plane pattern, and (b) horizontal plane tilted 30° toward the z-axis.

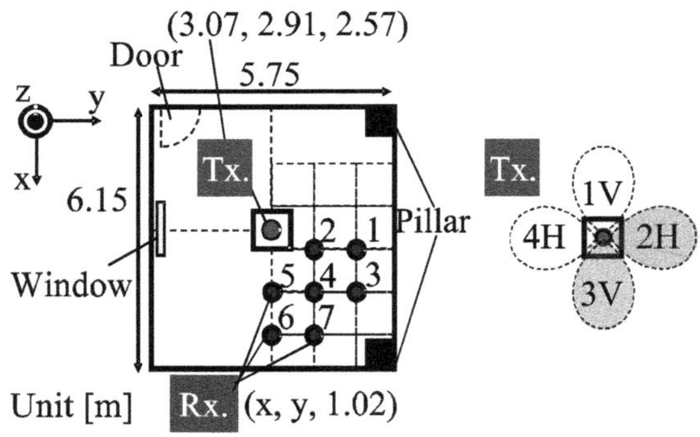

Figure 9.14 Measurement environment.

With regard to the BS, the antenna is fixed at the center of the ceiling, and two streams are radiated from one slot and one dipole. Channel measurements are also performed using sleeve arrays in BS as the reference. Here, sleeve antennas are used for vertical and horizontal polarization. For the MT, the sleeve antenna array whose elements are inclined parallel and arranged perpendicular to the ground is used. The channel matrices are obtained at seven room positions where no obstructions exist between the BS and MT and all the MT positions are in LOS. The measurement frequency is 2.45 GHz.

Figure 9.15 shows the measured SNR and channel capacities of the two antenna types. In five out of seven investigated positions,

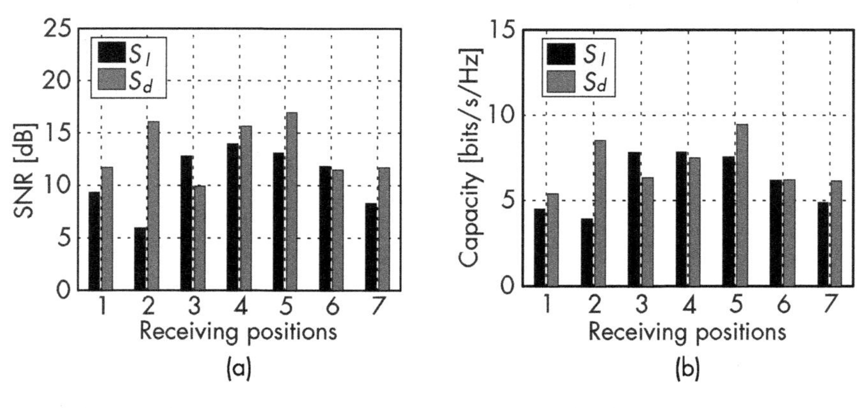

Figure 9.15 Measured SNR and channel capacity for the MIMO antennas constructed using either sleeve (S_l) or slot and dipole (S_d) antennas: (a) SNR, and (b) channel capacity.

SNR is enhanced by the slot and dipole antennas, and the channel capacity is then enhanced at the four positions. The place-averaged SNR of the slot and dipole antennas is enhanced by 2.6 dB and exhibits an improvement in the channel capacity of 16.2% compared with the sleeve antenna. A high channel capacity is obtained at positions with high SNR. Position 4 is an exception to this rule because of the degradation in the second eigenvalue, which is the minimum eigenvalue, in the spatial correlation matrix. Altogether, these results indicate that directive antennas enhance not only the high-order (maximum), but also the low-order (minimum), eigenvalues in the correlation matrix.

As discussed in Section 9.1, high SNR and low spatial correlation are effective in enhancing the channel capacity, which is verified by the results of the directive and dual-polarization antennas.

9.2.3 BS in Open Site

The channel capacity in an open site is estimated using the ray-tracing method, as presented in this section. Figure 9.16 shows that open site, A, above the earth's surface without any disturbing-ray object from a BS with height h_b.

Another open site, B, includes buildings inside and outside the observation area on earth, as shown in Figure 9.17. BS is mounted at the top of the building outside the observation area. The buildings are replaced with uniform dielectric blocks with electrical parameters of relative dielectric constant $\varepsilon_r = 6$ and conductivity $\sigma = 0.023$

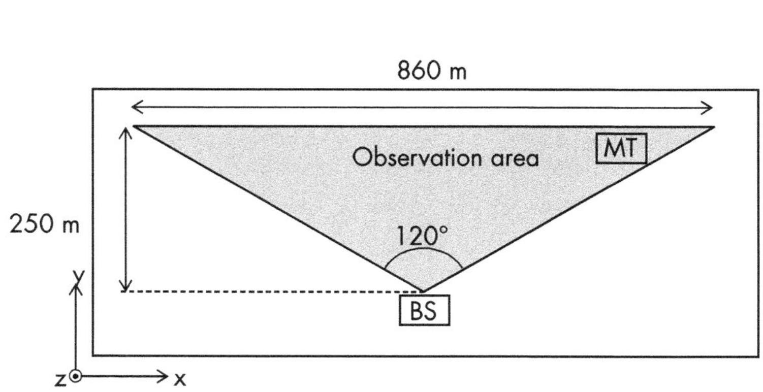

Figure 9.16 Open site A: $P_t = -5$ dBm, $N_f = -85$ dBm, $h_b = 30$m, $\theta_t = 10°$, $h_m = 1$m, $\varepsilon_g = 10$, $\sigma = 0.0023$ S/m.

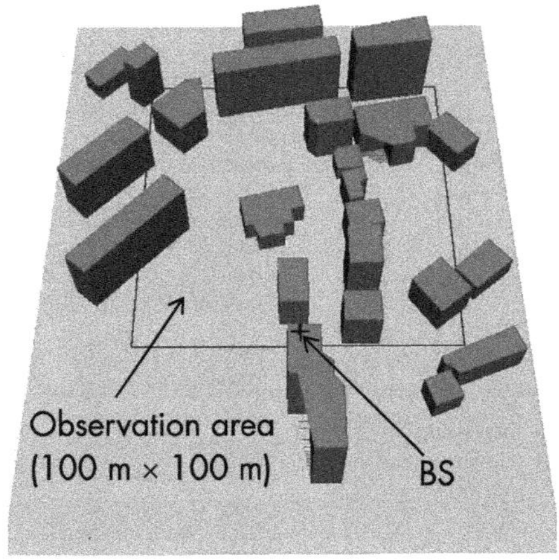

Figure 9.17 Open site B: $P_t = -5$ dBm, $N_f = -85$ dBm, $h_b = 32$m, $\theta_t = 10°$, $h_m = 1$m, $\varepsilon_g = 10$, $\varepsilon_b = 6$, $\sigma = 0.0023$ S/m.

S/m by assuming a dry concrete material. Additionally, the same electrical parameters $\varepsilon_r = 6$ and $\sigma = 0.023$ S/m represent the earth's parameters.

Two BS antennas are used in this ray-tracing simulation to determine the channel capacity in the observation area. Two identical antennas are arrayed to obtain three- and six-sector antennas with four branches, as shown in Figure 9.18. The element spacing is 0.9λ, and the array distance is 1.0λ. A crossed dipole antenna is used for the array element, and HPBW in the horizontal plane is controlled using the shape of the ground plane. The detailed shape of the ground plane is removed, as shown in Figure 9.18, to easily understand the geometry. One of the antennas has HPBW of $\theta_h = 85°$ to divide the horizontal plane into three sectors, and another antenna has HPBW of $\theta_h = 43°$ to emulate the six sectors. The HPBW value of both antennas in the vertical plane is $\theta_v = 7°$, which provides an antenna gain of 17 dBi for the three sectors and 20 dBi for the six sectors. The transmitting power is $P_t = -5$ dBm, and the noise level is set at $N_f = -85$ dBm.

The MT antenna contains a pair of ±45° polarization branches consisting of a sleeve and a slot element. The spacing of the two

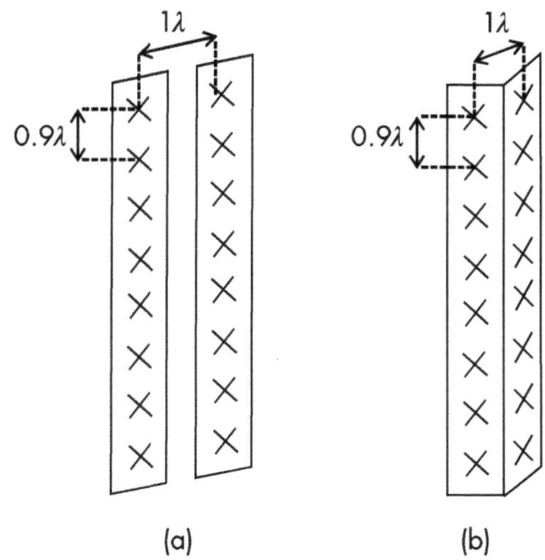

Figure 9.18 Geometry of base station antenna: (a) three-sector antenna, and (b) six-sector antenna.

branches is 1.0λ. The MT above the ground plane with a height of h_m = 1m moves in the observation area, and the electric field is calculated in a 1m grid to estimate the channel capacity. Figures 9.19 and 9.20 show the electric-field distribution in sites A and B, respectively. The radiation patterns of the two types of sector antennas are clearly projected on the observation area in site A. The field strength spreads over the center region by the six-sector antenna because of the tilted beams, whereas the field is focused around the center region by the three-sector antenna.

The average and peak channel capacities of both sites are shown in Figure 9.21, where the left character represents site A or B and the right character represents three or six sectors. The average capacity is increased by the six-sector antenna; however, the peak capacity is reduced by the six-sector antenna. The observed peak capacity around the overlapping area between two beams is enhanced by the three-sector antenna because two BS antennas face in the same direction. The channel capacity depends on the site environment and antenna configuration, which should be considered in the system design.

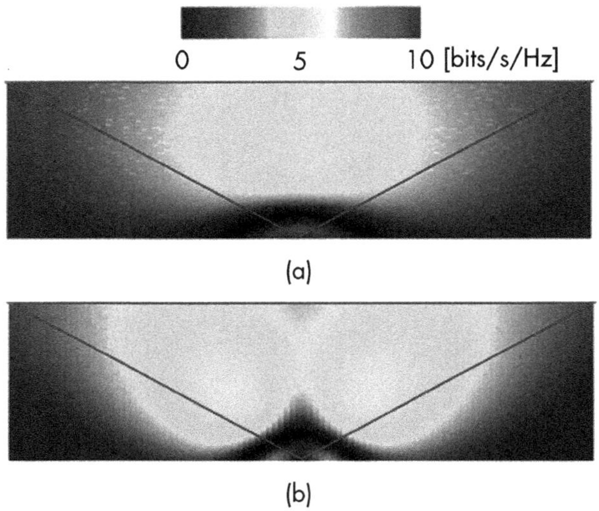

Figure 9.19 Channel capacity distribution of open site A: (a) three-sector antenna, and (b) six-sector antenna.

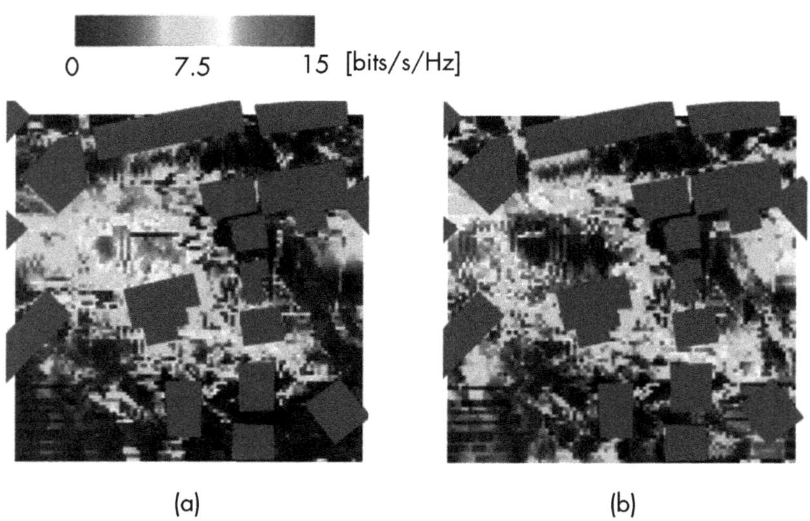

Figure 9.20 Channel capacity distribution of open site B: (a) three-sector antenna, and (b) six-sector antenna.

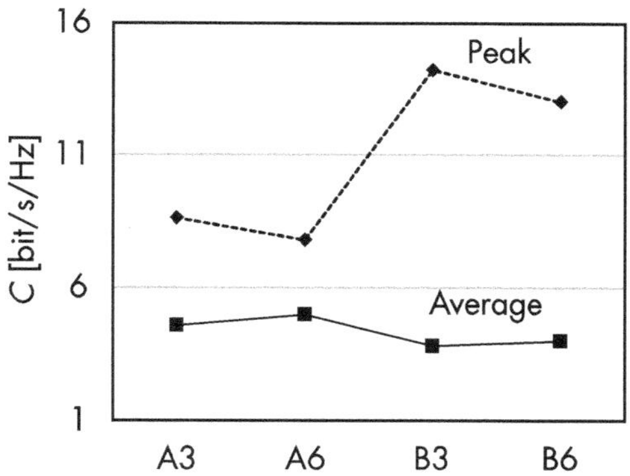

Figure 9.21 Average and peak channel capacity distribution of open site A, B.

9.3 TERMINAL ANTENNA DESIGN AND CHANNEL CAPACITY

This section describes the mutual coupling between the antennas installed in the MT to enhance the channel capacity, along with the influence of radiation efficiency and pattern distortion. The mutual coupling and radiation efficiency are related to the correlation and SNR parameters to estimate the channel capacity. Optimization of the antenna radiation pattern is considered under a realistic propagation model in the MT as well as the effects of the head, hands, and body of users by numerical simulations using a phantom. In this section, *correlation* refers to the antenna correlation because the correlation equation assumes that the distribution of an incoming wave is uniform. Correlation is derived according to the antenna characteristics only.

9.3.1 Mutual Coupling and Radiation Efficiency

Two major parameters, namely, SNR and correlation, are critical in enhancing the channel capacity in the BS antenna design. Obtaining high SNR by increasing the antenna gain is not easy for MT antennas because the available space in the MT is limited. Most built-in antennas are installed at the bottom of the MT to reduce the specific absorption ratio (i.e., the allowable exposure level of electromagnetic waves) when phones are held in the talking position. This location also restricts the spacing between antennas, which is necessary to obtain a low correlation. MT antennas are characterized by radiation efficiency, instead of SNR and correlation, because the antenna array distance is close. First, a simple simulation model based on two half-wavelength dipoles is considered as an example of MT antennas.

The effects of array spacing on the MIMO channel capacity are investigated for an array with and without mutual coupling. Figure 9.22 shows the geometry of two omnidirectional antennas on the x axis with spacing d under random incoming waves in the xy plane.

Assuming uniform incoming-wave distribution $A_i(\theta) = 1$, the spatial correlation of the two antenna elements is obtained by the array factor as follows:

$$\gamma_s(d) = \frac{\int_0^{2\pi} Ai(\theta)\exp\left(jk_o d\cos\theta\right)d\theta}{\int_0^{2\pi} Ai(\theta)\,d\theta} = J_0\left(k_o d\right) \tag{9.18}$$

where k_o is the wave number in free space, and $J_0(x)$ is the zeroth-order Bessel function. Small spatial correlation $\gamma_s < 0.5$ is obtained at element spacing $d > 0.2\lambda$, as shown in Figure 9.23; however, the mutual coupling and radiation efficiency between the antenna elements are not considered in (9.18). Built-in antennas in MTs are restricted in space to increase the mutual coupling and pattern distortion. To determine the effects of mutual coupling, radiation efficiency, and pattern distortion on the correlation, two $\lambda/2$ dipole antennas are used in the simulation for a simple model of MT antennas. The two dipoles are parallel in the z-axis direction. The center of one dipole

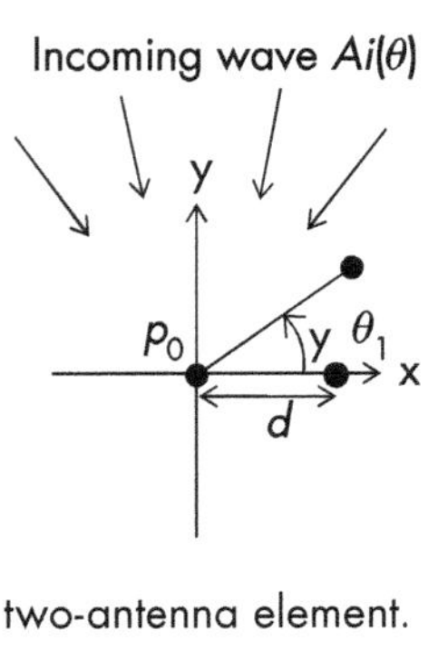

Figure 9.22 Geometry of a two-antenna element.

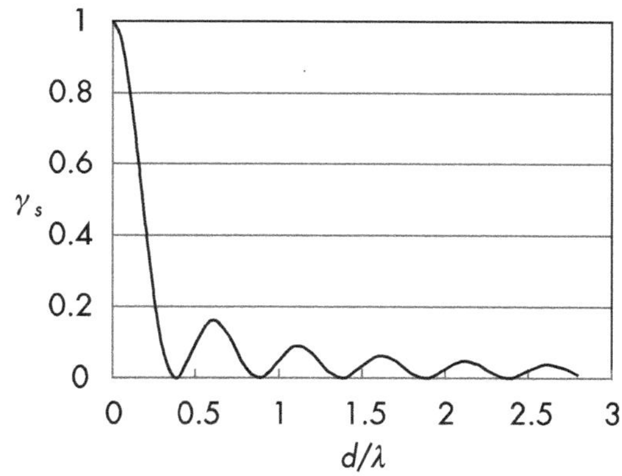

Figure 9.23 Correlation of two antenna element.

is at the origin, and the other dipole center is on the x axis at position $x = d$. The correlation between the two dipoles is given by the scattering parameters as [8]:

$$\gamma_d = \frac{\left|S_{11}^* S_{12} + S_{12}^* S_{22}\right|^2}{\left(1 - \left|S_{11}\right|^2 - \left|S_{21}\right|^2\right)\left(1 - \left|S_{12}\right|^2 - \left|S_{22}\right|^2\right)} \tag{9.19}$$

Assuming no losses in the antenna elements, the radiation efficiency of antenna 1 is given by S_{11} and S_{21} as follows:

$$\eta_r = 1 - \left|S_{11}\right|^2 - \left|S_{21}\right|^2 \tag{9.20}$$

Mutual coupling also causes a distortion in the radiation pattern, which is defined by the ratio of the maximum and minimum gains in the xy (E) plane as:

$$\Delta_g = \frac{G_{\max}}{G_{\min}} \tag{9.21}$$

The evaluation factors for the array spacing are shown in Figure 9.24. The correlation of γ_d is reduced by mutual coupling compared with

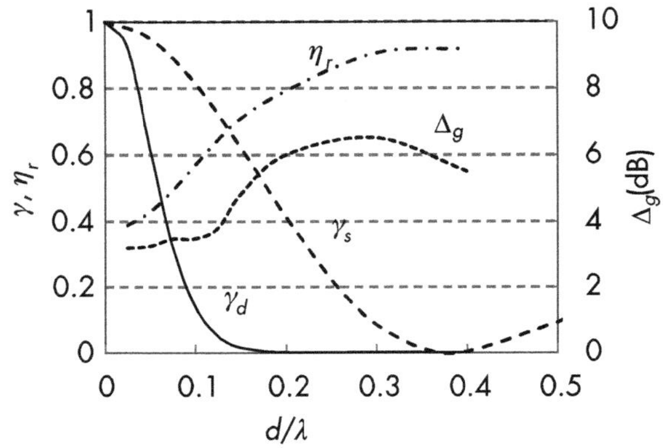

Figure 9.24 Correlation, radiation efficiency, and pattern distortion of two-element dipole antennas.

spatial correlation γ_s defined by (9.18), which is also explained by the observed pattern distortion at values of more than 3 dB in the *xy* plane. We should note that efficiency η_r is reduced by the small array spacing. The reduction in efficiency reduces the channel capacity.

As described in Section 9.1.1, the channel capacity is approximated using (9.13). In the case of the 2 × 2 MIMO systems, the ergodic channel capacity is approximated using the first term of (9.13) under a high-SNR regime. Thus, the channel capacity, including the efficiency, is approximated as:

$$C(\eta) \cong N_r \log_2\left(\frac{\eta_r \rho}{N_t}\right) \tag{9.22}$$

The channel capacity is normalized to limit the maximum radiation efficiency to $\eta_r = 1$, as shown in Figure 9.25. In small-spacing arrays, the radiation efficiency should be carefully considered in the design of MT antennas [6]. The saturation in the capacity increases at $\gamma_s < 0.5$ shown in Figure 9.4(b), and the effect of radiation efficiency are key parameters in the MT antenna design.

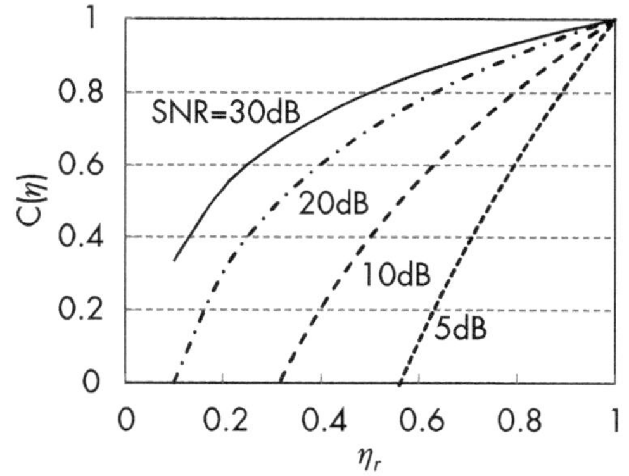

Figure 9.25 Normalized channel capacity as a function of radiation efficiency.

9.3.2 Channel Capacity Estimation of MT Antennas

For diversity application, orthogonal polarization is effective in reducing the correlation coefficient between two antenna ports. However, the channel capacity of MIMO also depends on the antenna gain. To characterize MT antennas for MIMO applications, this section presents a method for estimating the channel capacity of MTs. The propagation environment is based on a Nakagami-Rice model together with the Kronecker scattering assumption due to the surrounding objects, as shown in Figure 9.26 [9]. This model produces uncorrelated scattering by many scattering objects between the BS and MT, which can separate the transmitting and receiving correlation matrices into independent matrices.

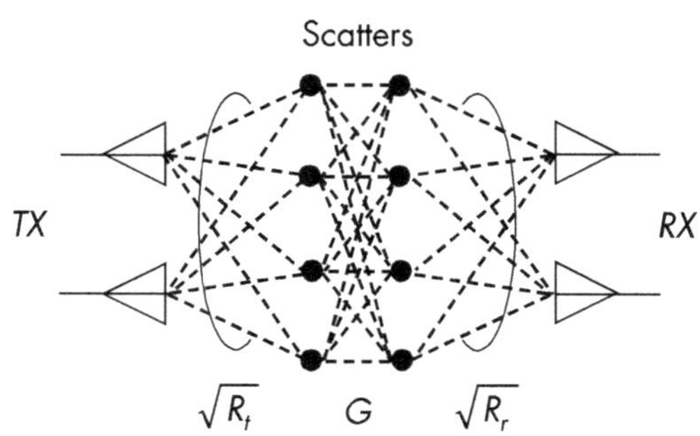

Figure 9.26 Kronecker model.

For simplicity in simulation, all incoming waves are concentrated in the *xy* plane, and the components of the channel matrix in (9.16) are expressed as follows:

$$\boldsymbol{H}_D(\phi) = \boldsymbol{\Gamma}(\phi) \otimes a(\phi),\ H_D(\phi) = \sqrt{\boldsymbol{R}_r}\boldsymbol{G}\left(\sqrt{\boldsymbol{R}_t}\right)^t \tag{9.23}$$

where $\boldsymbol{H}_D$ and $\boldsymbol{H}_S$ are the direct and scattering components, respectively, and ϕ is the angle of the incident wave in the *xy* plane [10]. $\boldsymbol{\Gamma}(\phi)$, $\boldsymbol{a}(\phi)$, and $\boldsymbol{G}$ are the complex radiation patterns of each antenna, an array factor, and the Gaussian distribution matrix, respectively, and $\otimes$ represents the Hadamard product. The correlation unit matrices at the transmitting and receiving sides ($\boldsymbol{R}_t$, $\boldsymbol{R}_r$) are given by assuming a 2×2 MIMO as follows:

$$\boldsymbol{R}_i = \begin{pmatrix} P_{i1} & \sqrt{P_{i1}P_{i2}}\gamma \\ \sqrt{P_{i1}P_{i2}}\gamma & P_{i1} \end{pmatrix}, \quad i = r,t \tag{9.24}$$

where P_{ij} (j = 1, 2) is the receiving power of each antenna, and γ is the correlation factor between the antennas. The uniform incoming-wave distribution in the *xy* plane, namely, γ (as shown in Figure 9.23), is obtained using the complex radiation pattern of the antennas as follows:

$$\gamma = \frac{|g_{12}|}{\sqrt{g_{11}(\phi)}\sqrt{g_{22}(\phi)}} \tag{9.25}$$

$$g_{ij} = \int_0^{2\pi} \left\{ XPR \times E_{\theta i}(\phi) E_{\theta j}^*(\phi) + E_{\phi i}(\phi) E_{\phi j}^*(\phi) \right\} d\phi \tag{9.26}$$

where $E_{ki}(\phi)$ ($k = \theta, \phi$) is the k component of the complex radiation pattern of antenna i in the *xy* plane, and *XPR* is the cross-polarization ratio of the incoming wave, as defined in Section 3.3.2. The primary component of E_θ is used in (9.26). Finally, by substituting (9.23) and (9.24) into (9.16), the channel matrix of 2×2 MIMO is obtained. Then, the channel capacity is estimated using (9.11).

9.3.3 Channel Capacity of MT Antennas

Figure 9.27 shows the geometries of a receiving antenna. The built-in antennas of MTs such as smartphones are installed at the bottom edge to reduce the power density of the transmitting power of MTs on a human body. The MT is approximated using a rectangular conductor plate, and an inverted-F antenna (IFA) is used as the built-in antenna. A nonradiating element of IFA is folded to reduce its volume ($w \times h \times l$). The length of nonradiating element l is adjusted for four geometries, as shown in Figure 9.27, to resonate at 2 GHz, and w and h are fixed at 5 mm.

IFA, denoted as R_1 in Figure 9.27(a), is a common antenna for other geometries, and the position of other IFAs is varied to find the optimum antenna location for a MIMO system. Figure 9.28 shows the correlation and radiation efficiency of each antenna geometry estimated using the formulas defined in Section 9.3.1. The simulation assumes no conducting loss. Therefore, the radiation efficiency dominated by S_{11} and S_{12} is equivalent to an average antenna gain in the xy plane. The small difference in radiation efficiency between the #1 and #2 antennas are caused by the asymmetrical antenna arrangement in the xy plane. The maximum efficiency is obtained in geometry C because the distance between the two antennas is the largest and the radiating elements of IFA are orthogonal to each

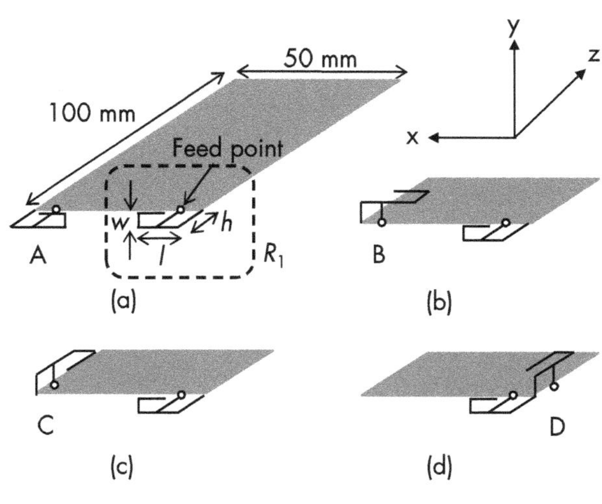

Figure 9.27 Inverted-F antennas on plate (category 1).

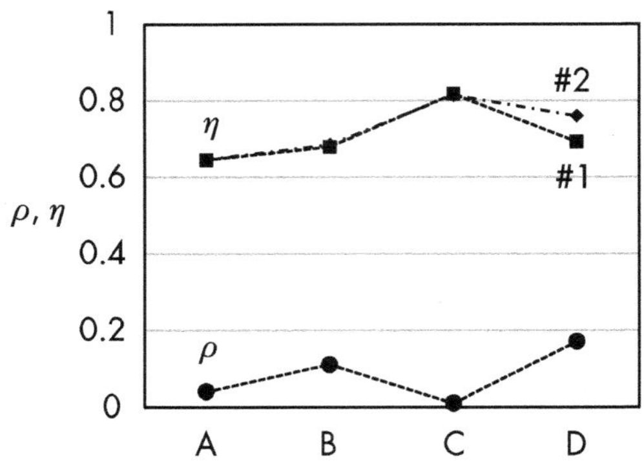

Figure 9.28 Correlation and radiation efficiency of category 1; #1 = R_1, #2 = A, B, C, D.

other. Figure 9.29 shows the channel-capacity estimation of the four antenna geometries. The maximum capacity is obtained by geometry C, which is almost the same as the radiation efficiency results shown in Figure 9.28. We conclude that the dominant factor that maximizes the channel capacity is the average antenna gain in the plane where the incoming wave exists. This condition is known as the maximum effective gain for a uniform incoming-wave distribution in an incident plane [11]. The correlation and radiation efficiency of the other antenna geometry shown in Figure 9.30 is shown in Figure 9.31, and those in Figure 9.27(c) are shown for reference. The efficiency is increased in antennas with a large distance, which is the same as that in Figure 9.28.

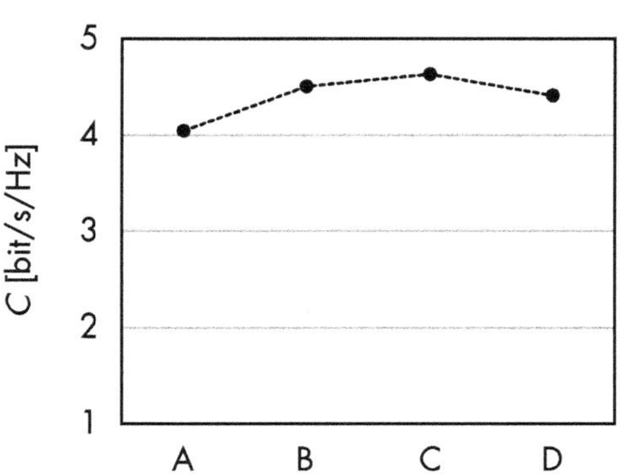

Figure 9.29 Channel capacity of category 1; K = 3 dB, SNR = 10 dB.

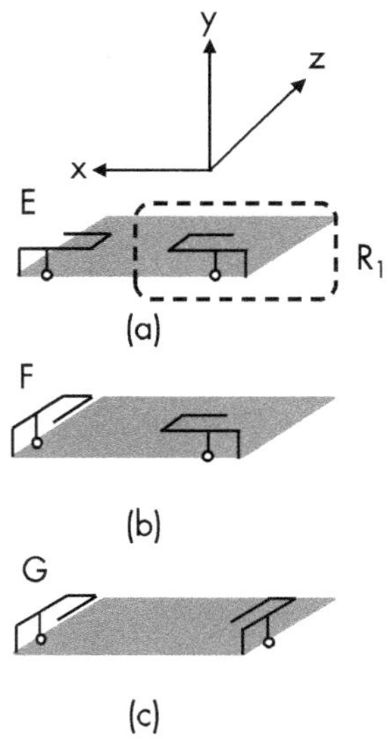

Figure 9.30 Inverted-F antennas on plate (category 2).

The channel capacity of the MT antennas shown in Figure 9.29 is estimated in which the effect of human body in using MTs is neglected. The radiation power from smartphone antennas is absorbed by the body, which degrades the channel capacity due to the decrease in radiation efficiency. The effect on the human body is emulated using a phantom with the same electrical parameters as

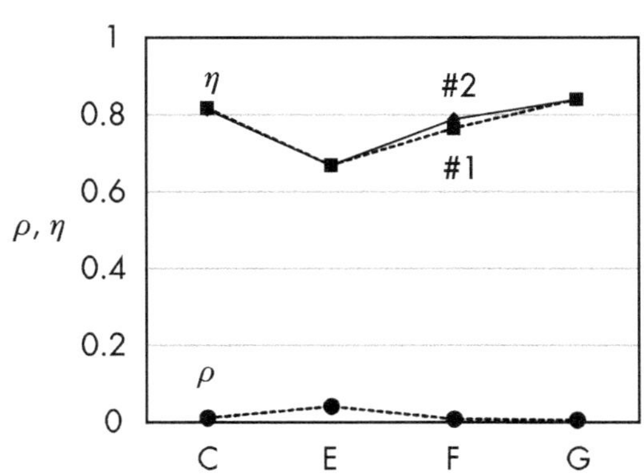

Figure 9.31 Correlation and radiation efficiency of category 1; #1 = R_1, #2 = E, F, G.

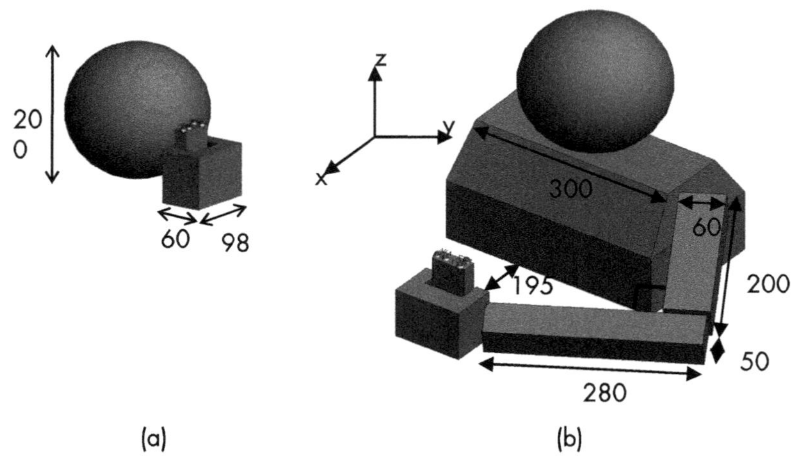

Figure 9.32 Phantom model for viewer mode, all units are in millimeters; head part ($\varepsilon_r = 40$, $\sigma = 1.4$ S/m), other parts ($\varepsilon_r = 54$, $\sigma = 1.45$ S/m). (a) Head and hand, and (b) head, chest, arm, and hand.

those in a human body. Figure 9.32 shows two phantom models used in the following simulation [12].

A head is represented by a dielectric sphere, and a hand is approximated by a U-shaped block, as shown in Figure 9.32(a). This condition is denoted as a talk position. A chest and an arm are shown in Figure 9.32(b) to emulate the smartphone used in the viewer mode. The channel capacity is not seriously affected by the MT position. However, the distance between the MT and the body affects the capacity, as shown in Figure 9.33. The reduction in the channel capacity in the talk position is larger than that in the viewer position.

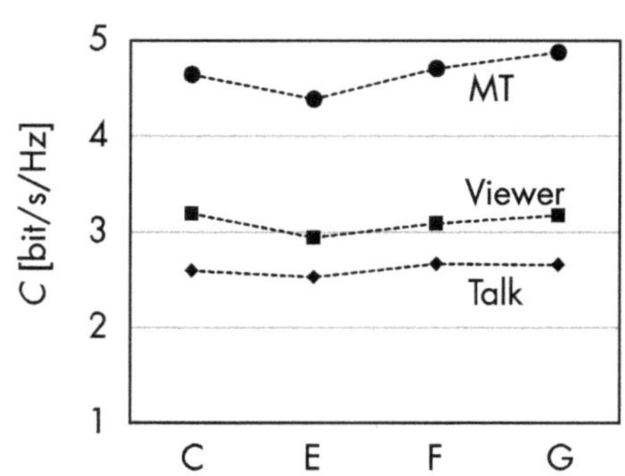

Figure 9.33 Channel capacity of MT in the presence of phantom.

References

[1] Foschini, G. J. "Layered Space-Time Architecture for Wireless Communication in Fading Environment Using Multi-Element Antennas," *Bell Labs Tech. J.*, Autumn 1996, pp. 41–59.

[2] McNamara, D. P., et al., "Capacity Variation of Indoor Multiple-Input Multiple-Output Channels," *Electron. Lett.*, Vol. 36, November 2000, pp. 2037–2038.

[3] Sakaguchi, K., et al., "MIMO Channel Capacity in an Indoor Line-of-Sight (LOS) Environment," *IEICE Trans. Commun.*, Vol. E88-B, No. 7, July 2005, pp. 3010–3019.

[4] Paulraj, A., et al., *Introduction to Space Time Wireless Communications*, Cambridge University Press, 2003.

[5] Uchida, D., et al., "Optimum Beam Direction and Width for Directional Antenna Indoor MIMO Systems," *Proc., Int. Symp. Antennas Propag.*, 1D5-2, August 2007, pp. 165–168.

[6] Uchida, D., et al., "Channel Capacity Enhancement by Polarization and Pattern of Access Point Antenna in Indoor MIMO Systems," *Proc., IEEE Antennas Propag. Soc. Int. Symp.*, July 2008.

[7] Uchida, D., et al., "A Low-Profile Dual-Polarized Directional Antenna for Enhancing Channel Capacity in Indoor MIMO Systems," *IEICE Trans. Commun.*, Vol. E93-B, No. 10, October 2010, pp. 2570–2577.

[8] Thaysen, J., and K. B. Jakobsen, "Envelope Correlation in (N, N) MIMO Antenna Array from Scattering Parameters," *Microwave and Optical Technology Letters*, May 2006, Vol. 48, No. 5.

[9] Kermoal, J. P., et al., "A Stochastic MIMO Radio Channel Model with Experimental Validation," *IEEE Journal on Selected Areas in Communications*, Vol. 20, No. 6, August 2002, pp. 1211–1226.

[10] Turuta, M., and Y. Karasawa, "Simplified Estimation Method of the Largest Eigenvalue distribution in Nakagami-Rice MIMO Channel," *IEICE Trans-B*, Vol. J87-B, No. 9, September 2004, pp. 1486–1495 (in Japanese).

[11] Taga, T., "Analysis for Mean Effective Gain of Mobile Antennas in Land Mobile Radio Environments," *IEEE Trans. Vehicular Technology*, Vol. 39, No. 2, May 1990, pp. 117–131.

[12] Arai, H., and J. Ohno, "Channel Capacity Enhancement by Pattern Controlled Handset Antenna," *Radioengineering*, Vol. 18, No. 4, December 2009, pp. 413–417.

10

PROPAGATION MEASUREMENTS

This chapter presents the measurement procedure to verify or adjust the propagation environment designed using the BS antenna radiation pattern. The propagation measurements are divided into narrowband and wideband measurements. The narrowband measurement provides the propagation-path-loss factor, and the wideband measurement provides the delay profile of the site. The two measurement procedures are described, and several examples are presented. On the basis of the path-loss estimation, the link budget is determined to design the cellular system parameters. The downlink and uplink budgets are explained using typical sample data. The performance test for the MTs is also described under indoor artificial-propagation environments, such as the field simulator for fading simulation and the MIMO emulator for throughput measurement.

10.1 PROPAGATION-LOSS AND DELAY-PROFILE MEASUREMENTS

Propagation loss is a dominant factor in the design of cellular systems to create a coverage area by adjusting the radiation pattern of the BS antenna. The propagation loss is predicted using the several methods described in Section 8.1, which needs to be verified in a real propagation environment. This section describes the measurement setup and parameters to characterize the narrowband propagation and presents examples that are measured in several sites. The delay-profile measurement is also demonstrated as a key process of digital mobile-communication systems.

10.1.1 Measurement System

The propagation measurement is used to determine the loss factor in a specific site and is required when the performance of the base station antenna and the corresponding coverage area are checked. This measurement is performed using a receiver installed on a vehicle to perform measurement in a wide coverage area. Figure 10.1 shows a typical onboard receiver configuration that is used to measure the received-signal strength. The system consists of the following subsystems: a location detector, a data processor, and an electric-field-strength measuring device. The location-detection subsystem estimates the precise position of the vehicle and consists of a directional sensor, a distance sensor, route-history data, and a road map to help generate the measured route history. Multiple measuring receivers are also installed on board to allow simultaneous measurements of multiple radio channels.

Two types of calibration procedures are generally used for precise propagation measurements. The first involves linearity compensation

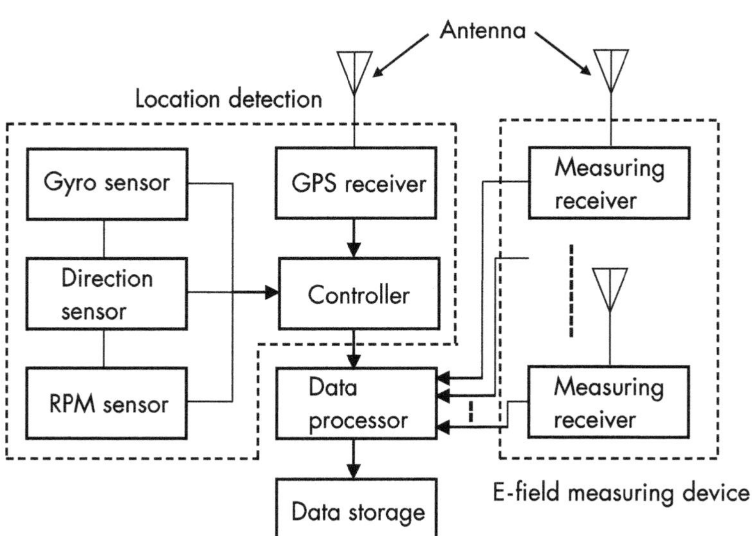

Figure 10.1 Measurement system.

of the receiver in systems that are used in short-range measurement. The second type of calibration procedure is the time standard calibration in the delay-measurement system described in Section 10.3. Calibration of the RF system is carried out by directly connecting the signal-generator output to the antenna input port of the RF receiver.

A problem is encountered in the propagation measurement (i.e., how to remove the electrical effect of the supporting structure of the mobile terminals). The receiving antennas are usually attached using low-dielectric-constant materials such as polyethylene foam, wood, or corrugated fiberboard. In the measurements, the receivers and data-recording devices are made of a metal frame. To minimize pattern distortions due to electrical obstacles, the equipment should be located away from the tested devices. An antenna needs some space to stably operate. This space is known as *antenna clearance* and is calculated by placing an electrical obstacle near the tested antenna [1]. This simple obstacle is a $\lambda/2$ conducting wire to disturb the antenna-input characteristics and radiation patterns (Figure 10.2). The input impedance is seriously affected by the wire distance from the test antenna less than λ, whereas the deviation radiation patterns do not converge in this spacing distance. Figure 10.3 shows that the H-plane radiation pattern of a $\lambda/2$ dipole antenna is seriously affected by the presence of the $\lambda/2$ conducting wires that are arrayed in parallel. The ripple of the H-plane radiation pattern is ±2.4 dB at distance $d = \lambda$ between the dipole and wire, and it is ±0.8 dB at $d = 3\lambda$, as shown in Figure 10.4. A pattern deviation of less than ±1 dB is

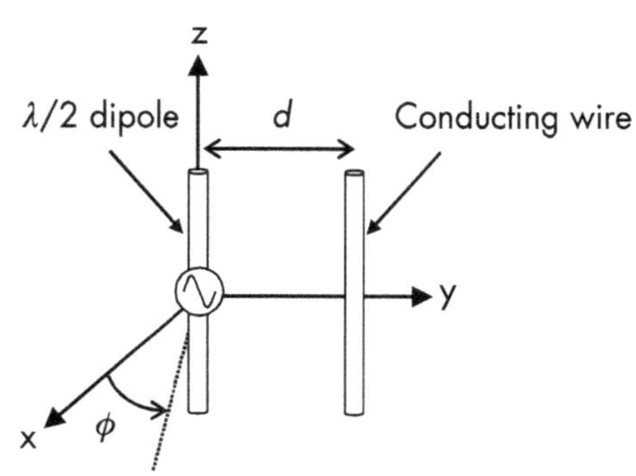

Figure 10.2 $\lambda/2$ dipole and conducting wire.

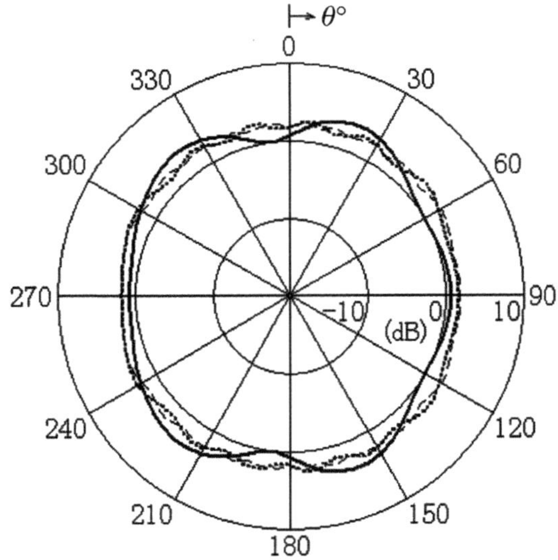

Figure 10.3 Radiation pattern of $\lambda/2$ dipole with conducting wire in length of $\lambda/2$; solid line is $d = 1\lambda$, dotted line is $d = 2\lambda$, dashed line is $d = 3\lambda$.

guaranteed for the tested devices that are at least 3λ away from the electrical obstacles.

10.1.2 Delay-Profile Measurement

Recent mobile-communication systems use ever-increasing higher speed data-transmission rates. The delay profile of a site limits the maximum data-transmission rate of the system. When the delay

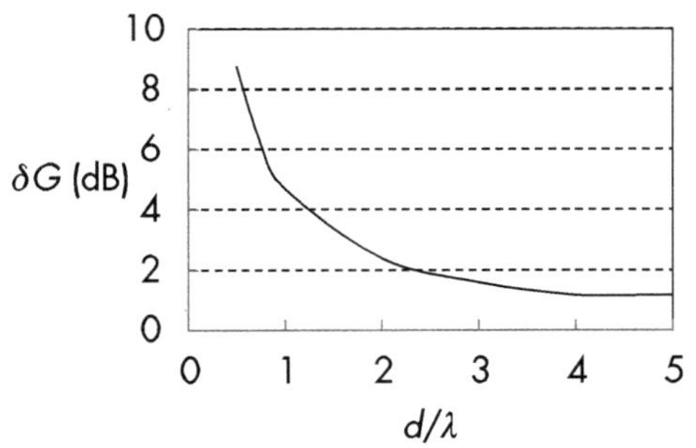

Figure 10.4 Pattern deviation of $\lambda/2$ dipole with conducting wire.

profile is known in advance, an equalizer circuit can be used to cancel the unwanted signal delay. A rake receiver can be used to lump several delayed signals using the direct propagation-path signal and thus increase the effective received-signal level [2]. Therefore, the delay profile at the site should be measured in advance to determine the rake-receiver parameters.

The delay-profile measurement system is shown in Figure 10.5. In this measurement type, a precise timing signal is needed for both the transmitter and receiver. When the receiving points are located close to the transmitter, a common signal-generator source can supply a timing signal to both the transmitter and receiver using a coaxial cable. However, the use of cables is not feasible for these measurements. Table 10.1 lists the stability of the oscillator [3].

The cesium-frequency standards are very stable and accurate, and they are used as frequency references in many institutions. Rubidium oscillators are available as small and cost-effective devices, and are used in a standard signal generator to synchronize the oscillators of the transmitter and receiver for delay-profile measurements. To synchronize the signal phase between the transmission and reception systems, two rubidium oscillators are synchronized before the actual measurement. However, this synchronization may last for a few days only; thus, for precise measurement, the synchronization process must be repeated every two to three hours.

When a delay-profile measurement is performed indoors, the transmitting and receiving antennas can both be connected to a vector network analyzer to obtain the received-signal time-domain characteristics. After the measurement has been performed over a wide frequency range, the transmission-coefficient (S_{21}) frequency data are transformed from the frequency domain to the time domain using Fourier transformation, which is the simplest method for obtaining the time-domain delay profile. However, we should note that the

Table 10.1
Stability of Oscillator

	Per One Second	Per One Day
TCXO	–	10^{-9} to 10^{-18}
Rubidium	10^{-12} to 10^{-11}	10^{-13}
Cesium	10^{-12} to 10^{-11}	10^{-14} to 10^{-13}

TCXO is temperature compensated crystal oscillator.

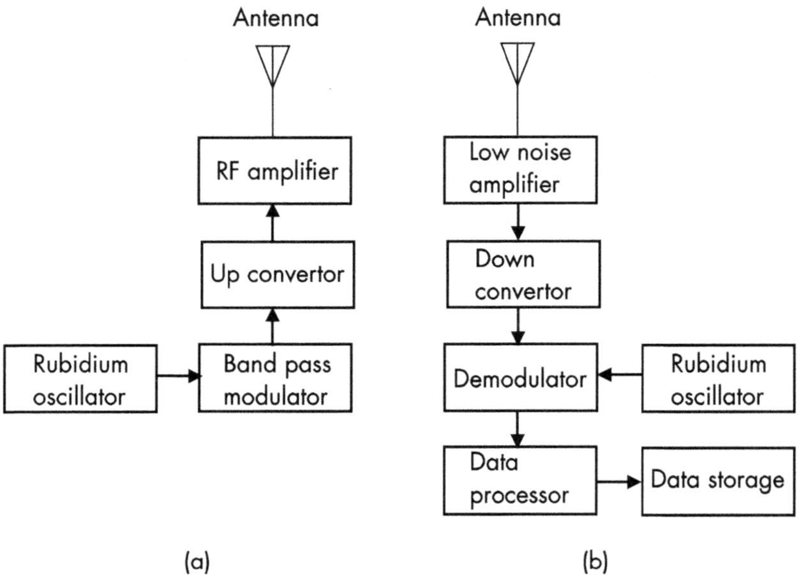

Figure 10.5 Relay profile measurement system: (a) transmitting system, and (b) receiving system.

measurement resolution depends on the frequency bandwidth of the employed antennas.

10.1.3 Examples of Propagation Measurement

Figure 10.6 shows the measured propagation loss in a city area, and the measurement parameters are described in the caption [4]. The free-space propagation loss expressed in (2.2) in each frequency band is indicated by the solid line. The approximation-loss curves are expressed as $(\lambda/4\pi d)^{\alpha}$, and the dotted line is obtained by fitting the measured data to the curve. Attenuation factor $\alpha \cong 2.5$ is a site-specific value. A precise loss-estimation curve is provided by the modified Walfisch-Bertoni model [5] by remodeling the propagation paths and adding multiple reflections between building [4].

Figure 10.7 shows another loss measurement obtained by Kokura [6]. The approximation curve is different from that shown in Figure 10.6 and is expressed as follows:

$$L = -23.0 + 56.3\log d \tag{10.1}$$

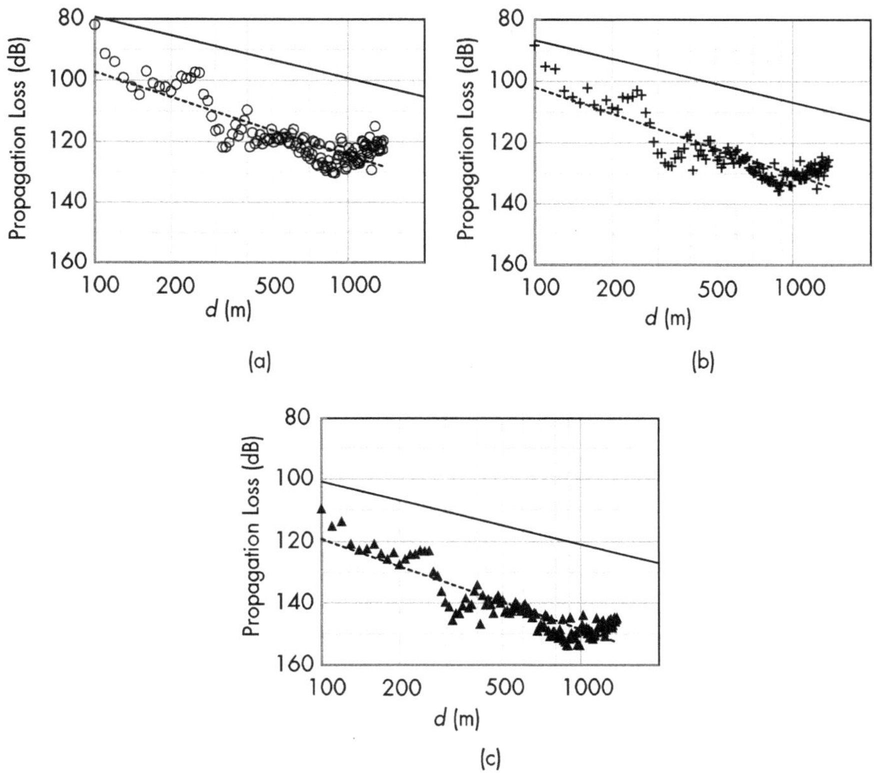

Figure 10.6 Propagation loss in city area (Tokyo). Solid line is propagation in free space, dotted line is approximation curve of measured data, TX antenna height is 42.5m, RX antenna height is 2.7m, each plot by median in 10m. (a) 2.2 GHz, (b) 5.2 GHz, and (c) 26.4 GHz.

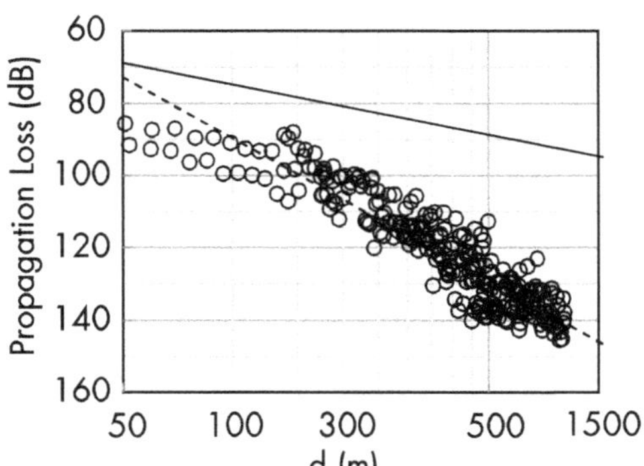

Figure 10.7 Propagation loss in city area (Kokura). Solid line is propagation in free space, f = 1.298 GHz, dotted line is approximation curve of measured data, TX antenna height is 12.5m, RX antenna height is 1.5m. Each plot by median is 10m.

where loss L is expressed in decibel units and distance d is expressed in meters. The propagation loss heavily depends on the site environment, and several parameters such as building height and road are included to increase the estimation accuracy described in Section 8.1.

The delay profile is measured at 8.45 GHz using the setup shown in Figure 10.5. The propagation environment shown in Figure 10.8 is a suburb area surrounded by mountains. Buildings A–I are located along the road, and the height parameters are listed in Table 10.2 [7].

An omnidirectional antenna is used as the BS antenna at the top of the building, and the receiving direction at the measurement position on the road is varied using a six-sector antenna [8]. The half-power beam width of the sector antenna is 60°, and the front-to-back ratio is more than 20 dB. The antenna consists of tapered slot antennas on a circular ground plane. Numbers 1 to 6 in Figure 10.8 denote the direction of each sector. The distance between BS and measurement points P_1 and P_2 are 240m and 110m, respectively.

P_1 is at the LOS of BS. Then, the typical propagation model of the direct wave exponentially decreases, as shown in Figure 10.9(a). Only a direct wave from BS is received at sector 3 (S3), and the delay spread defined by (2.40) is $\sigma = 0.02\ \mu s$. The delay waves shown in Figure 10.9(b) are observed at sector 6 (S6) in the opposite direction of the BS, and the delay spread is maximum ($\sigma = 0.273\ \mu s$).

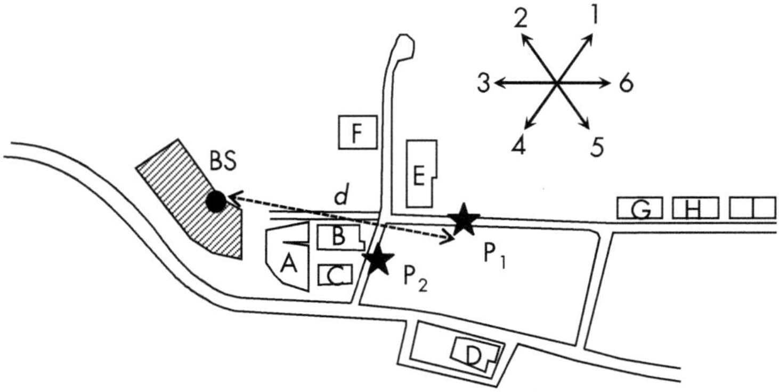

Figure 10.8 Delay profile measurement in suburb area (Yokosuka). $d_1 = 240$m, $d_2 = 110$m.

Table 10.2
Building Parameters

	BS	A	B	C	D	E	F	G	H	I
H [m]	25	31	34	6	21	30	14	24	30	33
Floor	6F	7F	7F	IF	5F	6F	3F	7F	9F	9F

A NLOS exists at point P_2, and the direct waves are low, as shown in Figure 10.10. A reflection wave at approximately $t = 3$ μs by the building is received at sector 1 (S1). However, it is not observed at sector 6. The delay spread at sector 6 (S6) is smaller than that at sector 1, which shows that few buildings cause delay waves. As indicated in these examples, the delay-spread measurements by the sector

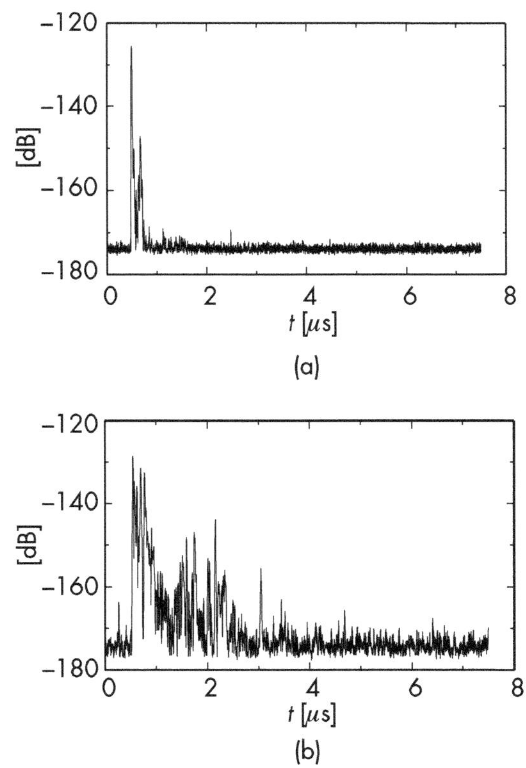

Figure 10.9 Measured delay profile at P_1: (a) S3, σ = 0.02 [μs], and (b) S6, σ = 0.273 [μs].

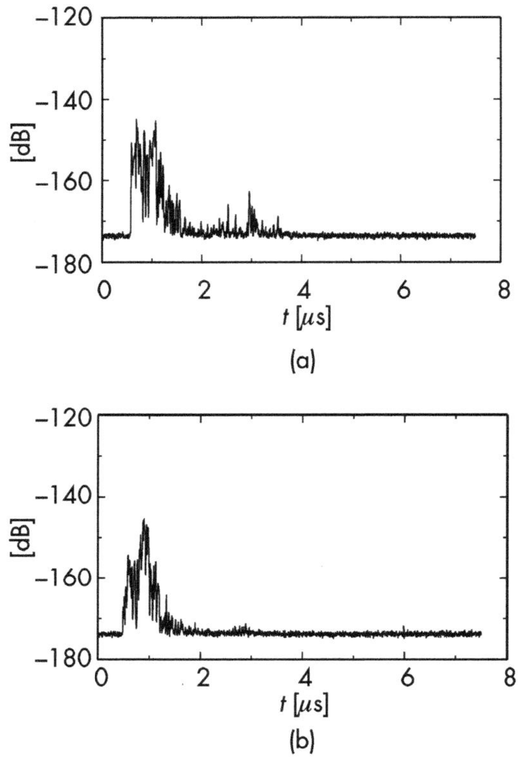

Figure 10.10 Measured delay profile at P_2: (a) S1, $\sigma = 0.274$ [μs], (b) S6, and $\sigma = 0.126$ [μs].

antenna provides the direction of arrival information from BS at the receiving points.

10.2 LINK BUDGET

In the design of cellular systems, a cell coverage area should be determined according to the BS antenna gain and transmitting power. The link-budget calculation provides the maximum signal attenuation between the BS and MT antennas. The uplink budget is provided by the MT, which transmits the maximum power level at the cell edge, whereas the downlink budget is provided by the BS, which transmits the maximum power level. The received power level must exceed the receiver sensitivity, which determines the minimum transmitted power level provided by the antenna gain and transmitter output.

The dominant loss factor in the cellular system is the propagation loss determined by the site environment, and another loss factor is added to the link-budget simulation.

The receiver sensitivity depends on the noise level. Thermal noise increases with temperature and is defined as follows [9]:

$$N_t(\text{dBm}) = -174 + 10\log\{\Delta f(\text{Hz})\} \tag{10.2}$$

where Δf is the bandwidth of the signal, and the noise obtained at the standard temperature is 290K. The receiver noise is defined by the noise figure (NF) as the difference between the noise output of the receiver in operation and that of an ideal receiver with the same overall gain and bandwidth. The ideal receiver is connected to corresponding sources at standard temperature. NF of the receiver is defined as follows:

$$NF = \frac{SNR_i}{SNR_o} \tag{10.3}$$

where SNR is the signal-to-noise ratio and subscripts i and o denote the input and output of the receiver, respectively. Finally, the sensitivity should be higher than $N_t + NF$.

Figure 10.11(a) shows the diagram of the downlink budget. The improved factors are the BS and MT antenna gains indicated by the dotted arrows, and the others are the loss factors. The transmit power is supplied to the BS antenna via coaxial cables, and cable loss L_c due to its length is the first loss factor in the downlink budget. Path loss L_p is determined by distance d between the BS and MT. The radius of the cell is estimated using the propagation model described in Section 8.1. L_p is the largest loss in the link budget and is the dominant factor. The two major margin factors are fading and interference, which are considered as a margin. The fading margin compensates the received-signal drop due to multipath fading, and the interference margin suppresses the signal level from the adjacent cell signals. The other losses include shadowing, indoor penetration, body effects, and those caused by the obstacles near MT.

An example of the downlink budget is listed in Table 10.3. The path loss is calculated using the Okumura-Hata model described in

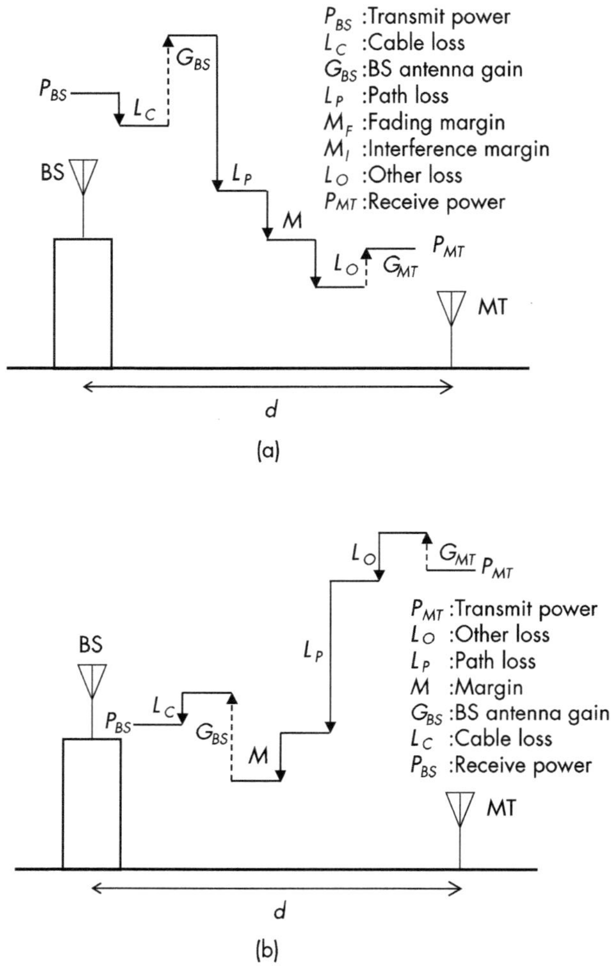

Figure 10.11 Diagram of budget: (a) downlink budget, and (b) unlink budget.

Section 8.1.1, where an urban area is assumed in the prediction. The transmit power of 40W is expressed in decibel milliwatts, and the antenna gain represents the isotropic gain expressed in (1.36). This example defines the receiving power level at the MT, which should exceed the receiver sensitivity. The thermal noise expressed in (10.2) is −103 dBm for a 15-MHz frequency bandwidth and is almost at the same level as P_{RX} listed in Table 10.3. The received power level is increased by the receiver NF, diversity gain, and factors due to the coding scheme. The approximately 20-dB total gain enhances

Table 10.3
Downlink Budget

f (MHz)	900	L_c (dB)	3
h_b (m)	40	L_p (dB)	132
h_m (m)	1.5	M (dB)	20
d (km)	1.5	L_O (dB)	5
P_{TX} (dBm)	46	G_{RX} (dBi)	0
G_{TX} (dBi)	17	P_{RX} (dBm)	−103

h_b is BS antenna height, h_m is MT antenna height, and $M = M_F + M_I$.

the sensitivity and provides the communication link in this budget design [10].

Figure 10.11(b) shows the uplink budget calculated using the same procedure as the uplink one [11]. The transmit power of MT is 24 dBm, which is much smaller than that of BS although it is compensated by the high sensitivity of the BS receiver.

10.3 INDOOR SIMULATOR MEASUREMENT

This section presents the indoor simulator measurement of MT. The field simulator presented in Section 8.3.3 is used for the MT test under Rayleigh and Nakagami-Rice fading environments. The cumulative-probability distributions of a suburb and countryside are reproduced inside the simulator to evaluate the sensitivity of the MTs. The MIMO performance, which is tested using MIMO described in Section 9.2.3, is also shown by measuring the downlink throughput of a 4 × 2 MIMO.

10.3.1 Fading Emulation

The field simulator consists of an $8m^3$ cube to measure the frequency of more than 800 MHz, and the MT under evaluation is mounted on the board hung from the ceiling. This size is sufficient to accommodate a human body to manipulate the MT inside the simulator, as shown in Figure 10.12. All outer walls that surround the measurement field consist of metals, three standard dipole antennas for transmission, and one dipole for reception. A block diagram of the field simulator is shown in Figure 10.13. The transmit signal is distributed

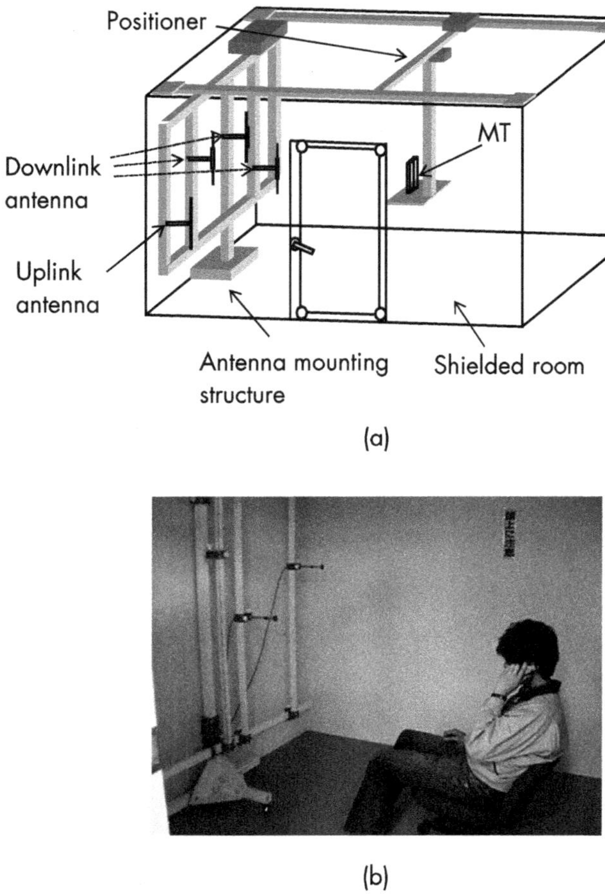

Figure 10.12 Field simulator: (a) outline of field simulator, and (b) photo of field simulator.

from the source to the three antennas at an arbitrary phase distribution controlled by a computer, which produces the required fading environment inside the measurement area.

Fading generators are used to randomly change the phase and independently produce multipass fading such as Rayleigh and Rician fading. Two fading generators are used to investigate the handover effect on the MT by receiving the electric fields using a directional antenna turned toward two different base stations. In general, the arrangement uses the base-station signal for RF input. However, in this case, a base-station simulator is used instead of an actual base station.

The fading generator consists of an eight-bit digital phase shifter to vary the antenna phase as a function of time. This simulator can also be used to investigate the MT of cellular systems using two RF signals from different BSs. The two RF signals are combined and input to the antennas inside the simulator. The variable attenuator inserted in front of the fading generator adjusts the level of transmitted RF signals and allows simulation of arbitrarily weak and medium-strength electric fields, which correspond to outdoor conditions. Although an actual base station is used in this example, a base-station simulator is substituted for the open-air antenna part. In this case, the field simulator is considered as an attenuator only in the propagation path. Therefore, the system is free from rules and regulations.

The received electric-field strength at the MT is picked up by a receiving dipole antenna installed inside the simulator, as shown in Figure 10.12(a). The reported signals from the handset are the received-signal-strength indicator level and bit-error ratio.

Figure 10.14 shows the cumulative probability distributions that are measured both inside the field simulator and outdoor in the weak electric field strength areas. Two examples of outdoor areas are the suburbs and countryside where the measured electric field strengths range from 5 to 15 dBm. The suburban area is located at the cell edge in a medium-size city, and the countryside is located at the cell edge at a lakeside. Both measured probability distributions inside the simulator become almost equal to those under outdoor

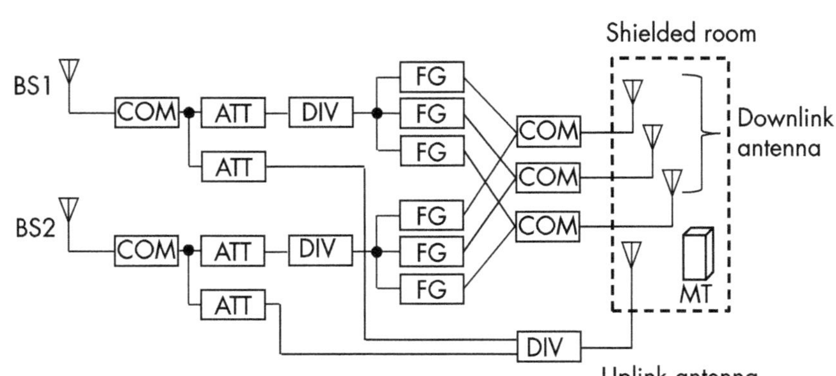

Figure 10.13 Block diagram of field simulator. COM: combiner, ATT: attenuator, DIV: divider, FG: fading generator.

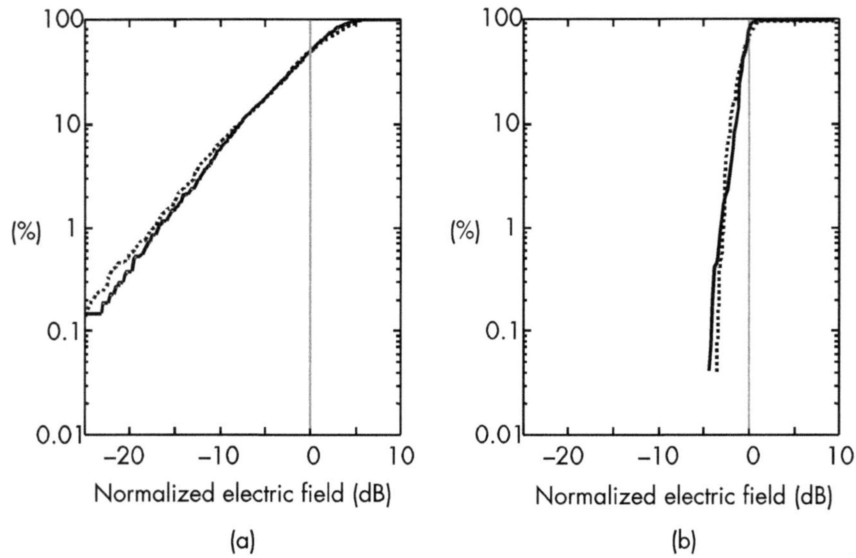

Figure 10.14 Cumulative probability distributions measured inside field simulator; solid line is measurement, dotted line is simulator: (a) data from suburb simulation, and (b) data from simulation of countryside.

conditions, which show the effectiveness of the field simulator from the viewpoint of fading distribution.

10.3.2 MIMO Emulation

The MIMO performance of the MT is measured using the MIMO emulator, and Figure 10.15(a) shows the block diagram of the measurement. A BS simulator is connected to a fading emulator to generate several cluster signals, and the RF outputs are amplified and transmitted from the dual-polarized antennas. A smartphone is tested as the MT inside the MIMO emulator, as shown in Figure 10.15(b), where four transmitting dual-polarization antennas are used. The generated signal is based on a single spatial-cluster model with multiple paths based on a spatial-channel model extension urban micromodel [12]. The fading generator generates a cluster signal consisting of three different delay profiles. Subsequently, a total of six clusters are transmitted to the MT. Four dual-polarization transmitting antennas and two MT antennas provide 4 × 2 MIMO downlink streams. The uplink signal from the MT is uploaded to the

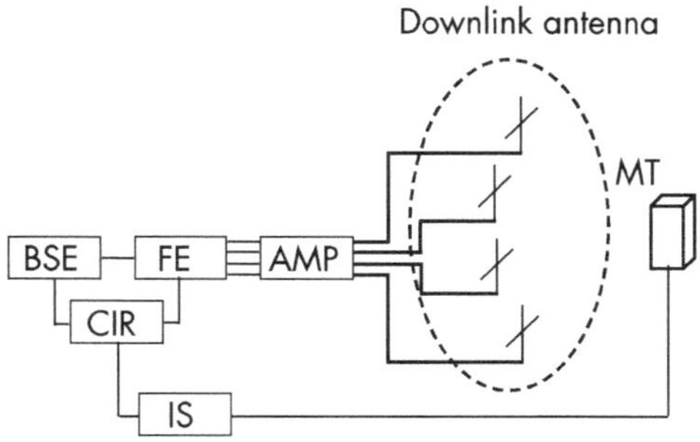

(a)

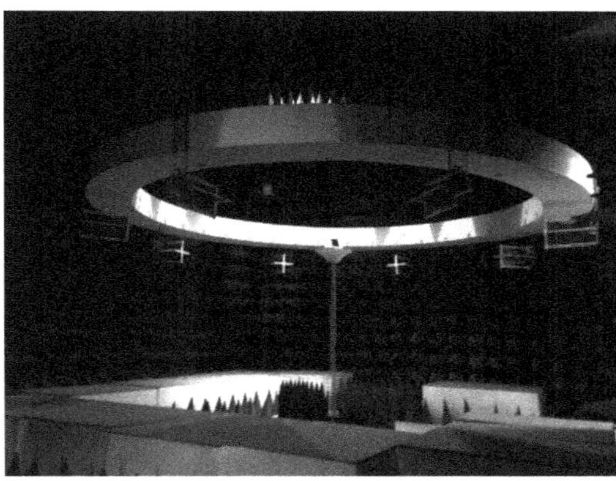

(b)

Figure 10.15 MIMO emulator: (a) block diagram of MIMO emulator (BSE: base station emulator, FE: fading generator, AMP: amplifier, CIR: circulator, IS: isolator), and (b) photo of MIMO emulator. (Courtesy of Microwave Factory.)

BS simulator through the circulator because the BS simulator ports are designed for the common use of the up- and downlinks.

Figure 10.16 shows the measured throughput by changing the input power level of the three different modulation schemes, where a SISO transmission and two different modulations are used for the 4 × 2 MIMO [13]. The throughput of the 4 × 2 MIMO is increased from 8 Mbps by QPSK to 20 Mbps by 16 QAM, which is much higher than that of a SISO that uses 16 QAM with two diversity receptions.

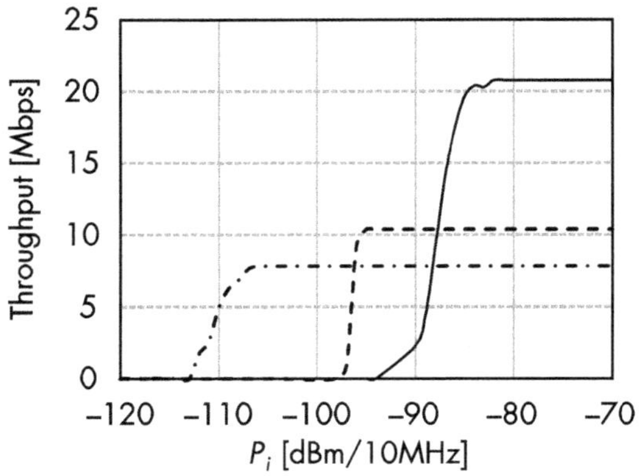

Figure 10.16 Throughput of MT. Dot and dashed line is QPSK MIMO, dashed line is 16 QAM SISO, solid line is 16 QAM MIMO.

References

[1] Arai, H., "Antenna Clearance, an Index Factor for Electrical Size of Small Antennas," *Conf. Dig., iWAT2008*, Chiba University, Japan, IT45, March 2008.

[2] Goldsmith, A., "Wireless Communications," Cambridge, U.K.: Cambridge University Press, 2005, pp. 419–420.

[3] Lewis, L. L., "An Introduction to Frequency Standard," *Proc. of IEEE*, Vol. 79, No. 7, July 1991, pp. 927–935.

[4] Yamada, Y., et, al., "Extended Walfisch-Bertoni Propagation Model to Cover Short Range and Millimeter-Wave Bands," *AGU Radio Science*, Vol. 56, Issue 3, March 2021.

[5] Walfisch, J., and H. L. Bertoni, "A Theoretical Model of UHF Propagation in Urban Environments," *IEEE Transactions on Antennas and Propagation*, Vol. 36, 1988, pp. 1788–1796.

[6] Imai, T., et al., "Radio Propagation Modeling Competition," *Technical Report IEICE*, Vol. 119, No. 28, AP2019-2, May 2019, pp. 7–12 (in Japanese).

[7] Mori, K., et al., "Delay Profile Measurement by the Sector Antenna at the Outdoor," *Antennas and Propagation Symposium*, 2P7, Davos, Switzerland, April 2000.

[8] Mori, K., et al., "Broadband Frequency Characteristics of 6-Sector Antenna USING PCTSA," *Society Conference of IEICE Japan*, B-1-105, October 1998 (in Japanese).

[9] Pierce, J. R., "Physical Sources of Noise," *Proceedings of the IRE*, Vol. 44, No. 5, May 1956, pp. 601–608.

[10] Coupechoux, M., "Link Budget 4G," https://marceaucoupechoux.wp.imt.fr/files/2018/02/BdL-4G-eng.pptx.pdf.

[11] Sharma, K. P., et al., "Cell Coverage Area and Link Budget Calculations in LTE System," *Indian Journal of Science and Technology*, Vol. 9 (SI), December 2016.

[12] 3GPP TR 37.976 V12.0.0, 2014-09.

[13] Cornes Technology, https://www.cornestech.co.jp/tech/webmagazine/webmagazine-153 (accessed February 13, 2022).

11

ANTENNA AND PROPAGATION CHALLENGES IN FUTURE MOBILE COMMUNICATION SYSTEMS

This chapter describes the strategies and challenges in antennas and propagation for beyond 5G mobile communication systems. A frequency spectrum of up to 100 GHz is allocated for 5G, and a 300-GHz band, known as the sub-terahertz (THz) band, is a candidate for the next-generation spectrum. Furthermore, fiber optical systems already use the infrared spectrum, which is the backbone of current cable communication. Laser diode and silicon photonics technology provide integrated optical circuits and are available for mobile communication systems. New spectrum development of the THz and optical bands examine novel high-gain antenna based on precise propagation prediction. This chapter describes a possible extension of the future mobile communication system and presents the channel capacity given by the frequency bandwidth and link budget simulation for the system design of THz and optical bands. Short-range optical wireless communication systems are also described.

11.1 OVERVIEW OF BEYOND 5G SYSTEM

The 5G mobile communication system enhances the channel capacity using sub-6 and millimeter frequency bands [1]. Low latency of the communication link is an advantage for self-driving vehicles and IoT applications, which are accelerated in the next-generation system. The 5G system primarily covers cellular services on terrestrial areas, underground areas, and inside large buildings using repeater systems or microcell systems. In future systems, this coverage area is extended to the atmosphere and outer space, which requires the development of a long-range communication link [2]. A high-gain antenna provides an optical link in outer space with a precise beam-tracking system, which is merged into the future system. Moreover, undersea communication is also expected to be used in the future by overcoming significant propagation loss in seawater. Figure 11.1 provides an overview of the coverage area of the future system. A recent drone development is expected to provide the possibility of a personal base station for each user by creating an in-line link with millimeter-wave communication [3]. Most wireless

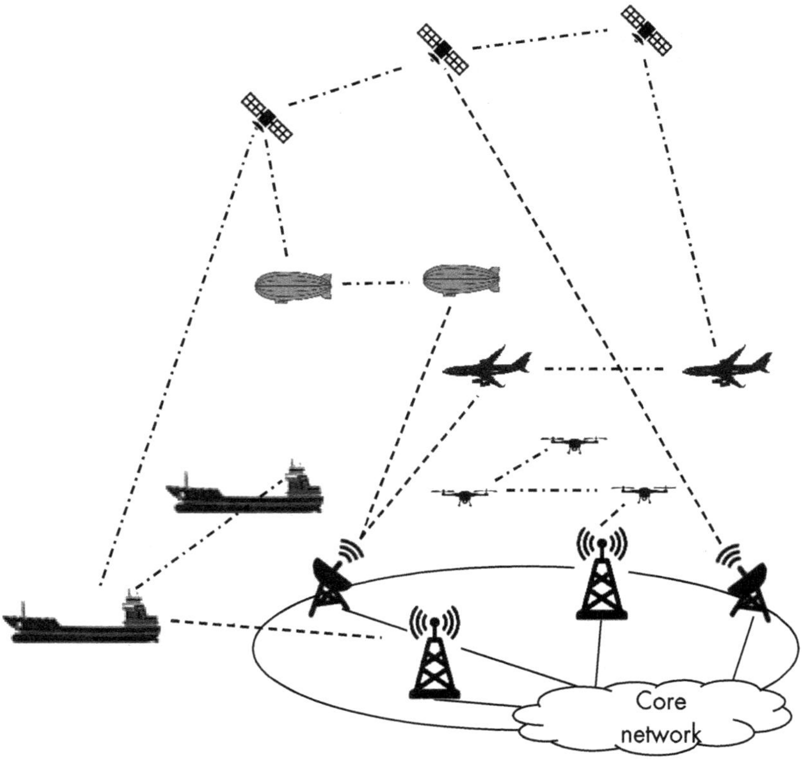

Figure 11.1 Future network; solid line is BS to core network links, dotted line is ground to moving object link, dot and dash line is wireless link for moving objects.

communication systems are merged with the future system to realize a seamless communication system.

Massive MIMO is introduced into 5G in addition to a new, unused spectrum. Ultra-high-speed data transmission is applied for short-range wireless systems, and the antenna is a key technology to cover small cells for each user. Antenna development for millimeter waves, terahertz, and optical spectra are required to extend the frequency bandwidth for realizing high-frequency systems. Millimeter and optical bands are already developed for automobile radars and fiber optics, respectively, and the terahertz band is currently evolving.

Cost-effective oscillators are realized for millimeter spectral bands and above the infrared spectrum. Terahertz oscillator development is still underway, which is known as the *terahertz gap*. The 324-/400-GHz CMOS oscillator was developed in 2008, and a 0.2–0.3 THz chip was fabricated in 2019. A large-scale, high-density terahertz SiGe radiator array has been introduced to compensate for the terahertz gap [4].

A high-gain and narrow-beam antenna with a limited coverage area is crucial to overcome large propagation loss in the optical band. Massive MIMO is equivalent to a high-gain multibeam former. The massive MIMO concept uses a large-scaled array, and beam formation via ultra-high-speed digital processing is mandatory. Further, a conventional frequency-scanned array provides high-speed beam scanning and is a promising antenna in the optical frequency range.

11.2 CHANNEL CAPACITY AND LINK BUDGET ESTIMATIONS

The maximum capacity of optical communication was limited to up to 100 Tbps around 2008 due to the fiber fuse propagation threshold of 1.2–1.5W [5]. Multiplexing transmission schemes using MIMO with multicore and multimode fibers enhanced the channel capacity by 1.05 Pbps, and the product of bandwidth and propagation distance achieved 1 Ebps/km in 2012 [6]. The increase in the capacity of optical communication is outstanding, whereas the terrestrial digital mobile communication capacity was only few kilobits per second in the early 1990s. International mobile telecommunications 2000 (IMT-2000) or 3G raised the capacity to 384 Kbps, 3.5G in 2006 achieved 14 Mbps, 3.9G in 2010 reached 100 Mpbs. The capacity in mobile communication is increased by 10^5 in 20 years. Long-term evolution or 4G provided 1 Gbps, and 5G achieved more than 10 Gbps [1]. The future 6G network is expected to have a capacity of around 1 Tbps.

The channel capacity of SISO defined in Section 9.1.2 is as follows:

$$C = \log_2(1 + \rho) \tag{11.1}$$

Then, the spectrum efficiency is given by multiplying the channel bandwidth in Hertz (B_c) with (11.1) as:

$$C' = B_c \log_2(1 + \rho) \tag{11.2}$$

Assuming SNR = 0 dB and $\rho = 1$ (11.1) gives $C = 1$ bps, suggesting that a bandwidth of $B_c > 1$ THz is required to achieve the 1 Tbps capacity in (11.2). The channel capacity of $N_t \times N_r$ MIMO is approximated as:

$$C \cong N_r \log_2\left(\frac{\rho}{N_t}\right) \tag{11.3}$$

where N_r and N_t are the number of receiving and transmitting antennas, respectively, under a constant total transmitting power. The use of 1000 receiving antennas reduced the required bandwidth to 1 GHz and obtain 1 Tbps; however, embedding several antennas into a smartphone using the microwave frequency band is difficult.

For example, the link budget of a short-range communication system is presented using the propagation loss defined in Section 2.2, and thermal noise is obtained using (10.2). As described in Section 10.2, several parameters should be considered to calculate the link budget. Figure 11.2 presents the receiving power level (P_r) as a function of propagation distance (d) for transmitting power ($P_t = 0$ dBm). The carrier frequency (f_c) is changed from 30 GHz to 300 THz, and the thermal noise (N_t) is calculated by changing the bandwidth (BW) from 1 kHz to 1 THz. The receiving power should be larger than the noise level defined by the bandwidth to obtain the communication link, and $P_r = -102$ dBm for $f_c = 300$ GHz at $d = 10$m, and $N_t = -84$ dBm is larger than P_r for BW = 1 GHz. This budget gap of 18

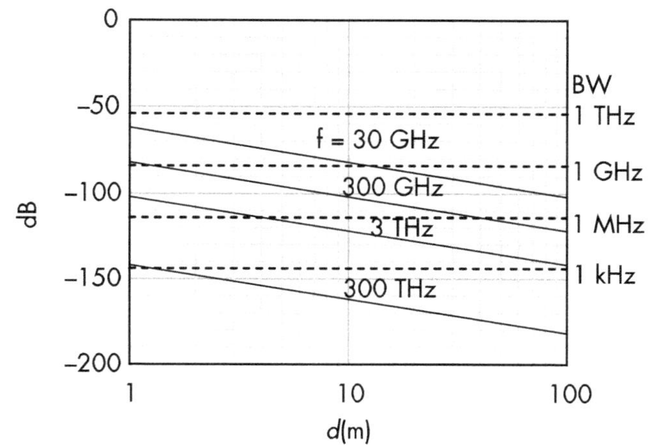

Figure 11.2 Propagation loss (solid line) and thermal noise (dotted line).

dB should be compensated by increasing the antenna gain or P_t. To achieve 1-THz capacity, BW > 1 is required for (11.2). Generally, the f_c is more than 100 times the required bandwidth in wireless communication systems; then, an f_c of more than 100 THz is recommended. Assuming $d = 1$m, $f_c = 300$ THz, and BW = 1 THz, the budget gap is 88 dB and the antenna gain is more than 44 dB for transmitting and recovering antennas, which should be developed for the ultra-high-speed communication system.

11.3 OPTICAL WIRELESS COMMUNICATION

A bandwidth of 1 THz is easily provided using an optical spectrum. Optical wireless communication (OWC) is a possible candidate for the future system to achieve a capacity of more than 1 Tbps after the development of millimeter waves and terahertz. An OWC is used as a backhaul between buildings and is known as free-space optics (FSO) using the atmospheric transmission window in the spectrum wavelength range of 0.3–14 μm [7]. The FSO depends on the presence of rain, fog, and haze, and the maximum data rate of 10 Gbps is achieved [8]. FSO can be applied in crosslinks to up-and-downlinks between satellites, aircraft, ships, and ground-standing or mobile terminals. Due to significant propagation loss, FSO requires high-gain transmitting and receiving antennas using high-power laser and high-precision beam tracking to obtain long-range communication links, more than a few hundred kilometers in space applications [9]. The optical spectrum is not disturbed during indoor usage and is appropriate for short-distance communication systems, as shown in Figure 11.3, because people tend to be indoors 80% of the time and more when they need to communicate [10].

Infrared data application (IrDA) provides 4 Mbps for a distance of 1m via DATA 1.4, 100 Mbps via ultra-fast IR (UFIR), and 1 Gbps via giga-IR [11], respectively. Wireless data transmission using visible light, known as visual light communication (VLC), has achieved a capacity of up to 100 Mbps and is expected to extend to 1 Gbps [12]. The transmitted light from LED is switched on and off in short intervals to modulate data into visible light. To increase the uplink capacity, a high-gain antenna or high-power emission is required at the user terminal in VLC. Additionally, to overcome the high propagation loss, an indoor FSO is realized using high-gain beam-steering

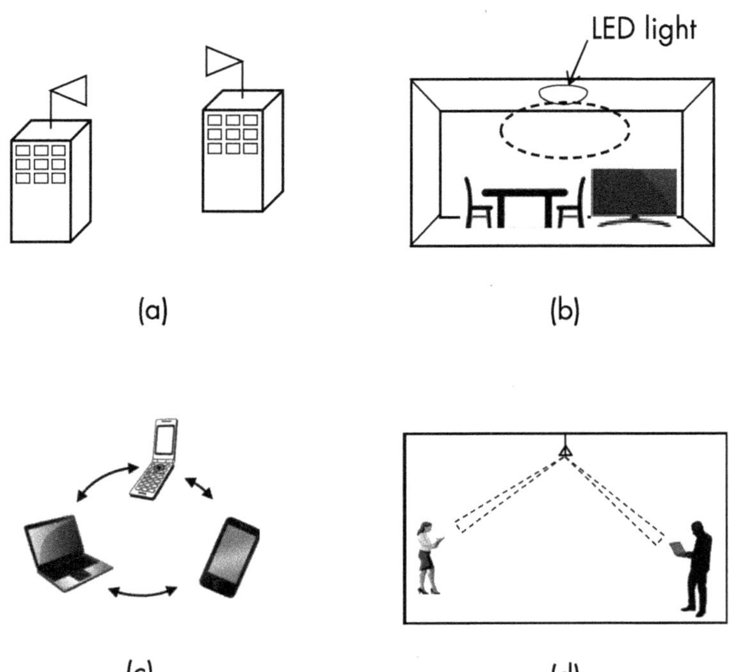

Figure 11.3 Optical wireless communication system: (a) free space optics, (b) visual light communication, (c) infrared ray, and (d) short range OWC.

antennas for transmitting and receiving systems as described in Section 11.2, which is another candidate for a 6G indoor system.

Optical beam scanning arrays with a combined tunable laser and phased array [13, 14] are used in sensor applications due to their limited scanning range. A two-axis beam scanner is also developed using a thermal switched phase shifter for display applications [15]. Moreover, a beam-scanning array with a switched leaky-wave waffled waveguide array was fabricated on a silicon wafer. Its beam scanning is based on wavelength sweep and beam switching with a Mach-Zehnder–type optical switch [16]. Optical leaky wave antennas are physically small but electrically large due to their short wavelength [17] and are expected to be used for OWC and light detection and ranging.

Issues associated with the short-range OWC include beam trucking, safety standards, and the development of high-speed electrical devices such as a modulator and demodulator. Under an axial symmetrical beam assumption, the antenna beamwidth B_w is approximately given for high gain antenna as [18]:

$$B_w = \sqrt{\frac{30000}{10^{\frac{G_t}{10}}}} \tag{11.4}$$

where B_w and antenna gain G_t are expressed in degree and decibel, respectively. Figure 11.4 shows the coverage area determined using B_w and the distance d from the antenna.

The coverage area of a circle with radius r is given by:

$$r = d\tan\left(\frac{\theta_w}{2}\right) \tag{11.5}$$

High-gain antennas are required to satisfy the link budget described in Section 11.2. An example of the coverage area radius of the 40-dBi antenna is less than 15 cm at the distance 3m from the BS antenna (Figure 11.4(b)) as a typical propagation distance under indoor environment. The coverage raidus r = 15 cm is enough to illuminate a mobile terminal used by a person; however, the coverage radius becomes small by increasing the antenna gain. When the antenna

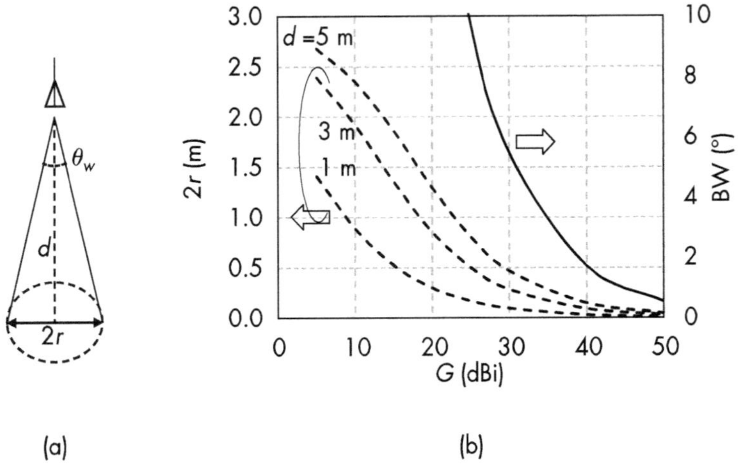

Figure 11.4 Coverage area of high gain antenna: (a) geometry of beam, (b) beam width and coverage area of beam.

gain is higher than this example, beam tracking with high precision is crucial. Efficient tracking methods are expected to develop for short-range communication systems.

Another challenge is laser safety. The OWC uses lasers for transmitting sources that are hazardous to the eye. According to the guidelines defined by the International Electrotechnical Commission, laser safety is categorized into several classes [19]. The class 2 limit corresponds to a 1 mW (0 dBm), which is the accessible laser output power (e.g., laser pointers) in the visible spectral range (λ = 400–700 nm). The red infrared spectrum is assumed for the OWC and is categorized as class 3R, which is five times the permissible optical power of class 2. Laser safety should be considered while designing short-range OWC in future systems.

References

[1] Medbo, J., et al., "Channel Modelling for the Fifth Generation Mobile Communications," CP06-1.1, EuCAP, 2014.

[2] Zhang, Z., et al., "6G Wireless network, Vision, Requirement, Architecture and Key Technologies," *VT Magazine*, Vol. 14, No. 3, September 2019, pp. 28–41.

[3] Nishimori, K., "A New Concept Using MIMO and Drone for Wide and High-Speed Are in Beyond 5G and 6G," *Proc., International Symposium of Antennas and Propagation*, Osaka, Japan, January 2021, pp. 461–462.

[4] Han, R., et al., "Filling the Gap: Silicon Terahertz Integrated Circuits Offer Our Best Bet," Microwave Magazine, Vol. 20, Issue 4, April 2019, pp. 80–93.

[5] Morioka, T., "New Generation Optical Infrastructure Technologies: 'EXAT initiative' Towards 2020 and Beyond," *Optoelectronics and Communication Conference*, FT4, July 2009.

[6] National Institute of Information and Communications Technology, "Going Beyond the Limits of Optical Fibers," https://www.nict.go.jp/en/press/2015/04/24-1.html.

[7] Henniger, H., and O. Wilfer, "An Introduction to Free-Space Optical Communications," *Radioengineering*, Vol. 19, No. 2, June 2010, pp. 203–211.

[8] Malik, A., and P. Singh, "Free Space Optics: Current Applications and Future Challenge," *International Journal of Optics*, Vol. 2015, Article ID 945483, 2015.

[9] Mukherjee, B., et al., *Springer Handbook of Optical Networks*, Cham, Switzerland: Springer, 2020, pp. 1057–1103.

[10] Davit, K., and H. Berndt, "6G Vision and Requirements," *IEEE Vehicular Technology Magazine*, September 2018, pp. 72–80.

[11] IRDA, https://www.irda.org.

[12] Strinati, E. C., et al., "The Next Frontier: From Holographic Messaging to Artificial Intelligence Using Subterahertz and Visible Light Communication," *Vehicular Technology Magazine*, Vol. 14, No. 3, September 2019, pp. 42–50.

[13] McManamon, P. F., et al., "A Review of Phased Array Steering for Narrow-Band Electro Optical Systems," *Proc. IEEE*, Vol. 97, No. 6, June 2009, pp. 1078–1096.

[14] Hutchison, D. N., et al., "High-Resolution Aliasing-Free Optical Beam Steering," *Optica 3*, 2016, pp. 887–890

[15] Acoleyen, K. V., et al., "Off-Chip Beam Steering with a One-Dimensional Optical Phased Array on Silicon-on-Insulator," *Opt. Lett.*, Vol. 34, No. 9, 2009, pp. 1477–1479.

[16] Morimoto, Y., and H. Arai, "Wavelength-Insensitive MZ Type Switch for Dielectric Leaky Wave Antenna for Optical Transmission," *IEICE Spring Conf.*, B-1-182, March 2014.

[17] Qi Song, Q., et al., "Silicon-Based Optical Leaky Wave Antenna with Narrow Beam Radiation," *Opt. Express 19*, 2011, pp. 8735–8749.

[18] Volakis, J. L., *Antenna Engineering Handbook*, Fourth Edition, New York: McGraw-Hill, 2007, pp. 10–14.

[19] RP Photonics, https://www.rp-photonics.com/laser_safety.html.

ABOUT THE AUTHOR

Hiroyuki Arai is a professor in the Department of Electrical and Computer Engineering at Yokohama National University. Dr. Arai is the author of six books, four research books chapters, more than 170 journal papers, and 1,400 conference papers. He is a fellow of the IEICE Japan and a fellow of IEEE. He holds a BE in electrical and electronic engineering and a ME and DE in physical electronics, all from Tokyo Institute of Technology.

INDEX

A

Active integration antenna, 143–46
Anechoic chamber, 72–77
Angular diversity, 52
Angular profile, 56–57
Angular spread, 57
Antenna elements (base station)
  about, 87–88
  dipole antenna, 88–93
  microstrip antenna, 93–95
  multiband dipole and microstrip antennas, 95–97
Antenna elements (mobile terminal)
  about, 104
  correlation between, 261
  electrically small antenna, 110–15
  geometry of, 260
  inverted-F shaped antenna, 107–10
  monopole on a finite ground plane, 104–7
Antenna factor (AF), 202
Antenna pattern
  correlation, 139–40
  half power beam width (HPBW), 11
  measurements in open sites, 180–82
  shape and size of antenna elements and, 23
  synthesis, 102–3
Antenna-radiation pattern measurement
  about, 70–71
  far-field, in anechoic chamber, 72
  far fields, 71, 73
  Fresnel region, 71–72, 73
  near fields, 71–72, 73, 77–80
Antennas
  basic elements of, 12–15
  basic theory of, 1–29
  effective height, 214
  gain, 15–16, 22–23, 76
  mutual coupling and, 28–29
  parameters, 16–23
  See also Base station antennas; Mobile terminal antennas
Antenna under test (AUT), 16, 182, 184–85
Antinode/node, 61
Aperture efficiency
  defined, 22
  illustrated, 22
  Taylor weighting distribution and, 100
Array antennas
  about, 23, 97
  advantage of, 25
  antenna pattern synthesis, 102–3
  colinear, 97–98
  design, 136–38
  linear, 24–28, 98–101
  patch, 250–51
  pattern, 26
  three-element, 27
Array factor, 24, 28, 136

Array spacing, 260
Attenuation coefficient, 80–81
Attenuation factor, 276
Attenuator (ATT), 74, 77
Autotransformer balun, 68
Axial ratio (AR), 74, 76

B

Babinet's principle, 7
Balanced feeding line, 67–70
Balun
  autotransformer, 68
  circuit, 68
  defined, 67
  sleeve, 68–69
Base station antenna measurements
  about, 179–80
  far-field, 181
  near-field, 182–86
  in open sites, 180–82
  overview, 184
  passive intermodulation, 187–90
  radiation pattern, 186
  specifications, 185
  very-near-field, 182–86
Base station antennas, 123–24
  antenna configuration of, 123–24
  array pattern design, 136–38
  beam coverage area, 145
  beam tilt of, 118–19, 124–25, 143–44
  beam width, 121
  categories of, 117–23
  at ceiling, 251–55
  circular coverage antennas, 118–19
  coaxial cables, 132–33
  components of, 125–36
  design of, 117–46
  diversity and MIMO systems, 139–46
  dual polarization, 124
  eight-element, structure of, 126
  feeding network of, 124, 125, 143
  geometry of, 257
  indoor, 123
  mechanical tilt, 127
  near walls, 248–51
  null-filled radiation pattern, 137–38
  omnidirectional, 130, 133, 134
  in open site, 255–59
  phase shifters, 132–33
  polarization diversity, 125
  reflector size, 129–30
  sector antennas, 119–21
  six-sector, 121
  vertical plane radiation pattern, 137
Base station antennas and channel capacity
  about, 247–48
  BS at ceiling, 251–55
  BS in open site, 255–59
  BS near walls, 248–51
Base station elements
  about, 87–88
  dipole antenna, 88–93
  microstrip antenna, 88–93
  multiband dipole and microstrip antennas, 95–97
Bazooka balun, 68–69
Beam forming network (BFN), 144
Beam tilt, 118–19, 124–25, 143–44
Beam width, 120–21, 131
Beyond 5G systems
  about, 288
  illustrated, 292
  overview, 291–93
Bidirectional pattern, 122
Bit error rate (BER), 192
Body effect
  about, 165
  antenna input characteristics and, 173
  antenna performance and, 171–76
  body effect, 171
  dielectric properties and, 170
  numerical phantom, 165–67
  phantom in measurement, 167–70
  user hand and, 174–76
  See also Mobile terminal antennas
Boundary value condition, 34
Breakpoint, 40, 41
Brewster's angle, 38–39
Built-in mobile terminal antennas
  about, 150
  chassis effect and, 150–52
  elements, 152–54
  geometry of, 152
  location, 150–52
  matching circuit, 154–57
  multielement element, 153
  radiation resistance, 157
  See also Mobile terminal antennas
Butler matrix, 141

C

Calibration procedures, 272–73
Ceiling, BS at, 251–55
Cesium-frequency standards, 275
Channel capacity
  about, 241
  average and peak, of open site, 259
  BS antenna design and, 247–49
  BS at ceiling, 251–55
  BS in open site, 255–59
  BS near walls, 248–51
  estimation, 241, 293–95
  estimation of MT antennas, 263–64
  as function of averaged spatial correlation coefficient, 247
  as function of SNR, 245
  large, radiation patterns leading to, 252
  MIMO systems, 241–47, 254
  of MT antennas, 265–68
  of MT in presence of phantom, 268
  Nakagami-Rice environment and, 246–47
  patch arrays, 250–51
  radiation efficiency and, 262–63
  SISO systems, 241–47, 294
  terminal antenna design and, 259–68
Channel matrix, 246, 264
Characteristic impedance, 225
Characteristics mode, 159
Chassis mode, 158–60
Circular coverage antennas, 118–19
Coaxial cable
  characteristic impedance of, 70
  colinear array, 97
  dipole antenna fed by, 68
  feeding circuits, 123
  illustrated, 14
  leaky current on surface, 88
  timing signal with, 85
  as unbalanced line, 67
Colinear array antennas, 97–98
Coordinate system
  current source and, 4
  knife-edge diffraction, 43
  MIMO performance measurement, 198, 199
  mobile terminal measurement, 191
  space diversity antenna, 139
Correction factor, 214
Correlation coefficient, 84, 85, 141, 234
Correlation factor, 264
COST-231 Walfisch-Ikegami model, 219
Coverage areas
  about, 117–18
  beam, 145
  bidirectional pattern, 122
  of circle, calculation, 297
  circular, 118–19
  multibeam, 122
  multisector, 122
  pattern shapes for, 118–23
  sector, 119
  spot beam, 121
Crossed dipole antenna, 92, 127–29, 132
Cross-polarization ratio (XPR), 74, 75, 83–84
Cumulative distribution function (CDF), 196, 197
Cumulative probability distributions, 285–86
Current distribution, 6, 10, 107
Current elements
  characteristics of, 4
  infinitesimal, radiation from, 3–6
  infinitesimal magnetic, radiation from, 7–8
Current source, 1, 3, 4, 7, 34
Cylindrical very-near-field measurement, 185

D

Decoupling methods, 162–65
Decoupling network, 163–64, 165
Deep fading, 52
Delay profile
  about, 53, 274–75
  distribution function of, 53–54
  double spike model, 55
  exponential function model, 56
  measured, 54, 279, 280
  model, 54
Delay-profile measurement
  illustrated, 86
  performing, 85–86
  propagation loss and, 274–76
  by sector antenna, 279–80
  in suburb area, 278
Delay spread, 55
Delay spread simulation, 234–35
Device under test (DUT), 66, 189, 190–91, 200–203, 236

Diagonal matrix, 243
Dielectric loss tangent, 167–68
Diffraction point, 225–27
Digital beam-forming antenna, 146
Digital phase-shifter, 238
Dipole antennas
  about, 88
  asymmetrical element, 155
  backed by reflector, 130
  base station elements, 87–93
  coaxial cable feeding and, 129
  crossed, 92, 127–29, 132
  current distribution of, 10
  dual resonant, 96
  elements, radiation pattern, 253
  geometry of, 19
  half-power beamwidth, 91–92
  hand effect and, 175
  illustrated, 88
  input characteristics of, 20
  Lecher line and, 8, 9, 13
  modified crossed, 129
  multiband, 95–97
  observation point and, 10
  principle of, 12
  printed, 89–90, 127–29
  radiation from, 8–12
  radiation pattern, 9–11, 88, 91
  reflection coefficient, 171–72
  short, radiation pattern, 27
  sleeve, 88, 89
  spacing of, 28
  spherical phantom and, 172
  twin, 92
Dipole mode, 160
Diversity and MIMO systems
  active integration antenna, 143–46
  antenna pattern correlation, 139–41
  base station configuration, 139–46
  Butler matrix and multibeam, 141–43
Diversity gain, 53
Diversity reception, 51–53, 234
Diversity systems, 52
Double sideband suppressed carrier modulation, 233
Double spike model, 55
Downlink budget, 281–82, 283
Dry phantom, 169
Dual polarization antenna, 124
Dyadic diffraction coefficient, 226–27

E

Effective area
  about, 19–22
  antenna gain and, 17, 22–23
  defined, 22
Effective height, 214
Effective isotropic radiated power (EIRP), 190–92, 196
Effective length, 19–22
Electrical equivalent phantom, 170
Electrically small antenna
  about, 110
  bent current path and, 111
  ground plane mode, 112
  high dielectric constant materials, 111–12
  matching current, 111
  miniaturization methods, 110–11
  minimum Q value and Q factors of, 114–15
  theory of, 113–15
Electric fields
  breakpoint and, 41
  complementarity of magnetic fields and, 7
  distribution, 51
  distributions by ray-tracing method, 230
  incident, 226
  multipath fading and, 51
  outgoing, 226
  time-varying, 47
Electromagnetic waves
  about, 1–2
  boundary value condition and, 34
  knife-edge diffraction and, 42–45
  multipath fading and, 46–47
  oblique incidence, 36–39
  See also Propagation
EMC measurements
  about, 199–200
  chamber, 201
  EMI assessment and, 200
  frequency range, 202–3
  open site EMI measurement and, 200–203
  reverberating enclosure, 203–5
  shield box size, 204
  site, 201
  site attenuation, 204
EM interference (EMI)

assessment, 199–200
open site measurement, 200–203
Equivalent current method, 183
Evaluation factor, 54
Exponential function model, 56

F

Fading emulation, 283–86
Fading-field simulator
fading measurement, 236
illustrated, 237
mobile terminal evaluation and, 235
Fading generators, 284–85
Fading margin, 281
Fading simulator, 232–34
Fading structure, 46–51
Far-field measurements
about, 71
in anechoic chamber, 72–77
with ground reflection, 181
with slant range, 182
with tall tower, 181
Feeding lines, 67–70
FEKO electromagnetic simulation software, 29
Fiberglass reinforced plastic (FRP), 123
Field components
decomposition of, 5
in far-field region, 6
magnetic current and, 8
radiation, 6
Field simulator
about, 283–84
block diagram, 285
outline, 284
photo, 284
Finite domain and time difference (FDTD) method, 223
Four-port arch shaped phase shifter, 135
Free-space optics (FSO), 295–96
Frequency-selective fading, 234–35
Fresnel zone, 45–46

G

Gain, antenna
about, 15–16
absolute, 76
deriving, 16
directive, 76
directivity, illustrated, 15
effective area and, 22–23
isotropic, 16
omnidirectional antennas, 130
Gain, diversity, 53
Gaussian distributions, 48, 247, 264
Gaussian noise, 232
Geometric theory of diffraction (GTD), 223
Global positioning system (GPS), 151
Goal, this book, *xii*
Ground
as perfect conductor, 40
reflection, effect of, 39–42
"Guidelines for Evaluation of Radio Transmission Technologies for IMT-2000 (M.1225)," 221

H

Half power beam width (HPBW), 120, 130, 248, 249, 251, 256
Halo antenna, 132
High-gain antennas, 297–98
Horn antennas, 14–15

I

Ikegami model, 218, 219
Image method, 227–28
Impedance mapping, 62
Impedance matching, 17–18
Impedance matrix, 66
Impedance measurement
about, 59–60
balanced and unbalanced feeding lines and, 67–70
reflection coefficient and, 60–63
Smith chart and, 63–64
S-parameter method, 64–67, 69–70
IMT-Advanced, 221
Indoor hotspot (InH) model, 222
Indoor simulator measurement
about, 283
fading emulation, 283–86
MIMO emulation, 286–88
Input impedance
about, 18–19
characteristic match break, 154
measurement, 69
monopole antenna on ground plane, 104, 105

Input impedance *(Cont.)*
normalized, of asymmetrical dipole, 155
Interference fading, 237–38
International Mobile Telecommunications-2000 (IMT-2000), 221
Inverted-F shaped antenna (IFA)
about, 107–8
as built-in antenna, 265
on chassis, 163
illustrated, 108
on plate, 265, 267
radiating elements of, 265
radiation pattern, 109–10
radiation resistance, 108
Inverted-L shaped antenna, 107–8
Isotropic antennas, power density of, 23
Isotropic gain, 16

K

K-factor, 50
Knife-edge diffraction
about, 42–43
coordinate system, 43
edge height and, 44
illustrated, 42
multiple edges and, 44–45
Vogler solution and, 44–45
Kronecker model, 263

L

Laser safety, 298
Layered-sphere phantom model, 166–67
Lecher line, 8, 9, 13, 127
Linear array antennas
binomial, 100
equally spaced, 24
feed network, 99
lobe suppression and, 99, 100
radiation pattern, 98, 100, 101
subarray, 99, 100
Line-of-sight (LOS) propagation, 32
Link budget
downlink, 281–82, 283
estimations, 293–95
gap, 294–95
short-range communication system, 294
uplink, 281–82
Low noise amplifier (LNA), 189–90

M

Mach-Zehnder-type optical switch, 297
Magnetic distribution, 6
Magnetic fields, 7
Massive MIMO, 293
Matching circuit
built-in mobile terminal antenna, 154–57
circuit topologies of, 155
loss, 157
reactance of, 156
Maxwell's equations, 1–3
Measurement(s)
about, 59
of antenna radiation pattern, 70–80
base station antenna, 179–90
delay-profile, 274–76, 278
EMC, 199–205
fading, 236
impedance, 59
indoor simulator, 283–88
input impedance, 69
mobile terminal, 190–99
phantom in measurement, 167–70
propagation, 80–86, 271–88
reverberating enclosure, 203–5
SAR/PD, 205
S-parameter, 64–67, 69–70
Measurement system, 272–74
Microstrip antennas
about, 93
base station elements, 93–95
dual polarization and, 95
dual resonant, 97
feeding structures, 94–95
illustrated, 13, 93
multiband, 95–97
radiation pattern, 94
Microstrip lines, 13
MIMO
about, 241
applications, 149
array spacing and, 260
channel capacity of, 241–47
channel measurements, 253
emulation, 286–88
independent transmission, 244
massive, 293

measured SNR and channel capacity for, 254
performance, 283, 286
photo, 253
QPSK, 288
system model, 242–43
MIMO emulator, 286–87
MIMO performance measurement
about, 196
coordinate system, 198, 199
emulator, 197–99
scatterer, 198, 199
Mobile terminal antenna elements
about, 104
electrically small antenna, 110–15
inverted-F shaped antenna, 107–10
monopole on a finite ground plane, 104–7
Mobile terminal antennas
about, 149
body effect, 165–76
built-in antenna, 150–57
channel capacity and, 259–68
design of, 149–76
multiantenna system, 157–65
mutual coupling and, 259–63
radiation efficiency and, 259–63
Mobile terminal design and channel capacity
about, 259
estimation of MT antennas, 263–64
of MT antennas, 265–68
mutual coupling and, 259–63
radiation efficiency and, 259–63
Mobile terminal measurement
about, 190
bit error rate (BER), 192
coordinate system, 191
effective isotropic radiated power (EIRP), 190–92, 196
MIMO performance, 196–99
OTA, 192–96
parameters, 190–92
total isotropic sensitivity (TIS), 192
total radiation power (TRP), 190–92
Monopole antenna on ground plane
about, 104
current distribution, 107
illustrated, 104
input impedance, 104, 105
radiation pattern, 105, 106
rectangular plate, 106
M-sequence generator, 234
Multiantenna system
about, 157–58
antenna position, 160–62
chassis mode, 158–60
decoupling method, 162–65
mutual coupling and, 160–62
See also Mobile terminal antennas
Multibeam, 122
Multipath fading
about, 46–47
delay spread simulation, 234–35
diversity reception and, 51–53
effects, reduction by diversity reception, 234
emulation, 232–38
fading-field simulator, 235–38
fading simulator, 232–34
frequency-selective, 234–35
interference, 237
measurement, 236
in outdoor environment, 236
Rayleigh distribution, 232
receiving electric field and, 51
Multiprobe measurement, 182
Multisector, 122
Mutual coupling
about, 28–29
antenna position and, 160–62
channel capacity and, 259–63
decoupling network and, 163–64
of $\lambda/2$ dipole array, 162
radiation patterns and, 261
between two short dipoles, 29

N

Nakagami-Rice distribution, 49, 50–51, 263
Nakagami-Rice environment, 246–47
Narrowband measurements, 271
Near-field measurements
about, 71–72, 182
cylindrical, 183
distance of, 77
multiprobe, 182
scanning area, 79
scans, 79
space requirement, 78–79
virtual boundary and, 78

Network analyzer, 59, 64, 66, 74, 275
Noise figure (NF), 281
Non-line-of-sight (NLOS) propagation, 32
Normalized site attenuation (NSA), 202

O

Observation point
  antenna size and, 71
  current source and, 3
  dipole antenna and, 10
Okumura-Hata model
  geographical features and, 212
  Okumura's curve, 212, 213, 216
  path loss calculation with, 281–82
  propagation loss by, 215
Omnidirectional antennas
  about, 130
  polarization diversity, 134
  slant polarization, 132, 133
  V/H-polarization, 133
Open site, channel capacity and, 255–59
Optical beam scanning arrays, 297
Optical leaky wave antennas, 297
Optical wireless communication (OWC), 295–98
Organization, this book, *xi*
Orthogonal polarization (transverse electric), 37
Oscillators, 86, 275, 293
OTA measurement
  about, 192–93
  conical cut method, 192–93
  conical cut method with probe array, 194–96
  great circle method, 194, 195
  of TIS, 196
  of TRP, 196
Overlaid multicell system, 141

P

Parallel polarization (transverse magnetic), 37, 38
Parameters, antenna
  about, 16–17
  cross sectional area and gain and, 22–23
  effective length and area and, 19–22
  impedance matching, 17–18
  radiation resistance and input impedance and, 18–19
Passive intermodulation
  about, 187
  frequency allocation for, 189
  k parameter and, 188
  in LNA, 189–90
  measurement, 188–89
  measurement diagram, 188–89
  odd order frequency, 188
  as phenomenon, 187
Patch arrays, 250–51
Phantom
  composition, 168
  dry, 169
  electrical equivalent, 170
  layered-sphere model, 166–67
  with loss tangent, 168
  in measurement, 167–70
  numerical, 165–67
  shell, 170
  spherical, 172
  wet, 168–69
Phase shifters
  about, 132–33
  digital, 238
  four-port arch shaped, 135
  with moveable dielectric plate, 135–36
  semifixed, 135
Physical model, 216–21
Polarization
  circular, 75
  dual, 95
  elliptical, 75
  TE, 36–37
  TM, 36–37
Polarization diversity antenna, 128, 134
Power density, 15, 23, 32–33, 34, 179, 265
Poynting vector, 34
Printed antennas
  dipole, 89–90, 127–29
  fabrication of, 13
Propagation
  classification of paths, 31–32
  distance, 33
  LOS, 32
  NLOS, 32
Propagation loss
  in city area (Kokura), 277
  in city area (Tokyo), 277
  COST-231 Walfisch-Ikegami model, 219
  in free space, 32
  as function of frequency, 215

measurement, 80–83, 271–74
models, 222
by Okumura-Hata model, 215
propagation distance and, 33
suburban/farmland, 215
thermal noise and, 295
of tilted beam, 119
in urban area, 213
Propagation loss estimation
about, 211–12
by big cities, 220
Okumura-Hata model, 211–16
physical model, 216–21
standardization of predictions, 221–22
Walfisch model, 218–19
Propagation measurement(s)
about, 80, 271
calibration procedures, 272–73
correlation coefficient, 84, 85
cross polarization, 83–85
delay profile, 85–86
examples of, 276–80
indoor simulator, 283–88
link budget and, 280–83
long-range field distribution, 81
loss, 80–83
loss factor and, 272
measurement system, 272–74
narrowband, 80, 271
sampling frequency and, 83
short-range field distribution, 81–82
standard system, 82–83
wideband, 271

R

Radiation
from dipole antenna, 8–12
efficiency, 262–63
from infinitesimal current element, 3–6
from infinitesimal magnetic current element, 7–8
resistance, 18–19, 157
Radiation patterns
base station antennas, 137–38
bidirectional, 122
by Butler matrix, 143
dipole antennas, 88, 91
field strength, 6
half-wavelength dipole antenna, 9–11
human body and, 175
illustrated, 12
inverted-F shaped antenna, 109–10
leading to large channel capacities, 252
linear array antenna, 24, 26, 98, 100, 101
measured BS, 186
measured spin linear, 76
microstrip antenna, 94
monopole antenna on ground plane, 105, 106
mutual coupling and, 261
sector antennas, 257
shape and size of elements and, 23
slot and dipole antenna elements, 253
three-element array, 27
Ray-launching method, 228
Rayleigh distribution, 48–49, 50, 246
Rayleigh fading distribution, 232
Ray-tracing methods
about, 223
basis of, 224
electric field distributions by, 230
FDTD method and, 223–24
image method, 227–28
ray-launching method, 228, 229
simulation examples, 228–31
simulation model, 229
Receiver sensitivity, 281
Receiving power, 20–21
Receiving voltage, 20
Rectangular waveguide, 14
Reflection and transmission at the boundary, 34–36
Reflection coefficient
dipole antenna, 171–72
at each port, 65
of electric field, 35
of ground, 40
impedance measurement and, 60–63
parallel polarization (transverse magnetic), 38
Reflection form load, 60
Reverberating enclosure, 237
Reverberating enclosure measurement, 203–5
Rician factor, 246
Rural macro (RMa) model, 222

S

SAR/PD measurement
about, 205

internal SAR and, 206
localized SAR, 205–6
by probe array, 207
by probe scanner, 207
Scientific and Technical Research 244 (COST 244), 165, 166
Sector antennas, 119–21, 257
Semifixed phase shifter, 135
Signal-to-noise ratio (SNR), 245, 247, 250, 254–55, 281
Sinusoidal function, 10
SISO systems
about, 241
channel capacity of, 241–47, 294
characterization, 242
system model, 242–43
See also MIMO
Sleeve antennas
about, 88
arrays, 251
fabrication of, 14
illustrated, 14, 89
Sleeve balun, 68–69
Slot antennas, 253
Smith chart
about, 63
illustrated, 64
impedance locus on, 64
in load impedance determination, 63
voltage measurement and, 61
Snell's law, 36
Source point vectors, 3
Space diversity antenna, 139, 141
Space diversity reception, 52
S-parameters, measurement, 64–67, 69–70
Spectrum efficiency, 294
Spherical coordinate system, unit vectors of, 4
Spherical phantom, 172
Spot beam, 121–22
Spreading factor, 224, 225, 226
Suburban macro (SMa) model, 222

T

Taylor weighting distribution, 100
Thermal noise, 294, 295
Total isotropic sensitivity (TIS), 192
Total radiation power (TRP), 190–92
Transpose matrix, 243
Transverse electric (TE) polarization, 36–37
Transverse magnetic (TM) polarization, 36–37
Twin dipole antenna, 92
Two-port network, 65

U

Ultra-fast IR (UFIR), 296
Unbalanced feeding line, 67–70
Uplink budget, 281–82
Urban macro (UMa) model, 222
Urban micro (UMi) model, 222

V

Vector formula, 2–3
Vertical polarized loop slot antenna, 131
Very-near-field measurements, 182–86
Visual light communication (VLC), 296
Voltage standing wave ratio (VSWR), 17–18, 63, 174

W

Walfisch-Bertoni model, 276
Walfisch model, 218–19
Walls, BS near, 248–51
Wavenumber
free-space, 136
of oblique incidence, 35
obtaining, 5
Wet phantom, 168–69
Wideband measurements, 271
Wireless local area network (WLAN), 151

Recent Titles in the Artech House Antennas and Propagation Library

Christos Christodoulou, Series Editor
Vince Rodriguez, Series Editor

Adaptive Array Measurements in Communications, M. A. Halim

Antenna-on-Chip: Design, Challenges, and Opportunities, Hammad M. Cheema, Fatima Khalid, and Atif Shamim

Antenna Design for Cognitive Radio, Youssef Tawk, Joseph Costantine, and Christos Christodoulou

Antenna Design for CubeSats, Reyhan Baktur

Antenna Design with Fiber Optics, A. Kumar

Antenna Engineering Using Physical Optics: Practical CAD Techniques and Software, Leo Diaz and Thomas Milligan

Antenna Measurement Techniques, Gary E. Evans.

Antennas and Propagation for Body-Centric Wireless Communications, Second Edition, Peter S. Hall and Yang Hao, editors

Antennas and Site Engineering for Mobile Radio Networks, Bruno Delorme

Antennas for Small Mobile Terminals, Kyohei Fujimoto and Koichi Ito, editors

Analysis, Design and Measurement of Small and Low-profile Antennas, Kazuhiro Hirasawa

Analysis of Radome-Enclosed Antennas, Second Edition, Dennis J. Kozakoff

Applications of Geographic Information Systems for Wireless Network Planning, Francisco Saez de Adana, Josefa Gómez Pérez, Abdelhamid Tayebi Tayebi, and Juan Casado Ballesteros

The Art and Science of Ultrawideband Antennas, Second Edition, Hans G. Schantz

AWAS for Windows Version 2.0: Analysis of Wire Antennas and Scatterers, Antonije R. Djordjević, et al.

Broadband Microstrip Antennas, Girsh Kumar and K. P. Ray

Broadband Patch Antennas, Jean-François Zürcher and Fred E. Gardiol

Circularly Polarized Dielectric Resonator Antennas, Raghvendra Kumar Chaudhary, Rajkishor Kumar, and Rakesh Chowdhury

Design and Applications of Active Integrated Antennas, Mohammad S. Sharawi and Oualid Hammi

Design and Measurement of Antennas and Propagation in Mobile Cellular Systems, Hiroyuki Arai

Designer Notes for Microwave Antennas, Richard C. Johnson

Dielectric Resonator Antenna Handbook, Aldo Petosa

Dielectric Resonator Antennas, Biswajeet Mukherjee and Monika Chauhan

Electromagnetics and Antenna Technology, Alan J. Fenn

Electromagnetics, Microwave Circuit and Antenna Design for Communications Engineering, Peter Russer

Engineering Applications of the Modulated Scatterer Technique, Jean-Charles Bolomey and Fred E. Gardiol

Fixed & Mobile Terminal Antennas, Akhileswar Kumar

Foundations of Antenna Engineering: A Unified Approach for Line-of -Sight and Multipath, Per-Simon Kildal

Four-Arm Spiral Antennas, Joseph A. Mosko

Frequency-Agile Antennas for Wireless Communications, Aldo Petosa

Fresnel Zones in Wireless Links, Zone Plate Lenses and Antennas, Hristo D. Hristov

Handbook of Antennas for EMC, Second Edition, Thereza Macnamara with John McAuley

Handbook of Reflector Antennas and Feed Systems, Volume I: Theory and Design of Reflectors, Satish Sharma, Sudhakar Rao, and Lotfollah Shafai, editors

Handbook of Reflector Antennas and Feed Systems, Volume II: Feed Systems, Lotfollah Shafai, Satish Sharma, and Sudhakar Rao, editors

Handbook of Reflector Antennas and Feed Systems, Volume III: Applications of Reflectors, Sudhakar Rao, Lotfollah Shafai, and Satish Sharma, editors

Interference Suppression Techniques for Microwave Antennas and Transmitters, Ernest Freeman

Introduction to Antenna Analysis Using EM Simulators, Hiroaki Kogure, Yoshie Kogure, and James C. Rautio

Introduction to Antennas and RF Propagation Analysis, Dean James Friesen

LONRS: Low-Noise Receiving Systems Performance and Analysis Toolkit, Charles T. Stelzried, Macgregor S. Reid, and Arthur J. Freiley

Measurement of Mobile Antenna Systems, Second Edition, Hiroyuki Arai

Microstrip Lines and Slotlines, Third Edition, Ramesh Garg, Inder J. Bahl, Maurizio Bozzi

Microstrip Antenna Design, K.C Gupta and Abdelaziz Benalla

Microwave and Millimeter-Wave Remote Sensing for Security Applications, Jeffrey A. Nanzer

Microwave Cavity Antennas, Akhileswar Kumar and Hristo D. Hristov

Millimeter Wave Microstrip and Printed Circuit Antennas, Prakash Bhartia

Mobile Antenna Systems Handbook, Third Edition, Kyohei Fujimoto, editor

Moment Methods in Antennas and Scattering, Robert C. Hansen

Multiband Integrated Antennas for 4G Terminals, David A. Sánchez-Hernández, editor

Near-Field Antenna Measurements, Dan Slater

Noise Temperature Theory and Applications for Deep Space Communications Antenna Systems, Tom Y. Otoshi

Phased Array Antenna Handbook, Third Edition, Robert J. Mailloux

Phased Array Antennas, Arthur A. Oliner

Phased Array Antennas with Optimized Element Patterns, Sergei P. Skobelev

Plasma Antennas, Second Edition, Theodore Anderson

Polarization in Electromagnetic Systems, Second Edition, Warren Stutzman

Practical Antenna Design for Wireless Products. Henry Lau

Practical Microstrip and Printed Antenna Design, Anil Pandey

Practical Phased Array Antenna Systems, Eli Brookner

Practical Simulation of Radar Antennas and Radomes, Herbert L. Hirsch

Printed MIMO Antenna Engineering, Mohammad S. Sharawi

Radiowave Propagation and Antennas for Personal Communications, Third Edition, Kazimierz Siwiak

Reconfigurable Antenna Design and Analysis, Mohammod Ali

Reflectarray Antennas: Analysis, Design, Fabrication and Measurement, Jafar Shaker, Mohammad Reza Chaharmir, and Jonathan Ethier

Reflector Lens Antennas: Analysis and Design Using Personal Computers, Carlyle J. Sletten, editor

RF Coaxial Slot Radiators: Modeling, Measurements, and Applications, Kok Yeow You

Shipboard Antennas, Second Edition, Preston Law

Significant Phased Array Papers, Robert C. Hansen

Smart Antenna Engineering, Ahmed El Zooghby

Solid Dielectric Horn Antennas, Carlos Salema, Carlos Fernandes, and Rama Kant Jha

Sparse Phased Array Antennas: Theory and Applications, Ashutosh Kedar

Ultrawideband Antennas for Microwave Imaging Systems, Tayeb A. Denidni and Gijo Augustin

Ultrawideband Short-Pulse Radio Systems, V. I. Koshelev, Yu. I. Buyanov, and V. P. Belichenko

Waveguide Components for Antenna Feed Systems: Theory and CAD, Jaroslaw Uher, Jens Bornemann, and Uwe Rosenberg

Wearable Antennas and Electronics, Asimina Kiourti and John L. Volakis, editors

For further information on these and other Artech House titles, including previously considered out-of-print books now available through our In-Print-Forever® (IPF®) program, contact:

Artech House
685 Canton Street
Norwood, MA 02062
Phone: 781-769-9750
Fax: 781-769-6334
e-mail: artech@artechhouse.com

Artech House
16 Sussex Street
London SW1V HRW UK
Phone: +44 (0)20 7596-8750
Fax: +44 (0)20 7630 0166
e-mail: artech-uk@artechhouse.com

Find us on the World Wide Web at: www.artechhouse.com